The Landscape of Theoretical Physics:
A Global View

Fundamental Theories of Physics

*An International Book Series on The Fundamental Theories of Physics:
Their Clarification, Development and Application*

Editor:
ALWYN VAN DER MERWE, *University of Denver, U.S.A.*

Editorial Advisory Board:
JAMES T. CUSHING, *University of Notre Dame, U.S.A.*
GIANCARLO GHIRARDI, *University of Trieste, Italy*
LAWRENCE P. HORWITZ, *Tel-Aviv University, Israel*
BRIAN D. JOSEPHSON, *University of Cambridge, U.K.*
CLIVE KILMISTER, *University of London, U.K.*
PEKKA J. LAHTI, *University of Turku, Finland*
ASHER PERES, *Israel Institute of Technology, Israel*
EDUARD PRUGOVECKI, *University of Toronto, Canada*
TONY SUDBURY, *University of York, U.K.*
HANS-JÜRGEN TREDER, *Zentralinstitut für Astrophysik der Akademie der
   Wissenschaften, Germany*

The Landscape of Theoretical Physics: A Global View

From Point Particles to the Brane World
and Beyond,
in Search of a Unifying Principle

by

Matej Pavšič

Department of Theoretical Physics,
Jožef Stefan Institute,
Ljubljana, Slovenia

KLUWER ACADEMIC PUBLISHERS
DORDRECHT / BOSTON / LONDON

A C.I.P. Catalogue record for this book is available from the Library of Congress.

ISBN 0-7923-7006-6

Published by Kluwer Academic Publishers,
P.O. Box 17, 3300 AA Dordrecht, The Netherlands.

Sold and distributed in North, Central and South America
by Kluwer Academic Publishers,
101 Philip Drive, Norwell, MA 02061, U.S.A.

In all other countries, sold and distributed
by Kluwer Academic Publishers,
P.O. Box 322, 3300 AH Dordrecht, The Netherlands.

Printed on acid-free paper

All Rights Reserved
© 2001 Kluwer Academic Publishers
No part of the material protected by this copyright notice may be reproduced or
utilized in any form or by any means, electronic or mechanical,
including photocopying, recording or by any information storage and
retrieval system, without written permission from the copyright owner.

Printed in the Netherlands.

QC
21
.3
P38
2001
PHYS

To my family

Contents

Preface	xi
Acknowledgments	xiii
Introduction	xv

Part I Point Particles

1.	**THE SPINLESS POINT PARTICLE**		3
	1.1	Point particles versus worldlines	3
	1.2	Classical theory	7
	1.3	First quantization	16
		Flat spacetime	16
		Curved spacetime	20
	1.4	Second quantization	31
		Classical field theory with invariant evolution parameter	31
		The canonical quantization	35
		Comparison with the conventional relativistic quantum field theory	45
2.	**POINT PARTICLES AND CLIFFORD ALGEBRA**		53
	2.1	Introduction to geometric calculus based on Clifford algebra	54
	2.2	Algebra of spacetime	59
	2.3	Physical quantities as polyvectors	62
	2.4	The unconstrained action from the polyvector action	69
		Free particle	69
		Particle in a fixed background field	73
	2.5	Quantization of the polyvector action	75
	2.6	On the second quantization of the polyvector action	83
	2.7	Some further important consequences of Clifford algebra	87

viii *THE LANDSCAPE OF THEORETICAL PHYSICS: A GLOBAL VIEW*

Relativity of signature	87
Grassmann numbers from Clifford numbers	90
2.8 The polyvector action and De Witt–Rovelli material reference system	92
3. HARMONIC OSCILLATOR IN PSEUDO-EUCLIDEAN SPACE	93
3.1 The 2-dimensional pseudo-Euclidean harmonic oscillator	94
3.2 Harmonic oscillator in d-dimensional pseudo-Euclidean space	96
3.3 A system of scalar fields	99
3.4 Conclusion	102

Part II Extended Objects

4. GENERAL PRINCIPLES OF MEMBRANE KINEMATICS AND DYNAMICS	107
4.1 Membrane space $\mathcal{M}$	108
4.2 Membrane dynamics	114
Membrane theory as a free fall in $\mathcal{M}$-space	114
Membrane theory as a minimal surface in an embedding space	126
Membrane theory based on the geometric calculus in $\mathcal{M}$-space	130
4.3 More about the interconnections among various membrane actions	138
5. MORE ABOUT PHYSICS IN $\mathcal{M}$-SPACE	145
5.1 Gauge fields in $\mathcal{M}$-space	146
General considerations	146
A specific case	147
$\mathcal{M}$-space point of view again	152
A system of many membranes	155
5.2 Dynamical metric field in $\mathcal{M}$-space	158
Metric of V_N from the metric of $\mathcal{M}$-space	163
6. EXTENDED OBJECTS AND CLIFFORD ALGEBRA	167
6.1 Mathematical preliminaries	168
Vectors in curved spaces	168
Vectors in an infinite-dimensional space	177
6.2 Dynamical vector field in $\mathcal{M}$-space	180
Description with the vector field in spacetime	183
6.3 Full covariance in the space ot parameters ϕ^A	188
Description in spacetime	188
Description in $\mathcal{M}$-space	195
Description in the enlarged $\mathcal{M}$-space	197

7. QUANTIZATION	203
7.1 The quantum theory of unconstrained membranes	203
The commutation relations and the Heisenberg equations of motion	204
The Schrödinger representation	205
The stationary Schrödinger equation for a membrane	210
Dimensional reduction of the Schrödinger equation	211
A particular solution to the covariant Schrödinger equation	212
The wave packet	217
The expectation values	220
Conclusion	222
7.2 Clifford algebra and quantization	223
Phase space	223
Wave function as a polyvector	225
Equations of motion for basis vectors	229
Quantization of the p-brane: a geometric approach	238

Part III Brane World

8. SPACETIME AS A MEMBRANE IN A HIGHER-DIMENSIONAL SPACE	249
8.1 The brane in a curved embedding space	249
8.2 A system of many intersecting branes	256
The brane interacting with itself	258
A system of many branes creates the bulk and its metric	260
8.3 The origin of matter in the brane world	261
Matter from the intersection of our brane with other branes	261
Matter from the intersection of our brane with itself	261
8.4 Comparison with the Randall–Sundrum model	264
The metric around a brane in a higher-dimensional bulk	267
9. THE EINSTEIN–HILBERT ACTION ON THE BRANE AS THE EFFECTIVE ACTION	271
9.1 The classical model	272
9.2 The quantum model	274
9.3 Conclusion	281
10. ON THE RESOLUTION OF TIME PROBLEM IN QUANTUM GRAVITY	283
10.1 Space as a moving 3-dimensional membrane in V_N	285
10.2 Spacetime as a moving 4-dimensional membrane in V_N	287
General consideration	287
A physically interesting solution	289
Inclusion of sources	294

Part IV Beyond the Horizon

11. THE LANDSCAPE OF THEORETICAL PHYSICS:
    A GLOBAL VIEW   303

12. NOBODY REALLY UNDERSTANDS QUANTUM
    MECHANICS   315
    12.1 The 'I' intuitively understands quantum mechanics   318
    12.2 Decoherence   323
    12.3 On the problem of basis in the Everett interpretation   326
    12.4 Brane world and brain world   328
    12.5 Final discussion on quantum mechanics, and conclusion   333

13. FINAL DISCUSSION   339
    13.1 What is wrong with tachyons?   339
    13.2 Is the electron indeed an event moving in spacetime?   340
    13.3 Is our world indeed a single huge 4-dimensional membrane? 342
    13.4 How many dimensions are there?   343
    13.5 Will it ever be possible to find solutions to the classical and
         quantum brane equations of motion and make predictions? 344
    13.6 Have we found a unifying principle?   345

Appendices   347
    The dilatationally invariant system of units   347

References   353

Index
                                                                 363

Preface

Today many important directions of research are being pursued more or less independently of each other. These are, for instance, strings and membranes, induced gravity, embedding of spacetime into a higher-dimensional space, the brane world scenario, the quantum theory in curved spaces, Fock–Schwinger proper time formalism, parametrized relativistic quantum theory, quantum gravity, wormholes and the problem of "time machines", spin and supersymmetry, geometric calculus based on Clifford algebra, various interpretations of quantum mechanics including the Everett interpretation, and the recent important approach known as "decoherence".

A big problem, as I see it, is that various people thoroughly investigate their narrow field without being aware of certain very close relations to other fields of research. What we need now is not only to see the trees but also the forest. In the present book I intend to do just that: to carry out a first approximation to a synthesis of the related fundamental theories of physics. I sincerely hope that such a book will be useful to physicists.

From a certain viewpoint the book could be considered as a course in theoretical physics in which the foundations of all those relevant fundamental theories and concepts are attempted to be thoroughly reviewed. Unsolved problems and paradoxes are pointed out. I show that most of those approaches have a common basis in the theory of unconstrained membranes. The very interesting and important concept of membrane space, $\mathcal{M}$, the tensor calculus in $\mathcal{M}$ and functional transformations in $\mathcal{M}$ are discussed. Next I present a theory in which spacetime is considered as a 4-dimensional unconstrained membrane and discuss how the usual classical gravity, together with sources, emerges as an effective theory. Finally, I point out that the Everett interpretation of quantum mechanics is the natural one in that theory. Various interpretational issues will be discussed and the relation to the modern "decoherence" will be pointed out.

If we look at the detailed structure of a landscape we are unable to see the connections at a larger scale. We see mountains, but we do not see the mountain range. A view from afar is as important as a view from nearby. Every position illuminates reality from its own perspective. It is analogously so, in my opinion, in theoretical physics also. Detailed investigations of a certain fundamental theory are made at the expense of seeing at the same time the connections with other theories. What we need today is some kind of atlas of the many theoretical approaches currently under investigation. During many years of effort I can claim that I do see a picture which has escaped from attention of other researchers. They certainly might profit if they could become aware of such a more global, though not as detailed, view of fundamental theoretical physics.

MATEJ PAVŠIČ

Acknowledgments

This book is a result of many years of thorough study and research. I am very grateful to my parents, who supported me and encouraged me to persist on my chosen path. Also I thank my beloved wife Mojca for all she has done for me and for her interest in my work. I have profited a lot from discussions and collaboration with E. Recami, P. Caldirola, A.O. Barut, W.A. Rodrigues, Jr., V. Tapia, M. Maia, M. Blagojević, R. W. Tucker, A. Zheltukhin, I. Kanatchikov, L. Horwitz, J. Fanchi, C. Castro, and many others whom I have met on various occasions. The work was supported by the Slovenian Ministry of Science and Technology.

Introduction

The unification of various branches of theoretical physics is a joint project of many researchers, and everyone contributes as much as he can. So far we have accumulated a great deal of knowledge and insight encoded in such marvelous theories as general relativity, quantum mechanics, quantum field theory, the standard model of electroweak interaction, and chromodynamics. In order to obtain a more unified view, various promising theories have emerged, such as those of strings and "branes", induced gravity, the embedding models of gravity, and the "brane world" models, to mention just a few. The very powerful Clifford algebra as a useful tool for geometry and physics is becoming more and more popular. Fascinating are the ever increasing successes in understanding the foundations of quantum mechanics and their experimental verification, together with actual and potential practical applications in cryptography, teleportation, and quantum computing.

In this book I intend to discuss the conceptual and technical foundations of those approaches which, in my opinion, are most relevant for unification of general relativity and quantum mechanics on the one hand, and fundamental interactions on the other hand. After many years of active research I have arrived at a certain level of insight into the possible interrelationship between those theories. Emphases will be on the exposition and understanding of concepts and basic techniques, at the expense of detailed and rigorous mathematical development. Theoretical physics is considered here as a beautiful landscape. A global view of the landscape will be taken. This will enable us to see forests and mountain ranges as a whole, at the cost of seeing trees and rocks.

Physicists interested in the foundations of physics, conceptual issues, and the unification program, as well as those working in a special field and desiring to broaden their knowledge and see their speciality from a wider perspective, are expected to profit the most from the present book. They are assumed to possess a solid knowledge at least of quantum mechanics, and special and general relativity.

As indicated in the subtitle, I will start from point particles. They move along geodesics which are the lines of minimal, or, more generally, extremal length, in spacetime. The corresponding action from which the equations of motion are derived is invariant with respect to reparametrizations of an arbitrary parameter denoting position on the worldline swept by the particle. There are several different, but equivalent, reparametrization invariant point particle actions. A common feature of such an approach is that actually there is no dynamics in spacetime, but only in space. A particle's worldline is frozen in spacetime, but from the 3-dimensional point of view we have a point particle moving in 3-space. This fact is at the roots of all the difficulties we face when trying to quantize the theory: either we have a covariant quantum theory but no evolution in spacetime, or we have evolution in 3-space at the expense of losing manifest covariance in spacetime. In the case of a point particle this problem is not considered to be fatal, since it is quite satisfactorily resolved in relativistic quantum field theory. But when we attempt to quantize extended objects such as branes of arbitrary dimension, or spacetime itself, the above problem emerges in its full power: after so many decades of intensive research we have still not yet arrived at a generally accepted consistent theory of quantum gravity.

There is an alternative to the usual relativistic point particle action proposed by Fock [1] and subsequently investigated by Stueckelberg [2], Feynman [3], Schwinger [4], Davidon [5], Horwitz [6, 7] and many others [8]–[20]. In such a theory a particle or "event" in spacetime obeys a law of motion analogous to that of a nonrelativistic particle in 3-space. The difference is in the dimensionality and signature of the space in which the particle moves. None of the coordinates x^0, x^1, x^2, x^3 which parametrize spacetime has the role of evolution parameter. The latter is separately postulated and is Lorentz invariant. Usually it is denoted as τ and evolution goes along τ. There are no constraints in the theory, which can therefore be called *the unconstrained theory*. First and second quantizations of the unconstrained theory are straightforward, very elegant, and manifestly Lorentz covariant. Since τ can be made to be related to proper time such a theory is often called a *Fock–Schwinger proper time formalism*. The value and elegance of the latter formalism is widely recognized, and it is often used, especially when considering quantum fields in curved spaces [21]. There are two main interpretations of the formalism:

(i) According to the first interpretation, it is considered merely as a useful calculational tool, without any physical significance. Evolution in τ and the absence of any constraint is assumed to be fictitious and unphysical. In order to make contact with physics one has to get rid of τ in all the

INTRODUCTION xvii

expressions considered by integrating them over τ. By doing so one projects unphysical expressions onto the physical ones, and in particular one projects unphysical states onto physical states.

(ii) According to the second interpretation, evolution in τ is genuine and physical. There is, indeed, dynamics in spacetime. Mass is a constant of motion and not a fixed constant in the Lagrangian.

Personally, I am inclined to the interpretation (ii). In the history of physics it has often happened that a good new formalism also contained good new physics waiting to be discovered and identified in suitable experiments. It is one of the purposes of this book to show a series of arguments in favor of the interpretation (ii). The first has roots in geometric calculus based on Clifford algebra [22]

Clifford numbers can be used to represent vectors, multivectors, and, in general, polyvectors (which are Clifford aggregates). They form a very useful tool for geometry. The well known equations of physics can be cast into elegant compact forms by using the geometric calculus based on Clifford algebra.

These compact forms suggest the generalization that every physical quantity is a polyvector [23, 24]. For instance, the momentum polyvector in 4-dimensional spacetime has not only a vector part, but also a scalar, bivector, pseudovector and pseudoscalar part. Similarly for the velocity polyvector. Now we can straightforwardly generalize the conventional constrained action by rewriting it in terms of polyvectors. By doing so we obtain in the action also a term which corresponds to the pseudoscalar part of the velocity polyvector. A consequence of this extra term is that, when confining ourselves, for simplicity, to polyvectors with pseudoscalar and vector part only, the variables corresponding to 4-vector components can all be taken as independent. After a straightforward procedure in which we omit the extra term in the action (since it turns out to be just the total derivative), we obtain Stueckelberg's unconstrained action! This is certainly a remarkable result. The original, constrained action is equivalent to the unconstrained action. Later in the book (Sec. 4.2) I show that the analogous procedure can also be applied to extended objects such as strings, membranes, or branes in general.

When studying the problem of how to identify points in a generic curved spacetime, several authors [25], and, especially recently Rovelli [26], have recognized that one must fill spacetime with a reference fluid. Rovelli considers such a fluid as being composed of a bunch of particles, each particle carrying a clock on it. Besides the variables denoting positions of particles there is also a variable denoting the clock. This extra, clock, variable must enter the action, and the expression Rovelli obtains is formally the same as the expression we obtain from the polyvector action (in which we neglect the bivector, pseudovector, and scalar parts). We may therefore identify the pseudoscalar part of the velocity polyvector with the speed of the clock variable. Thus have a relation between the polyvector generalization of the usual constrained relativistic point particle, the Stueckleberg particle, and the DeWitt–Rovelli particle with clock.

A relativistic particle is known to posses spin, in general. We show how spin arises from the polyvector generalization of the point particle and how the quantized theory contains the Dirac spinors together with the Dirac equation as a particular case. Namely, in the quantized theory a state is naturally assumed to be represented as a polyvector wave function Φ, which, in particular, can be a spinor. That spinors are just a special kind of polyvectors (Clifford aggregates), namely the elements of the minimal left or right ideals of the Clifford algebra, is an old observation [27]. Now, scalars, vectors, spinors, etc., can be reshuffled by the elements of the Clifford algebra. This means that scalars, vectors, etc., can be transformed into spinors, and *vice versa*. Within Clifford algebra thus we have transformations which change bosons into fermions. In Secs. 2.5 and 2.7 I discuss the possible relation between the Clifford algebra formulation of the spinning particle and a more widely used formulation in terms of Grassmann variables.

A very interesting feature of Clifford algebra concerns the signature of the space defined by basis vectors which are generators of the Clifford algebra. In principle we are not confined to choosing just a particular set of elements as basis vectors; we may choose some other set. For instance, if e^0, e^1, e^2, e^3 are the basis vectors of a space M_e with signature $(+ + + +)$, then we may declare the set $(e^0, e^0e^1, e^0e^2, e^0e^3)$ as basis vectors γ^0, γ^1, γ^2, γ^3 of some other space M_γ with signature $(+ - - -)$. That is, by suitable choice of basis vectors we can obtain within the same Clifford algebra a space of arbitrary signature. This has far reaching implications. For instance, in the case of even-dimensional space we can always take a signature with an equal number of pluses and minuses. A harmonic oscillator in such a space has vanishing zero point energy, provided that we define the vacuum state in a very natural way as proposed in refs. [28]. An immediate consequence is that there are no central terms and anomalies in string theory living in spacetime with signature $(+ + + ... - - -)$, even if the dimension of such a space is not critical. In other words, spacetime with such a 'symmetric' signature need not have 26 dimensions [28].

The principle of such a harmonic oscillator in a pseudo-Euclidean space is applied in Chapter 3 to a system of scalar fields. The metric in the space of fields is assumed to have signature $(+ + + ... - - -)$ and it is shown that the vacuum energy, and consequently the cosmological constant, are then exactly zero. However, the theory contains some negative energy fields ("exotic matter") which couple to the gravitational field in a different way than the usual, positive energy, fields: the sign of coupling is reversed, which implies a repulsive gravitational field around such a source. This is the price to be paid if one wants to obtain a small cosmological constant in a straightforward way. One can consider this as a prediction of the theory to be tested by suitably designed experiments.

The problem of the cosmological constant is one of the toughest problems in theoretical physics. Its resolution would open the door to further understanding of the relation between quantum theory and general relativity. Since all more conventional approaches seem to have been more or less exploited without unambiguous success, the time is right for a more drastic novel approach. Such is the one which relies on the properties of the harmonic oscillator in a pseudo-Euclidean space.

In Part II I discuss the theory of extended objects, now known as "branes" which are membranes of any dimension and are generalizations of point particles and strings. As in the case of point particles I pay much attention to the unconstrained theory of membranes. The latter theory is a generalization of the Stueckelberg point particle theory. It turns out to be very convenient to introduce the concept of the infinite-dimensional *membrane space* $\mathcal{M}$. Every point in $\mathcal{M}$ represents an unconstrained membrane. In $\mathcal{M}$ we can define distance, metric, covariant derivative, etc., in an analogous way as in a finite-dimensional curved space. A membrane action, the corresponding equations of motion, and other relevant expressions acquire very simple forms, quite similar to those in the point particle theory. We may say that a membrane is a point particle in an infinite dimensional space!

Again we may proceed in two different interpretations of the theory:

(i) We may consider the formalism of the membrane space as a useful calculational tool (a generalization of the Fock–Schwinger proper time formalism) without any genuine physical significance. Physical quantities are obtained after performing a suitable projection.

(ii) The points in $\mathcal{M}$-space are physically distinguishable, that is, a membrane can be physically deformed in various ways and such a deformation may change with evolution in τ.

If we take the interpretation (ii) then we have a marvelous connection (discussed in Sec. 2.8) with the Clifford algebra generalization of the conventional constrained membrane on the one hand, and the concept of DeWitt–Rovelli reference fluid with clocks on the other hand.

Clifford algebra in the infinite-dimensional membrane space $\mathcal{M}$ is described in Sec. 6.1. When quantizing the theory of the unconstrained membrane one may represent states by wave functionals which are polyvectors in $\mathcal{M}$-space. A remarkable connection with quantum field theory is shown in Sec. 7.2

When studying the $\mathcal{M}$-space formulation of the membrane theory we find that in such an approach one cannot postulate the existence of a background embedding space independent from a membrane configuration. By "membrane configuration" I understand a system of (many) membranes, and the membrane configuration is identified with the embedding space. There is no embedding space without the membranes. This suggests that our spacetime is nothing but a membrane configuration. In particular, our spacetime could be just one 4-dimensional membrane (4-brane) amongst many other membranes within the configuration. Such a model is discussed in Part III. The 4-dimensional gravity is due to the induced metric on our 4-brane V_4, whilst matter comes from the self-intersections of V_4, or the intersections of V_4 with other branes. As the intersections there can occur manifolds of various dimensionalities. In particular, the intersection or the self-intersection can be a 1-dimensional worldline. It is shown that such a worldline is a geodesic on V_4. So we obtain in a natural way four-dimensional gravity with sources. The quantized version of such a model is also discussed, and it is argued that the kinetic term for the 4-dimensional metric $g_{\mu\nu}$ is induced by quantum fluctuations of the 4-brane embedding functions.

In the last part I discuss mainly the problems related to the foundations and interpretation of quantum mechanics. I show how the brane world view sheds new light on our understanding of quantum mechanics and the role of the observer.

I

POINT PARTICLES

Chapter 1

THE SPINLESS POINT PARTICLE

1.1. POINT PARTICLES VERSUS WORLDLINES

The simplest objects treated by physics are point particles. Any object, if viewed from a sufficiently large scale, is approximately a point particle. The concept of an exact point particle is an idealization which holds in the limit of infinitely large scale from which it is observed. Equivalently, if observed from a finite scale its size is infinitely small. According to the special and general theory of relativity the arena in which physics takes place is not a 3-dimensional space but the 4-dimensional space with the pseudo–Euclidean signature, say (+ - - -), called *spacetime*. In the latter space the object of question is actually a worldline, a 1-dimensional object, and it appears as a point particle only from the point of view of a space-like 3-surface which intersects the worldline.

Kinematically, any worldline (i.e., a curve in spacetime) is possible, but not all of them can be realized in a dynamical situation. An obvious question arises of which amongst all kinematically possible worldlines in spacetime are actually possible. The latter worldlines are solutions of certain differential equations.

An action from which one can derive the equations of a worldline is proportional to the length of the worldline:

$$I[X^\mu] = m \int d\tau (\dot{X}^\mu \dot{X}^\nu g_{\mu\nu})^{1/2}. \tag{1.1}$$

Here τ is an arbitrary parameter associated with a point on the worldline, the variables X^μ denote the position of that point in spacetime, the fixed constant m is its mass, and $g_{\mu\nu}(x)$ is the metric tensor which depends on the position $x \equiv x^\mu$ in spacetime.

Variation of the action (1.1) with respect to X^μ gives *the geodesic equation*

$$E^\mu \equiv \frac{1}{\sqrt{\dot{X}^2}} \frac{d}{d\tau}\left(\frac{\dot{X}^\mu}{\sqrt{\dot{X}^2}}\right) + \Gamma^\mu_{\alpha\beta} \frac{\dot{X}^\alpha \dot{X}^\beta}{\dot{X}^2} = 0 \qquad (1.2)$$

where

$$\Gamma^\mu_{\alpha\beta} \equiv \tfrac{1}{2} g^{\mu\rho}(g_{\rho\alpha,\beta} + g_{\rho\beta,\alpha} - g_{\alpha\beta,\rho}) \qquad (1.3)$$

is the affinity of spacetime. We often use the shorthand notation $\dot{X}^2 \equiv \dot{X}^\mu \dot{X}_\mu$, where indices are lowered by the metric tensor $g_{\mu\nu}$.

The action (1.1) is invariant with respect to transformations of spacetime coordinates $x^\mu \to x'^\mu = f^\mu(x)$ (diffeomorphisms) and with respect to reparametrizations $\tau \to f(\tau)$. A consequence of the latter invariance is that the equations of motion (1.2) are not all independent, since they satisfy the identity

$$E^\mu \dot{X}_\mu = 0 \qquad (1.4)$$

Therefore the system is under-determined: there are more variables X^μ than available equations of motion.

From (1.1) we have

$$p_\mu \dot{X}^\mu - m(\dot{X}^\mu \dot{X}_\mu)^{1/2} = 0 \qquad (1.5)$$

or

$$H \equiv p_\mu p^\mu - m^2 = 0, \qquad (1.6)$$

where

$$p_\mu = \frac{\partial L}{\partial \dot{X}^\mu} = \frac{m \dot{X}_\mu}{(\dot{X}^\nu \dot{X}_\nu)^{1/2}}. \qquad (1.7)$$

This demonstrates that the canonical momenta p_μ are not all independent, but are subjected to the constraint (1.6). A general theory of constrained systems is developed by Dirac [29]. An alternative formulation is owed to Rund [30].

We see that the Hamiltonian for our system —which is a worldline in spacetime— is zero. This is often interpreted as there being no evolution in spacetime: worldlines are frozen and they exist 'unchanged' in spacetime.

Since our system is under-determined we are free to choose a relation between the X^μ. In particular, we may choose $X^0 = \tau$, then the action

(1.1) becomes a functional of the reduced number of variables[1]

$$I[X^i] = m \int d\tau (g_{00} + 2\dot{X}^i g_{0i} + \dot{X}^i \dot{X}^i g_{ij})^{1/2}. \tag{1.8}$$

All the variables X^i, $i = 1, 2, 3$, are now independent and they represent *motion* of a point particle in 3-space. From the 3-dimensional point of view the action (1.1) describes a dynamical system, whereas from the 4-dimensional point of view there is no dynamics.

Although the reduced action (1.8) is good as far as dynamics is concerned, it is not good from the point of view of the theory of relativity: it is not manifestly covariant with respect to the coordinate transformations of x^μ, and in particular, with respect to Lorentz transformations.

There are several well known classically equivalent forms of the point particle action. One of them is the second order or Howe–Tucker action [31]

$$I[X^\mu, \lambda] = \frac{1}{2} \int d\tau \left(\frac{\dot{X}^\mu \dot{X}_\mu}{\lambda} + \lambda m^2 \right) \tag{1.9}$$

which is a functional not only of the variables $X^\mu(\tau)$ but also of λ which is the Lagrange multiplier giving the relation

$$\dot{X}^\mu \dot{X}_\mu = \lambda^2 m^2. \tag{1.10}$$

Inserting (1.10) back into the Howe–Tucker action (1.9) we obtain the minimal length action (1.1). The canonical momentum is $p_\mu = \dot{X}_\mu / \lambda$ so that (1.10) gives the constraint (1.6).

Another action is the first order, or phase space action,

$$I[X^\mu, p_\mu, \lambda] = \int d\tau \left(p_\mu \dot{X}^\mu - \frac{\lambda}{2}(p_\mu p^\mu - m^2) \right) \tag{1.11}$$

which is also a functional of the canonical momenta p_μ. Varying (1.11) with respect to p_μ we have the relation between p^μ, $\dot{X}^\mu$, and λ:

$$p^\mu = \frac{\dot{X}^\mu}{\lambda} \tag{1.12}$$

which, because of (1.10), is equivalent to (1.7).

The actions (1.9) and (1.11) are invariant under reparametrizations $\tau' = f(\tau)$, provided λ is assumed to transform according to $\lambda' = (d\tau/d\tau') \lambda$.

---

[1] In flat spacetime we may have $g_{00} = 1$, $g_{rs} = -\delta_{rs}$, and the action (1.8) takes the usual special relativistic form $I = m \int d\tau \sqrt{1 - \dot{\mathbf{X}}^2}$, where $\dot{\mathbf{X}} \equiv \dot{X}^i$.

The Hamiltonian corresponding to the action (1.9),

$$H = p_\mu \dot{X}^\mu - L, \tag{1.13}$$

is identically zero, and therefore it does not generate any genuine evolution in τ.

Quantization of the theory goes along several possible lines [32]. Here let me mention the Gupta–Bleuler quantization. Coordinates and momenta become operators satisfying the commutation relations

$$[X^\mu, p_\nu] = i\delta^\mu{}_\nu, \quad [X^\mu, X^\nu] = 0, \quad [p_\mu, p_\nu] = 0. \tag{1.14}$$

The constraint (1.6) is imposed on state vectors

$$(p^\mu p_\mu - m^2)|\psi\rangle = 0. \tag{1.15}$$

Representation of the operators is quite straightforward if spacetime is *flat*, so we can use a coordinate system in which the metric tensor is everywhere of the Minkowski type, $g_{\mu\nu} = \eta_{\mu\nu}$ with signature $(+---)$. A useful representation is that in which the coordinates x^μ are diagonal and $p_\mu = -i\partial_\mu$, where $\partial_\mu \equiv \partial/\partial x^\mu$. Then the constraint relation (1.15) becomes the Klein–Gordon equation

$$(\partial_\mu \partial^\mu + m^2)\psi = 0. \tag{1.16}$$

Interpretation of the wave function $\psi(x) \equiv \langle x|\psi\rangle$, $x \equiv x^\mu$, is not straightforward. It cannot be interpreted as the one particle probability amplitude. Namely, if ψ is complex valued then to (1.16) there corresponds the non-vanishing current

$$j_\mu = \frac{i}{2m}(\psi^* \partial_\mu \psi - \psi \partial_\mu \psi^*). \tag{1.17}$$

The components j_μ can all be positive or negative, and there is the problem of which quantity then serves as the probability density. Conventionally the problem is resolved by switching directly into the second quantized theory and considering j_μ as the charge–current operator. I shall not review in this book the conventional relativistic quantum field theory, since I am searching for an alternative approach, much better in my opinion, which has actually already been proposed and considered in the literature [1]–[20] under various names such as the Stueckelberg theory, the unconstrained theory, the parametrized theory, etc.. In the rest of this chapter I shall discuss various aspects of the *unconstrained point particle theory*.

1.2. CLASSICAL THEORY

It had been early realized that instead of the action (1.1) one can use the action

$$I[X^\mu] = \frac{1}{2}\int d\tau \frac{\dot{X}^\mu \dot{X}_\mu}{\Lambda}, \tag{1.18}$$

where $\Lambda(\tau)$ is a fixed function of τ, and can be a constant. The latter action, since being quadratic in velocities, is much more suitable to manage, especially in the quantized version of the theory.

There are two principal ways of interpreting (1.18).

Interpretation (a) We may consider (1.18) as a gauge fixed action (i.e., an action in which reparametrization of τ is fixed), equivalent to the constrained action (1.1).

Interpretation (b) Alternatively, we may consider (1.18) as an unconstrained action.

The equations of motion obtained by varying (1.18) with respect to X^μ are

$$\frac{1}{\Lambda}\frac{d}{d\tau}\left(\frac{\dot{X}^\mu}{\Lambda}\right) + \Gamma^\mu_{\alpha\beta}\frac{\dot{X}^\alpha \dot{X}^\beta}{\Lambda^2} = 0. \tag{1.19}$$

This can be rewritten as

$$\frac{\sqrt{\dot{X}^2}}{\Lambda^2}\frac{d}{d\tau}\left(\frac{\dot{X}^\mu}{\sqrt{\dot{X}^2}}\right) + \Gamma^\mu_{\alpha\beta}\frac{\dot{X}^\alpha \dot{X}^\beta}{\Lambda^2} + \frac{\dot{X}^\mu}{\sqrt{\dot{X}^2}}\frac{d}{d\tau}\left(\frac{\sqrt{\dot{X}^2}}{\Lambda}\right) = 0. \tag{1.20}$$

Multiplying the latter equation by $\dot{X}^\mu$, summing over μ and assuming $\dot{X}^2 \neq 0$ we have

$$\frac{\sqrt{\dot{X}^2}}{\Lambda}\frac{d}{d\tau}\left(\frac{\sqrt{\dot{X}^2}}{\Lambda}\right) = \frac{1}{2}\frac{d}{d\tau}\left(\frac{\dot{X}^2}{\Lambda^2}\right) = 0. \tag{1.21}$$

Inserting (1.21) into (1.20) we obtain that (1.19) is equivalent to the geodetic equation (1.1).

According to *Interpretation (a)* the parameter τ in (1.19) is not arbitrary, but it satisfies eq. (1.21). In other words, eq. (1.21) is *a gauge fixing equation* telling us the $\dot{X}^2 = constant \times \Lambda^2$, which means that $(dX^\mu dX_\mu)^{1/2} \equiv ds = constant \times \Lambda d\tau$, or, when $\dot{\Lambda} = 0$ that $s = constant \times \tau$; the parameter τ is thus proportional to the proper time s.

According to *Interpretation (b)* eq. (1.21) tells us that mass is a constant of motion. Namely, the canonical momentum belonging to the action (1.18) is

$$p_\mu = \frac{\partial L}{\partial \dot{X}^\mu} = \frac{\dot{X}_\mu}{\Lambda} \tag{1.22}$$

and its square is

$$p^\mu p_\mu \equiv M^2 = \frac{\dot X^\mu \dot X_\mu}{\Lambda^2}. \tag{1.23}$$

From now on I shall assume *Interpretation (b)* and consider (1.18) as *the unconstrained action*. It is very convenient to take Λ as a constant. Then τ is just proportional to the proper time s; if $\Lambda = 1$, then it is the proper time. However, instead of τ, we may use another parameter $\tau' = f(\tau)$. In terms of the new parameter the action (1.18) reads

$$\frac{1}{2}\int d\tau' \frac{dX^\mu}{d\tau'} \frac{dX_\mu}{d\tau'} \frac{1}{\Lambda'} = I'[X^\mu], \tag{1.24}$$

where

$$\Lambda' = \frac{d\tau}{d\tau'} \Lambda. \tag{1.25}$$

The form of the transformed action is the same as that of the 'original' action (1.18): our unconstrained action is *covariant* under reparametrizations. But it is not *invariant* under reparametrizations, since the transformed action is a different functional of the variables X^μ than the original action. The difference comes from Λ', which, according to the assumed eq. (1.25), does not transform as a scalar, but as a 1-dimensional vector field. The latter vector field $\Lambda(\tau)$ is taken to be a fixed, background, field; it is not a dynamical field in the action (1.18).

At this point it is important to stress that in a given parametrization the "background" field $\Lambda(\tau)$ can be arbitrary in principle. If the parametrizations changes then $\Lambda(\tau)$ also changes according to (1.25). This is in contrast with the constrained action (1.9) where λ is a 'dynamical' field (actually the Lagrange multiplier) giving the constraint (1.10). In eq. (1.10) λ is arbitrary, it can be freely chosen. This is intimately connected with the choice of parametrization (gauge): choice of λ means choice of gauge. Any change of λ automatically means change of gauge. On the contrary, in the action (1.18) and in eq. (1.23), since M^2 is not prescribed but it is a constant of motion, $\Lambda(\tau)$ is not automatically connected to choice of parametrization. It does change under a reparametrization, but in a given, fixed parametrization (gauge), $\Lambda(\tau)$ can still be different in principle. This reflects that Λ is assumed here to be a physical field associated to the particle. Later we shall find out (as announced already in Introduction) that physically Λ is a result either of (i) *the "clock variable"* sitting on the particle, or (ii) of the *scalar part of the velocity polyvector* occurring in the Clifford algebra generalization of the theory.

I shall call (1.18) the *Stueckelberg action*, since Stueckelberg was one of the first protagonists of its usefulness. The Stueckelberg action is not invariant under reparametrizations, therefore it does not imply any constraint.

The spinless point particle

All X^μ are independent dynamical variables. The situation is quite analogous to the one of a non-relativistic particle in 3-space. The difference is only in the number of dimensions and in the signature, otherwise the mathematical expressions are the same. Analogously to the evolution in 3-space, we have, in the unconstrained theory, evolution in 4-space, and τ is the evolution parameter. Similarly to 3-space, where a particle's trajectory is "built up" point by point while time proceeds, so in 4-space a worldline is built up point by point (or event by event) while τ proceeds. The Stueckelberg theory thus implies genuine *dynamics* in spacetime: a particle is a point-like object not only from the 3-dimensional but also from the 4-dimensional point of view, and it *moves* in spacetime.[2]

A trajectory in V_4 has a status analogous that of the usual trajectory in E_3. The latter trajectory is not an existing object in E_3; it is a mathematical line obtained after having collected all the points in E_3 through which the particle has passed during its motion. The same is true for a moving particle in V_4. The unconstrained relativity is thus just a theory of *relativistic dynamics*. In the conventional, constrained, relativity there is no dynamics in V_4; events in V_4 are considered as frozen.

Non-relativistic limit. We have seen that the unconstrained equation of motion (1.19) is equivalent to the equation of geodesic. The trajectory of a point particle or "event" [6] moving in spacetime is a geodesic. In this respect the unconstrained theory gives the same predictions as the constrained theory. Now let us assume that spacetime is flat and that the particle "spatial" speed $\dot{X}^r$, $r = 1, 2, 3$ is small in comparison to its "time" speed $\dot{X}^0$, i.e.,

$$\dot{X}^r \dot{X}_r \ll (\dot{X}^0)^2. \tag{1.26}$$

Then the constant of motion $M = \sqrt{\dot{X}^\mu \dot{X}_\mu}/\Lambda$ (eq. (1.23)) is approximately equal to the time component of 4-momentum:

$$M = \frac{\sqrt{(\dot{X}^0)^2 + \dot{X}^r \dot{X}_r}}{\Lambda} \approx \frac{\dot{X}^0}{\Lambda} \equiv p^0. \tag{1.27}$$

Using (1.25) the action (1.18) becomes

$$I = \frac{1}{2} \int d\tau \int \left[\left(\frac{\dot{X}^0}{\Lambda} \right)^2 + \frac{\dot{X}^r \dot{X}_r}{\Lambda} \right] \approx \frac{1}{2} \int d\tau \left[M^2 + \frac{M}{\dot{X}^0} \dot{X}^r \dot{X}_r \right] \tag{1.28}$$

---

[2]Later, we shall see that realistic particles obeying the usual electromagnetic interactions should actually be modeled by time-like strings evolving in spacetime. But at the moment, in order to set the concepts and develop the theory we work with the idealized, point-like objects or *events* moving in spacetime.

Since M is a constant the first term in (1.28) has no influence on the equations of motion and can be omitted. So we have

$$I \approx \frac{M}{2} \int d\tau \, \frac{\dot{X}^r \dot{X}_r}{\dot{X}^0}. \tag{1.29}$$

which can be written as

$$I \approx \frac{M}{2} \int dt \, \frac{dX^r}{dt} \frac{dX_r}{dt}, \qquad t \equiv X^0 \tag{1.30}$$

This is the usual non-relativistic free particle action.

We see that the non-relativistic theory examines changes of the particle coordinates X^r, $r = 1, 2, 3$, with respect to $t \equiv X^0$. The fact that t is yet another coordinate of the particle (and not an "evolution parameter"), that the space is actually not three- but four-dimensional, and that t itself can change during evolution is obscured in the non relativistic theory.

The constrained theory within the unconstrained theory. Let us assume that the basic theory is the unconstrained theory[3]. A particle's mass is a constant of motion given in (1.23). Instead of (1.18), it is often more convenient to introduce an arbitrary fixed constant κ and use the action

$$I[X^\mu] = \frac{1}{2} \int d\tau \left(\frac{\dot{X}^\mu \dot{X}_\mu}{\Lambda} + \Lambda \kappa^2 \right), \tag{1.31}$$

which is equivalent to (1.18) since $\Lambda(\tau)$ can be written as a total derivative and thus the second term in (1.31) has no influence on the equations of motion for the variables X^μ. We see that (1.31) is analogous to the unconstrained action (1.9) except for the fact that Λ in (1.31) is not the Lagrange multiplier.

Let us now choose a constant $M = \kappa$ and write

$$\Lambda M = \sqrt{\dot{X}^\mu \dot{X}_\mu}. \tag{1.32}$$

Inserting the latter expression into (1.31) we have

$$I[X^\mu] = M \int d\tau \, (\dot{X}^\mu \dot{X}_\mu)^{1/2}. \tag{1.33}$$

This is just the constrained action (1.1), proportional to the length of particle's trajectory (worldline) in V_4. In other words, fixing or choosing a value

---

[3] Later we shall see that there are more fundamental theories which contain the unconstrained theory formulated in this section.

The spinless point particle 11

for the constant of motion M yields the theory which looks like the constrained theory. The fact that mass M is actually not a prescribed constant, but results dynamically as a constant of motion in an unconstrained theory is obscured within the formalism of the constrained theory. This can be shown also by considering the expression for the 4-momentum $p^\mu = \dot{X}^\mu/\Lambda$ (eq. (1.22)). Using (1.32) we find

$$p^\mu = \frac{M\dot{X}^\mu}{(\dot{X}^\nu \dot{X}_\nu)^{1/2}}, \qquad (1.34)$$

which is the expression for 4-momentum of the usual, constrained relativity. For a chosen M it appears as if p^μ are constrained to a mass shell:

$$p^\mu p_\mu = M^2 \qquad (1.35)$$

and are thus not independent. This is shown here to be just an illusion: within the unconstrained relativity all 4 components p^μ are independent, and M^2 can be arbitrary; which particular value of M^2 a given particle possesses depends on its initial 4-velocity $\dot{X}^\mu(0)$.

Now a question arises. If all X^μ are considered as independent dynamical variables which evolve in τ, what is then the meaning of the equation $x^0 = X^0(\tau)$? If it is not a gauge fixing equation as it is in the constrained relativity, what is it then? In order to understand this, one has to abandon certain deep rooted concepts learned when studying the constrained relativity and to consider spacetime V_4 as a higher-dimensional analog of the usual 3-space E_3. A particle moves in V_4 in the analogous way as it moves in E_3. In E_3 all its three coordinates are independent variables; in V_4 all its four coordinates are independent variables. Motion along x^0 is assumed here as a physical fact. Just think about the well known observation that today we experience a different value of x^0 from yesterday and from what we will tomorrow. The quantity x^0 is just a coordinate (a suitable number) given by our calendar and our clock. The value of x^0 was different yesterday, is different today, and will be different tomorrow. But wait a minute! What is 'yesterday', 'today' and 'tomorrow'? Does it mean that we need some additional parameter which is related to those concepts. Yes, indeed! The additional parameter is just τ which we use in the unconstrained theory. After thinking hard enough for a while about the fact that we experience yesterday, today and tomorrow, we become gradually accustomed to the idea of motion along x^0. Once we are prepared to accept this idea intuitively we are ready to put it into a more precise mathematical formulation of the theory, and in return we shall then get an even better intuitive understanding of 'motion along x^0'. In this book we will return several times to this idea and formulate it at subsequently higher levels of sophistication.

Point particle in an electromagnetic field. We have seen that a free particle dynamics in V_4 is given by the action (1.18) or by (1.31). There is yet another, often very suitable, form, namely *the phase space* or *the first order* action which is the unconstrained version of (1.11):

$$I[X^\mu, p_\mu] = \int d\tau \left(p_\mu \dot{X}^\mu - \frac{\Lambda}{2}(p_\mu p^\mu - \kappa^2) \right). \tag{1.36}$$

Variation of the latter action with respect to X^μ and p_μ gives:

$$\delta X^\mu : \qquad \dot{p}_\mu = 0, \tag{1.37}$$

$$\delta p_\mu : \qquad p_\mu = \frac{\dot{X}_\mu}{\Lambda}. \tag{1.38}$$

We may add a total derivative term to (1.36):

$$I[X^\mu, p_\mu] = \int d\tau \left(p_\mu \dot{X}^\mu - \frac{\Lambda}{2}(p_\mu p^\mu - \kappa^2) + \frac{d\phi}{d\tau} \right) \tag{1.39}$$

where $\phi = \phi(\tau, x)$ so that $d\phi/d\tau = \partial\phi/\partial\tau + \partial_\mu \phi \dot{X}^\mu$. The equations of motion derived from (1.39) remain the same, since

$$\delta X^\mu : \quad \frac{d}{d\tau}\frac{\partial L'}{\partial \dot{X}^\mu} - \frac{\partial L'}{\partial X^\mu} = \frac{d}{d\tau}(p_\mu + \partial_\mu \phi) - \partial_\mu \partial_\nu \phi \dot{X}^\nu - \frac{\partial}{\partial\tau}\partial_\mu \phi$$

$$= \dot{p}_\mu = 0, \tag{1.40}$$

$$\delta p_\mu : \quad p_\mu = \frac{\dot{X}_\mu}{\Lambda}. \tag{1.41}$$

But the canonical momentum is now different:

$$p'_\mu = \frac{\partial L}{\partial \dot{X}^\mu} = p_\mu + \partial_\mu \phi. \tag{1.42}$$

The action (1.36) is not covariant under such a transformation (i.e., under addition of the term $d\phi/d\tau$).

In order to obtain a covariant action we introduce compensating vector and scalar fields A_μ, V which transform according to

$$eA'_\mu = eA_\mu + \partial_\mu \phi, \tag{1.43}$$

$$eV' = eV + \frac{\partial \phi}{\partial \tau}, \tag{1.44}$$

and define

$$I[X^\mu, p_\mu] = \int d\tau \left(p_\mu \dot{X}^\mu - \frac{\Lambda}{2}(\pi_\mu \pi^\mu - \kappa^2) + eV \right), \tag{1.45}$$

The spinless point particle

where $\pi_\mu \equiv p_\mu - eA_\mu$ is *the kinetic momentum*. If we add the term $d\phi/d\tau = \partial\phi/\partial\tau + \partial_\mu\phi\,\dot{X}^\mu$ to the latter action we obtain

$$I[X^\mu, p_\mu] = \int d\tau \left(p'_\mu \dot{X}^\mu - \frac{\Lambda}{2}(\pi_\mu \pi^\mu - \kappa^2) + eV'\right), \qquad (1.46)$$

where $p'_\mu = p_\mu + \partial_\mu\phi$. The transformation (1.43) is called *gauge transformation*, A_μ is a gauge field, identified with the electromagnetic field potential, and e the electric charge. In addition we have also a scalar gauge field V.

In eq. (1.45) Λ is a fixed function and there is no constraint. Only variations of p_μ and X^μ are allowed to be performed. The resulting equations of motion in flat spacetime are

$$\dot{\pi}^\mu = eF^\mu{}_\nu \dot{X}^\nu + e\left(\partial_\mu V - \frac{\partial A_\mu}{\partial \tau}\right), \qquad (1.47)$$

$$\pi_\mu = \frac{\dot{X}_\mu}{\Lambda}, \qquad \pi^\mu \pi_\mu = \frac{\dot{X}^\mu \dot{X}_\mu}{\Lambda^2}, \qquad (1.48)$$

where $F_{\mu\nu} = \partial_\mu A_\nu - \partial_\nu A_\mu$ is the electromagnetic field tensor.

All components p_μ and π_μ are independent. Multiplying (1.47) by π_μ (and summing over μ) we find

$$\dot{\pi}^\mu \pi_\mu = \tfrac{1}{2}\frac{d}{d\tau}(\pi^\mu \pi_\mu) = e\left(\partial_\mu V - \frac{\partial A_\mu}{\partial \tau}\right)\frac{\dot{X}^\mu}{\Lambda}, \qquad (1.49)$$

where we have used $F_{\mu\nu}\dot{X}^\mu\dot{X}^\nu = 0$. In general, V and A_μ depend on τ. In a special case when they do not depend on τ, the right hand side of eq. (1.49) becomes $e\partial_\mu V \dot{X}^\mu = edV/d\tau$. Then eq. (1.49) implies

$$\frac{\pi^\mu \pi_\mu}{2} - eV = \text{constant}. \qquad (1.50)$$

In the last equation V depends on the spacetime point x^μ. In particular, it can be independent of x^μ. Then eqs.(1.47), (1.49) become

$$\dot{\pi}^\mu = eF^\mu{}_\nu \dot{X}^\nu, \qquad (1.51)$$

$$\dot{\pi}^\mu \pi_\mu = \tfrac{1}{2}\frac{d}{d\tau}(\pi^\mu \pi_\mu) = 0. \qquad (1.52)$$

The last result implies

$$\pi^\mu \pi_\mu = \text{constant} = M^2. \qquad (1.53)$$

We see that in such a particular case when $\partial_\mu V = 0$ and $\partial A_\mu/\partial \tau = 0$, the square of the kinetic momentum is a constant of motion. As before, let this this constant of motion be denoted M^2. It can be positive, zero or negative. We restrict ourselves in this section to the case of bradyons ($M^2 > 0$) and leave a discussion of tachyons ($M^2 < 0$) to Sec. 13.1. By using (1.53) and (1.48) we can rewrite the equation of motion (1.51) in the form

$$\frac{1}{\sqrt{\dot{X}^2}} \frac{\mathrm{d}}{\mathrm{d}\tau} \left(\frac{\dot{X}^\mu}{\sqrt{\dot{X}^2}} \right) = \frac{e}{M} F^\mu{}_\nu \frac{\dot{X}^\nu}{\sqrt{\dot{X}^2}}. \tag{1.54}$$

In eq. (1.54) we recognize the familiar Lorentz force law of motion for a point particle in an electromagnetic field. A worldline which is a solution to the usual equations of motion for a charged particle is also a solution to the unconstrained equations of motion (1.51).

From (1.54) and the expression for the kinetic momentum

$$\pi^\mu = \frac{M \dot{X}^\mu}{(\dot{X}^\alpha \dot{X}_\alpha)^{1/2}} \tag{1.55}$$

it is clear that, since mass M is a constant of motion, a particle cannot be accelerated by the electromagnetic field beyond the speed of light. The latter speed is a limiting speed, at which the particle's energy and momentum become infinite.

On the contrary, when $\partial_\mu V \neq 0$ and $\partial A_\mu/\partial \tau \neq 0$ the right hand side of eq. (1.49) is not zero and consequently $\pi^\mu \pi_\mu$ is not a constant of motion. Then each of the components of the kinetic momentum $\pi^\mu = \dot{X}^\mu/\Lambda$ could be independently accelerated to arbitrary value; there would be no speed limit. A particle's trajectory in V_4 could even turn backwards in x^0, as shown in Fig. 1.1. In other words, a particle would first overcome the speed of light, acquire infinite speed and finally start traveling "backwards" in the coordinate time x^0. However, such a scenario is classically not possible by the familiar electromagnetic force alone.

To sum up, we have formulated a classical unconstrained theory of a point particle in the presence of a gravitational and electromagnetic field. This theory encompasses the main requirements of the usual constrained relativity. It is covariant under general coordinate transformations, and locally under the Lorentz transformations. Mass normally remains constant during the motion and particles cannot be accelerated faster than the speed of light. However, the unconstrained theory goes beyond the usual theory of relativity. In principle, *the particle is not constrained to a mass shell*. It only appears to be constrained, since normally its mass does not change; this is true even if the particle moves in a gravitational and/or

The spinless point particle 15

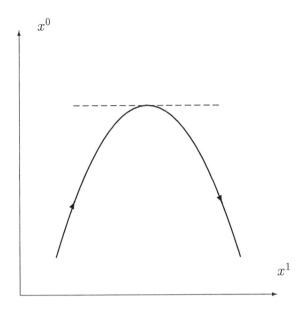

Figure 1.1. A possible trajectory of a particle accelerated by an exotic 4-force f^μ which does not satisfy $f^\mu \dot{X}_\mu = 0$.

electromagnetic field. Release of the mass shell constraint has far reaching consequences for the quantization.

Before going to the quantized theory let us briefly mention that the Hamiltonian belonging to the unconstrained action (1.31)

$$H = p_\mu \dot{X}^\mu - L = \frac{\Lambda}{2}(p^\mu p_\mu - \kappa^2) \qquad (1.56)$$

is different from zero and it is the generator of the genuine τ-evolution. As in the non-relativistic mechanics one can straightforwardly derive the Hamilton equations of motion, Poisson brackets, etc.. We shall not proceed here with such a development.

1.3. FIRST QUANTIZATION

Quantization of the unconstrained relativistic particle is straightforward. Coordinates X^μ and momenta p_μ become operators satisfying

$$[x^\mu, p_\nu] = i\delta^\mu{}_\nu , \quad [x^\mu, x^\nu] = 0 , \quad [p_\mu, p_\nu] = 0, \qquad (1.57)$$

which replace the corresponding Poisson brackets of the classical unconstrained theory.

FLAT SPACETIME

Let us first consider the situation in flat spacetime. The operators x^μ, p_μ act on the state vectors which are vectors in Hilbert space. For the basis vectors we can choose $|x'\rangle$ which are eigenvectors of the operators x^μ:

$$x^\mu |x'\rangle = x'^\mu |x'\rangle. \qquad (1.58)$$

They are normalized according to

$$\langle x'|x''\rangle = \delta(x' - x'), \qquad (1.59)$$

where

$$\delta(x' - x'') \equiv \delta^4(x' - x'') \equiv \prod_\mu \delta(x'^\mu - x''^\mu).$$

The eigenvalues of x^μ are spacetime coordinates. A generic vector $|\psi\rangle$ can be expanded in terms of $|x\rangle$:

$$|\psi\rangle = \int |x\rangle \mathrm{d}\langle x|\psi\rangle, \qquad (1.60)$$

where $\langle x|\psi\rangle \equiv \psi(x)$ is *the wave function*.

The matrix elements of the operators x^μ in the coordinate representation are

$$\langle x'|x^\mu|x''\rangle = x'^\mu \delta(x' - x'') = x''^\mu \delta(x' - x''). \qquad (1.61)$$

The matrix elements $\langle x'|p_\mu|x''\rangle$ of the momentum operator p_μ can be calculated from the commutation relations (1.57):

$$\langle x'|[x^\mu, p_\nu]|x''\rangle = (x'^\mu - x''^\mu)\langle x'|p_\nu|x''\rangle = i\delta^\mu{}_\nu \delta(x' - x''). \qquad (1.62)$$

Using

$$(x'^\mu - x''^\mu)\partial_\alpha \delta(x' - x'') = -\delta^\mu{}_\alpha \delta(x' - x'')$$

(a 4-dimensional analog of $f(x)\mathrm{d}\delta(x)/\mathrm{d}x = -(\mathrm{d}f(x)/\mathrm{d}x)\delta(x)$) we have

$$\langle x'|p_\mu|x''\rangle = -i\partial'_\mu \delta(x' - x''). \qquad (1.63)$$

Alternatively,

$$\begin{aligned}
\langle x'|p_\mu|x''\rangle &= \int \langle x'|p'\rangle \mathrm{d}p' \langle p'|p|p''\rangle \mathrm{d}p'' \langle p''|x''\rangle \\
&= \int \left(\frac{1}{2\pi}\right)^{D/2} e^{ip'_\mu x'^\mu} \mathrm{d}p' \delta(p'-p'') p'' \mathrm{d}p'' \left(\frac{1}{2\pi}\right)^{D/2} e^{-ip''_\mu x''^\mu} \\
&= \left(\frac{1}{2\pi}\right)^D \mathrm{d}p' \mathrm{d}p' \, e^{ip'_\mu (x'^\mu - x''^\mu)} \\
&= -i \frac{\partial}{\partial x'} \delta(x'-x'')
\end{aligned} \quad (1.64)$$

This enables us to calculate

$$\begin{aligned}
\langle x'|p_\mu|\psi\rangle &= \int \langle x'|p_\mu|x''\rangle \mathrm{d}x'' \langle x''|\psi\rangle = \int -i\partial'_\mu \delta(x'-x'')\mathrm{d}x'' \, \psi(x'') \\
&= \int i\partial''_\mu \delta(x'-x'')\mathrm{d}x'' \psi(x'') \\
&= -\int i\, \delta(x'-x'')\mathrm{d}x'' \, \partial''_\mu \psi(x'') \\
&= -i\partial'_\mu \psi(x')
\end{aligned} \quad (1.65)$$

where $\mathrm{d}x'' \equiv \mathrm{d}^4 x'' \equiv \prod_\mu \mathrm{d}x''^\mu$. The last equation can be rewritten as

$$p_\mu \psi(x) = -i\,\partial_\mu \psi(x) \quad (1.66)$$

which implies that p_μ acts as the differential operator $-i\partial_\mu$ on the wave function.

We work in the Schrödinger picture in which operators are independent of the evolution parameter τ, while the actual evolution is described by a τ-dependent state vector $|\psi\rangle$. A state can be represented by the wave function $\psi(\tau, x^\mu)$ which depends on four spacetime coordinates and the evolution parameter τ.

The quantity $\psi^*\psi$ is *the probability density* in spacetime. The probability of finding a particle within a spacetime region Ω is $\int \mathrm{d}^4 x\, \psi^*(\tau,x)\psi(\tau,x)$ where the integration is performed in spacetime (and not in space).

A wave function is assumed to satisfy the evolution equation

$$i\frac{\partial \psi}{\partial \tau} = H\psi, \quad H = \frac{\Lambda}{2}(-i)^2(\partial_\mu \partial^\mu - \kappa^2), \quad (1.67)$$

which is just the *Schrödinger equation* in spacetime.

For a constant Λ a general solution is a superposition

$$\psi(\tau, x) = \int \mathrm{d}^4 p\, c(p)\exp\left[ip_\mu x^\mu - i\frac{\Lambda}{2}(p^2 - \kappa^2)\tau\right], \quad (1.68)$$

where p_μ are now the eigenvalues of the momentum operator. In general (1.68) represents a particle with *indefinite p^2*, that is with *indefinite mass*. A wave packet (1.68) is localized in spacetime (Fig. 1.2a) and it moves in spacetime as the parameter τ increases (see Fig. 1.2b).

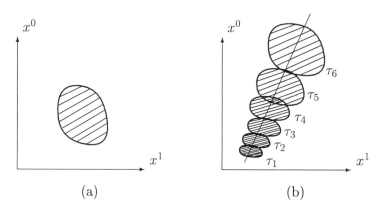

Figure 1.2. (a) Illustration of a wave packet which is localized in spacetime. (b) As the parameter τ increases, position of the wave packet in spacetime is moving so that its centre describes a world line.

The continuity equation which is admitted by the Schrödinger equation (1.67) is

$$\frac{\partial \rho}{\partial \tau} + \partial_\mu J^\mu = 0, \tag{1.69}$$

$$\rho \equiv \psi^*\psi \; ; \qquad J^\mu \equiv -\frac{i}{2}(\psi^*\partial_\mu \psi - \psi \partial_\mu \psi^*). \tag{1.70}$$

The probability density ρ is positive by definition, whilst components of the probability current J^μ (including the time-like component J^0) can be positive or negative. On the contrary, during the development of the constrained theory (without the parameter τ) J^0 had been expected to have the meaning of the probability density in 3-space. The occurrence of negative J^0 had been a nuisance; it was concluded that J^0 could not be the probability density but a charge density, and J^μ a charge current. In *the unconstrained, or parametrized theory*, the probability density is $\rho \equiv \psi^*\psi$ and there is no problem with J^μ being the probability current. A wave function which has positive J^0 evolves into the future (as τ increases, the time coordinate $t \equiv x^0$ of the center of the wave packet also increases), whilst a wave function with negative J^0 evolves into the past (as τ increases, t decreases). The sign of J^0 thus distinguishes a particle from its

The spinless point particle

antiparticle, as in the usual theory. Since particles and antiparticles have different charges —the electric or any other charge— it is obvious that J^0 is proportional to a charge, and J^μ to a charge current.

We normalized the wave function in *spacetime* so that for arbitrary τ we have
$$\int d^4x\, \psi^*\psi = 1. \tag{1.71}$$

Such a normalization is consistent with the continuity equation (1.70) and consequently the evolution operator $U \equiv \exp[iH\tau]$ which brings $\psi(\tau) \to \psi(\tau') = U\,\psi(\tau)$ is *unitary*. The generator H of the unitary transformation $\tau \to \tau + \delta\tau$ is the Hamiltonian (1.56).

Instead of a function $\psi(\tau, x)$, we can consider its Fourier transform[4]
$$\phi(\mu, x) = \int_{-\infty}^{\infty} d\tau\, e^{i\mu^2 \Lambda \tau}\, \psi(\tau, x) = \int d^4p\, c(p)\, e^{ip_\mu x^\mu}\, \delta(p^2 - \kappa^2 - \mu^2) \tag{1.72}$$

which, for a chosen value of μ, has *definite mass* $p^2 = M^2 = \kappa^2 + \mu^2$. A general function $\psi(\mu, x)$ solves the Klein-Gordon equation with mass square $\kappa^2 + \mu^2$:
$$\partial_\mu \partial^\mu \phi + (\kappa^2 + \mu^2)\phi = 0. \tag{1.73}$$

Here κ^2 is arbitrary fixed constant, while μ is a variable in the Fourier transformed function $\phi(\mu, x)$. Keeping κ^2 in our equations is only a matter of notation. We could have set $\kappa^2 = 0$ in the action (1.31) and the Hamiltonian (1.56). Then the Fourier transform (1.72) would simply tell us that mass is given by $p^2 = \mu^2$.

If the limits of integration over τ in (1.72) are from $-\infty$ to ∞, we obtain a *wave function with definite mass*. But is we take the limits from zero to infinity, then instead of (1.72) we obtain
$$\Delta(\mu, x) = \int_0^{\infty} d\tau\, e^{i\mu^2 \Lambda \tau}\, \psi(\tau, x) = i\int d^4p\, c(p)\, \frac{\exp[ip_\mu x^\mu]}{p^2 - \kappa^2 - \mu^2}. \tag{1.74}$$

For the choice $c(p) = 1$ the above equation becomes the well known *Feynman propagator*.

Quantization can be performed also in terms of the Feynman path integral. By using (1.31) with fixed Λ we can calculate the transition amplitude or propagator
$$K(\tau, x; \tau', x') = \langle x, \tau | x', \tau' \rangle = \int \mathcal{D}X(\tau)\, e^{iI[X(\tau)]}$$

---
[4] We assume here that the coefficient $c(p)$ is suitably redefined so that it absorbs the constant resulting from the integration over τ.

$$= \left(\frac{1}{2\pi i \Lambda (\tau' - \tau)}\right)^{D/2} \exp\left[\frac{i}{2\Lambda} \frac{(x'-x)^2}{\tau' - \tau}\right], \tag{1.75}$$

where D is the dimension of spacetime ($D = 4$ in our case). In performing the above functional integration there is no need for ghosts. The latter are necessary in a constrained theory in order to cancel out the unphysical states related to reparametrizations. There are no unphysical states of such a kind in the unconstrained theory.

In eq. (1.75) $K(\tau, x; \tau', x')$ is the probability amplitude to find the particle in a spacetime point x^μ at a value of the evolution parameter τ when it was previously found in the point x'^μ at τ'. The propagator has also the role of the Green's function satisfying the equation

$$\left(i\frac{\partial}{\partial \tau} - H\right) K(\tau, x; \tau', x') = \delta(\tau - \tau')\delta^D(x - x'). \tag{1.76}$$

Instead of $K(\tau, x; \tau', x')$ we can use its Fourier transform

$$\Delta(\mu, x - x') = \int d\tau\, e^{i\mu^2 \Lambda \tau} K(\tau, x; \tau', x'), \tag{1.77}$$

which is just (1.74) with $c(p) = 1$, x^μ and τ being replaced by $x^\mu - x'^\mu$ and $\tau - \tau'$, respectively.

CURVED SPACETIME

In order to take into account the gravitational field one has to consider curved spacetime. Dynamical theory in curved spaces has been developed by DeWitt [33]. Dimension and signature of the space has not been specified; it was only assumed that the metric is symmetric and nonsingular. Therefore DeWitt's theory holds also for a 4-dimensional spacetime, and its flat space limit is identical to the unconstrained theory discussed in Secs. 1.2 and 1.4.

Coordinates x^μ and momenta p_μ are Hermitian operators satisfying the commutation relations which are the same as those in flat spacetime (see eq. (1.57)). Eigenvectors $|x'\rangle$ satisfy

$$x^\mu |x'\rangle = x'^\mu |x'\rangle \tag{1.78}$$

and they form a complete eigenbasis. A state vector $|\psi\rangle$ can be expanded in terms of $|x'\rangle$ as follows:

$$|\psi\rangle = \int |x'\rangle \sqrt{|g(x')|}\, dx'\, \langle x'|\psi\rangle. \tag{1.79}$$

The quantity $\langle x'|\psi\rangle \equiv \psi(\tau, x')$ is the wave function and $|\psi|^2 \sqrt{|g|}\, dx'$, $dx' \equiv d^n x$, is the probability that the coordinates of the system will be found in

The spinless point particle

the volume $\sqrt{|g|}\mathrm{d}x'$ in the neighborhood of the point x' at the evolution time τ. The probability to find the particle anywhere in spacetime is 1, hence[5]

$$\int \sqrt{|g|}\mathrm{d}x\, \psi^*(\tau,x)\psi(\tau,x) = 1. \quad (1.80)$$

Multiplying (1.79) by $\langle x''|$ (that is projecting the state $|\psi\rangle$ into an eigenstate $\langle x''|$) we have

$$\langle x''|\psi\rangle = \psi(\tau, x'') = \int \langle x''|x'\rangle \sqrt{|g(x')|}\mathrm{d}x\, \langle x'|\psi\rangle. \quad (1.81)$$

This requires the following normalization condition for the eigenvectors

$$\langle x'|x''\rangle = \frac{\delta(x'-x'')}{\sqrt{|g(x')|}} = \frac{\delta(x'-x'')}{\sqrt{|g(x'')|}} \equiv \delta(x',x''). \quad (1.82)$$

From (1.82) we derive

$$\partial'_\mu \delta(x',x'') = -\partial''_\mu \delta(x',x'') - \frac{1}{\sqrt{|g(x'')|}} \partial''_\mu \sqrt{|g(x'')|}\, \delta(x',x''), \quad (1.83)$$

$$(x'^\mu - x''^\mu)\partial'_\alpha \delta(x',x'') = -\delta^\mu{}_\alpha \delta(x',x''). \quad (1.84)$$

We can now calculate the matrix elements $\langle x'|p_\mu|x''\rangle$ according to

$$\langle x'|[x^\mu, p_\nu]|x''\rangle = (x'^\mu - x''^\mu)\langle x'|p_\nu|x''\rangle = i\delta^\mu{}_\nu \delta(x',x''). \quad (1.85)$$

Using the identities (1.83), (1.84) we find

$$\langle x'|p_\mu|x''\rangle = (-i\partial'_\mu + F_\mu(x')),\delta(x',x'') \quad (1.86)$$

where $F_\mu(x')$ is an arbitrary function. If we take into account the commutation relations $[p_\mu, p_\nu] = 0$ and the Hermitian condition

$$\langle x'|p_\mu|x''\rangle = \langle x''|p_\mu|x'\rangle^* \quad (1.87)$$

we find that F_μ is not entirely arbitrary, but can be of the form

$$F_\mu(x) = -i|g|^{-1/4}\partial_\mu |g|^{1/4}. \quad (1.88)$$

Therefore

$$p_\mu \psi(x') \equiv \langle x'|p_\mu|\psi\rangle = \int \langle x'|p_\mu|x''\rangle \sqrt{|g(x'')|}\, \mathrm{d}x'' \langle x''|\psi\rangle$$

$$= -i(\partial_\mu + |g|^{-1/4}\partial_\mu |g|^{1/4})\psi(x') \quad (1.89)$$

---

[5] We shall often omit the prime when it is clear that a symbol without a prime denotes the eigenvalues of the corresponding operator.

or
$$p_\mu = -i(\partial_\mu + |g|^{-1/4}\partial_\mu |g|^{1/4}) \tag{1.90}$$

$$p_\mu \psi = -i|g|^{-1/4}\partial_\mu(|g|^{1/4}\psi). \tag{1.91}$$

The expectation value of the momentum operator is

$$\langle p_\mu \rangle = \int \psi^*(x')\sqrt{|g(x')|}\mathrm{d}x'\, \langle x'|p_\mu|x''\rangle \sqrt{|g(x'')|}\,\mathrm{d}x''\, \psi(x'')$$

$$= -i \int \sqrt{|g|}\,\mathrm{d}x\, \psi^*(\partial_\mu + |g|^{-1/4}\partial_\mu |g|^{1/4})\psi. \tag{1.92}$$

Because of the Hermitian condition (1.87) the expectation value of p_μ is real. This can be also directly verified from (1.92):

$$\langle p_\mu \rangle^* = i\int \sqrt{|g|}\,\mathrm{d}x\, \psi(\partial_\mu + |g|^{-1/4}\partial_\mu |g|^{1/4})\psi^*$$

$$= -i\int \sqrt{|g|}\,\mathrm{d}x\, \psi^*(\partial_\mu + |g|^{-1/4}\partial_\mu |g|^{1/4})\psi$$

$$+ i\int \mathrm{d}x\, \partial_\mu(\sqrt{|g|}\psi^*\psi). \tag{1.93}$$

The surface term in (1.93) can be omitted and we find $\langle p_\mu \rangle^* = \langle p_\mu \rangle$.

Point transformations   In classical mechanics we can transform the generalized coordinates according to

$$x'^\mu = x'^\mu(x). \tag{1.94}$$

The conjugate momenta then transform as covariant components of a vector:

$$p'_\mu = \frac{\partial x^\nu}{\partial x'^\mu} p_\nu. \tag{1.95}$$

The transformations (1.94), (1.95) preserve the canonical nature of x^μ and p_μ, and define what is called *a point transformation*.

According to DeWitt, point transformations may also be defined in quantum mechanics in an unambiguous manner. The quantum analog of eq. (1.94) retains the same form. But in eq. (1.95) the right hand side has to be symmetrized so as to make it Hermitian:

$$p'_\mu = \tfrac{1}{2}\left(\frac{\partial x^\nu}{\partial x'^\mu}p_\nu + p_\nu \frac{\partial x^\nu}{\partial x'^\mu}\right). \tag{1.96}$$

Using the definition (1.90) we obtain by explicit calculation

$$p'_\mu = -i\left(\partial'_\mu + |g'|^{-1/4}\partial'_\mu|g'|^{1/4}\right). \tag{1.97}$$

The spinless point particle

Expression (1.90) is therefore covariant under point transformations.

Quantum dynamical theory is based on the following postulate:

The temporal behavior of the operators representing the observables of a physical system is determined by the unfolding-in-time of a unitary transformation.

'Time' in the above postulate stands for the evolution time (the evolution parameter τ). As in flat spacetime, the unitary transformation is

$$U = e^{iH\tau} \tag{1.98}$$

where the generator of evolution is the Hamiltonian

$$H = \frac{\Lambda}{2}(p^\mu p_\mu - \kappa^2). \tag{1.99}$$

We shall consider the case when Λ and the metric $g_{\mu\nu}$ are independent of τ.

Instead of using the Heisenberg picture in which operators evolve, we can use the Schrödinger picture in which the states evolve:

$$|\psi(\tau)\rangle = e^{-iH\tau}|\psi(0)\rangle. \tag{1.100}$$

From (1.98) and (1.100) we have

$$i\frac{\partial|\psi\rangle}{\partial\tau} = H|\psi\rangle, \quad i\frac{\partial\psi}{\partial\tau} = H\psi, \tag{1.101}$$

which is *the Schrödinger equation*. It governs the state or the wave function defined over the entire spacetime.

In the expression (1.99) for the Hamiltonian there is an ordering ambiguity in the definition of $p^\mu p_\mu = g_{\mu\nu}p^\mu p^\nu$. Since p_μ is a differential operator, it matters at which place one puts $g_{\mu\nu}(x)$; the expression $g_{\mu\nu}p^\mu p^\nu$ is not the same as $p^\mu g_{\mu\nu} p^\nu$ or $p^\mu p^\nu g_{\mu\nu}$.

Let us use the identity

$$|g|^{1/4} p_\mu |g|^{-1/4} = -i\,\partial_\mu \tag{1.102}$$

which follows immediately from the definition (1.90), and define

$$\begin{aligned}
p^\mu p_\nu \psi &= |g|^{-1/2}\left(|g|^{1/4}p_\mu|g|^{-1/4}\right)|g|^{1/2}g^{\mu\nu}\left(|g|^{1/4}p_\nu|g|^{-1/4}\right)\psi \\
&= -|g|^{-1/2}\partial_\mu(|g|^{1/2}g^{\mu\nu}\,\partial_\nu\psi) \\
&= -\mathrm{D}_\mu \mathrm{D}^\mu \psi,
\end{aligned} \tag{1.103}$$

where D_μ is covariant derivative with respect to the metric $g_{\mu\nu}$. Expression (1.103) is nothing but an ordering prescription: if one could neglect that p_μ and $g_{\mu\nu}$ (or $|g|$) do not commute, then the factors $|g|^{-1/2}$, $|g|^{1/4}$, etc., in eq. (1.103) altogether would give 1.

One possible definition of the square of the momentum operator is thus

$$p^2\psi = |g|^{-1/4} p_\mu g^{\mu\nu} |g|^{1/2} p_\nu |g|^{-1/4} \psi = -D_\mu D^\mu \psi. \tag{1.104}$$

Another well known definition is

$$p^2\psi = \tfrac{1}{4}(p_\mu p_\nu g^{\mu\nu} + 2 p_\mu g^{\mu\nu} p_\nu + g^{\mu\nu} p_\mu p_\nu)\psi$$

$$= (-D_\mu D^\mu + \tfrac{1}{4} R + g^{\mu\nu} \Gamma^\alpha_{\beta\mu} \Gamma^\beta_{\alpha\nu})\psi. \tag{1.105}$$

Definitions (1.104), (1.105) can be combined according to

$$p^2\psi = \tfrac{1}{6}(2|g|^{-1/4} p_\mu g^{\mu\nu} |g|^{1/2} p_\nu |g|^{-1/4} + p_\mu p_\nu g^{\mu\nu} + 2 p_\mu g^{\mu\nu} p_\nu + g^{\mu\nu} p_\mu p_\nu)\psi$$

$$= (-D_\mu D^\mu + \tfrac{1}{6} R + \tfrac{2}{3} g^{\mu\nu} \Gamma^\alpha_{\beta\mu} \Gamma^\beta_{\alpha\nu})\psi. \tag{1.106}$$

One can verify that the above definitions all give the Hermitian operators p^2. Other, presumably infinitely many, Hermitian combinations are possible. Because of such an ordering ambiguity the quantum Hamiltonian

$$H = \frac{\Lambda}{2} p^2 \tag{1.107}$$

is undetermined up to the terms like $(\hbar^2 \kappa/2)(R + 4 g^{\mu\nu} \Gamma^\alpha_{\beta\mu} \Gamma^\beta_{\alpha\nu})$, (with $\kappa = 0$, $\tfrac{1}{4}$, $\tfrac{1}{6}$, etc., which disappear in the classical approximation where $\hbar \to 0$).

The Schrödinger equation (1.101) admits the following continuity equation

$$\frac{\partial \rho}{\partial \tau} + D_\mu j^\mu = 0 \tag{1.108}$$

where $\rho = \psi^* \psi$ and

$$j^\mu = \frac{\Lambda}{2}[\psi^* p^\mu \psi + (\psi^* p^\mu \psi)^*] = -i\frac{\Lambda}{2}(\psi^* \partial^\mu \psi - \psi \partial^\mu \psi^*). \tag{1.109}$$

The stationary Schrödinger equation.   If we take the ansatz

$$\psi = e^{-iE\tau} \phi(x), \tag{1.110}$$

where E is a constant, then we obtain the stationary Schrödinger equation

$$\frac{\Lambda}{2}(-D_\mu D^\mu - \kappa^2)\phi = E\phi, \tag{1.111}$$

The spinless point particle

which has the form of the Klein–Gordon equation in curved spacetime with squared mass $M^2 = \kappa^2 + 2E/\Lambda$.

Let us derive *the classical limit* of eqs. (1.91) and (1.101). For this purpose we rewrite those equations by including the Planck constant $\hbar$[6]:

$$\hat{p}_\mu \psi = -i\hbar |g|^{-1/4} \partial_\mu (|g|^{1/4} \psi), \tag{1.112}$$

$$\hat{H}\psi = i\hbar \frac{\partial \psi}{\partial \tau}. \tag{1.113}$$

For the wave function we take the expression

$$\psi = A(\tau, x) \exp\left[\frac{i}{\hbar} S(\tau, x)\right] \tag{1.114}$$

with real A and S.

Assuming (1.114) and taking the limit $\hbar \to 0$, eq. (1.112) becomes

$$\hat{p}_\mu \psi = \partial_\mu S \, \psi. \tag{1.115}$$

Inserting (1.114) into eq. (1.113), taking the limit $\hbar \to 0$ and writing separately the real and imaginary part of the equation, we obtain

$$-\frac{\partial S}{\partial \tau} = \frac{\Lambda}{2}(\partial_\mu S \partial^\mu S - \kappa^2), \tag{1.116}$$

$$\frac{\partial A^2}{\partial \tau} + D_\mu(A^2 \partial^\mu S) = 0. \tag{1.117}$$

Equation (1.116) is just the Hamilton–Jacobi equation of the classical point particle theory in curved spacetime, where $E = -\partial S/\partial \tau$ is "energy" (the spacetime analog of the non-relativistic energy) and $p_\mu = \partial_\mu S$ the classical momentum.

Equation (1.117) is the continuity equation, where $\psi^* \psi = A^2$ is *the probability density* and

$$A^2 \partial^\mu S = j^\mu \tag{1.118}$$

is *the probability current*.

---

[6] Occasionally we use the hat sign over the symbol in order to distinguish operators from their eigenvalues.

The expectation value. If we calculate the expectation value of the momentum operator p_μ we face a problem, since the quantity $\langle p_\mu \rangle$ obtained from (1.92) is not, in general, a vector in spacetime. Namely, if we insert the expression (1.114) into eq. (1.92) we obtain

$$\int \sqrt{|g|}\,\mathrm{d}x\, \psi^* p_\mu \psi = \int \sqrt{|g|}\,\mathrm{d}x\, A^2 \partial_\mu S - i\hbar \int \sqrt{|g|}\,\mathrm{d}x\, (A\partial_\mu A$$

$$+ A^2 |g|^{-1/4} \partial_\mu |g|^{1/4}$$

$$= \int \sqrt{|g|}\,\mathrm{d}x\, A^2 \partial_\mu S - \frac{i\hbar}{2} \int \mathrm{d}x\, \partial_\mu (\sqrt{|g|} A^2)$$

$$= \int \sqrt{|g|}\,\mathrm{d}x\, A^2 \partial_\mu S, \qquad (1.119)$$

where in the last step we have omitted the surface term.

If, in particular, we choose a wave packet localized around a classical trajectory $x^\mu = X_c^\mu(\tau)$, then for a certain period of τ (until the wave packet spreads too much), the amplitude is approximately

$$A^2 = \frac{\delta(x - X_c)}{\sqrt{|g|}}. \qquad (1.120)$$

Inserting (1.120) into (1.119) we have

$$\langle p_\mu \rangle = \partial_\mu S |_{X_c} = p_\mu(\tau) \qquad (1.121)$$

That is, the expectation value is equal to the classical momentum 4-vector of the center of the wave packet.

But in general there is a problem. In the expression

$$\langle p_\mu \rangle = \int \sqrt{|g|}\,\mathrm{d}x\, A^2 \partial_\mu S \qquad (1.122)$$

we integrate a vector field over spacetime. Since one cannot covariantly sum vectors at different points of spacetime, $\langle p_\mu \rangle$ is not a geometric object: it does not transform as a 4-vector. The result of integration depends on which coordinate system (parametrization) one chooses. It is true that the expression (1.119) is covariant under the point transformations (1.94), (1.96), but the expectation value is not a classical geometric object: it is neither a vector nor a scalar.

One way to resolve the problem is in considering $\langle p_\mu \rangle$ as a generalized geometric object which is a functional of parametrization. If the parametrization changes, then also $\langle p_\mu \rangle$ changes in a well defined manner. For the

The spinless point particle 27

ansatz (1.114) we have

$$\langle p'_\mu \rangle = \int \sqrt{|g'|}\, dx'\, A'^2(x')\, \partial'_\mu S = \int \sqrt{|g|}\, dx\, A^2(x) \frac{\partial x^\nu}{\partial x'^\mu} \partial_\nu S. \qquad (1.123)$$

The expectation value is thus a parametrization dependent generalized geometric object. The usual classical geometric object p_μ is also parametrization dependent, since it transforms according to (1.95).

Product of operators.. Let us calculate the product of two operators. We have

$$\begin{aligned}
\langle x|p_\mu p_\nu|\psi\rangle &= \int \langle x|p_\mu|x'\rangle \sqrt{|g(x')|}\, dx'\, \langle x'|p_\nu|\psi\rangle \\
&= \int (-i)(\partial_\mu + \tfrac{1}{2}\Gamma^\alpha_{\mu\alpha}) \frac{\delta(x-x')}{\sqrt{|g(x')|}} (-i)\partial'_\nu \psi \sqrt{|g(x')|}\, dx' \\
&= -\partial_\mu \partial_\nu \psi - \tfrac{1}{2}\Gamma^\alpha_{\mu\alpha} \partial_\nu \psi,
\end{aligned} \qquad (1.124)$$

where we have used the relation

$$\frac{1}{|g|^{1/4}} \partial_\mu |g|^{1/4} = \frac{1}{2\sqrt{|g|}} \partial_\mu \sqrt{|g|} = \tfrac{1}{2}\Gamma^\alpha_{\mu\alpha}. \qquad (1.125)$$

Expression (1.124) is covariant with respect to point transformations, but it is not a geometric object in the usual sense. If we contract (1.124) by $g^{\mu\nu}$ we obtain $g^{\mu\nu}(-\partial_\mu\partial_\nu\psi - \tfrac{1}{2}\Gamma^\alpha_{\mu\alpha}\partial_\nu\psi)$ which is not a scalar under reparametrizations (1.94). Moreover, as already mentioned, there is an ordering ambiguity where to insert $g^{\mu\nu}$.

FORMULATION IN TERMS OF A LOCAL LORENTZ FRAME

Components p_μ are projections of the vector p into basic vectors γ_μ of a coordinate frame. A vector can be expanded as $p = p_\nu \gamma^\nu$ and the components are given by $p_\mu = p \cdot \gamma_\mu$ (where the dot means the scalar product[7]). Instead of a *coordinate frame* in which

$$\gamma_\mu \cdot \gamma_\nu = g_{\mu\nu} \qquad (1.126)$$

we can use a local Lorentz frame $\gamma_a(x)$ in which

$$\gamma_a \cdot \gamma_b = \eta_{ab} \qquad (1.127)$$

where η_{ab} is the Minkowski tensor. The scalar product

$$\gamma^\mu \cdot \gamma_a = e_a{}^\mu \qquad (1.128)$$

---
[7] For a more complete treatment see the section on Clifford algebra.

is *the tetrad*, or *vierbein*, *field*.

With the aid of the vierbein we may define the operators

$$p_a = \tfrac{1}{2}(e_a{}^\mu p_\mu + p_\mu e_a{}^\mu) \tag{1.129}$$

with matrix elements

$$\langle x|p_a|x'\rangle = -i\left(e_a{}^\mu \partial_\mu + \frac{1}{2\sqrt{|g|}}\partial_\mu(\sqrt{|g|}\,e_a{}^\mu)\right)\delta(x,x') \tag{1.130}$$

and

$$\langle x|p_a|\psi\rangle = -i\left(e_a{}^\mu \partial_\mu + \frac{1}{2\sqrt{|g|}}\partial_\mu(\sqrt{|g|}\,e_a{}^\mu)\right).\psi \tag{1.131}$$

In the coordinate representation we thus have

$$p_a = -i(e_a{}^\mu \partial_\mu + \Gamma_a), \tag{1.132}$$

where

$$\Gamma_a \equiv \frac{1}{2\sqrt{|g|}}\partial_\mu(\sqrt{|g|}\,e_a{}^\mu). \tag{1.133}$$

One can easily verify that the operator p_a is invariant under the point transformations (1.95).

The expectation value

$$\langle p_a \rangle = \int \mathrm{d}x \sqrt{|g|}\,\psi^* p_a \psi \tag{1.134}$$

has some desirable properties. First of all, it is *real*, $\langle p_a\rangle^* = \langle p_a\rangle$, which reflects that p_a is a Hermitian operator. Secondly, it is invariant under coordinate transformations (i.e., it transforms as a scalar).

On the other hand, since p_a depends on a chosen local Lorentz frame, the integral $\langle p_a \rangle$ is a functional of the frame field $e_a{}^\mu(x)$. In other words, the expectation value is an object which is defined with respect to a chosen local Lorentz frame field.

In *flat spacetime* we can choose a constant frame field γ_a, and the components $e_a{}^\mu(x)$ are then the transformation coefficients into a curved coordinate system. Eq. (1.134) then means that after having started from the momentum p_μ in arbitrary coordinates, we have have calculated its expectation value in a Lorentz frame. Instead of a constant frame field γ_a, we can choose an x-dependent frame field $\gamma_a(x)$; the expectation value of momentum will then be different from the one calculated with respect to the constant frame field. In curved spacetime there is no constant frame field. First one has to choose a frame field $\gamma_a(x)$, then define components $p_a(x)$ in this frame field, and finally calculate the expectation value $\langle p_a \rangle$.

The spinless point particle

Let us now again consider the ansatz (1.114) for the wave function. For the latter ansatz we have

$$\langle p_a \rangle = \int dx \sqrt{|g|}\, A^2\, e_a{}^\mu \partial_\mu S, \tag{1.135}$$

where the term with total divergence has been omitted. If the wave packet is localized around a classical world line $X_c(\tau)$, so that for a certain τ-period $A^2 = \delta(x - X_c)/\sqrt{|g|}$ is a sufficiently good approximation, eq. (1.135) gives

$$\langle \hat{p}_a \rangle = e_a{}^\mu \partial_\mu S|_{X_c(\tau)} = p_a(\tau), \tag{1.136}$$

where $p_a(\tau)$ is the classical momentum along the world line X_c^μ in a local Lorentz frame.

Now we may contract (1.136) by $e^a{}_\mu$ and we obtain $p_\mu(\tau) \equiv e^a{}_\mu p_a(\tau)$, which are components of momentum along X_c^μ in a coordinate frame.

Product of operators. The product of operators in a local Lorentz frame is given by

$$\begin{aligned}\langle x|p_a p_b|\psi\rangle &= \int \langle x|p_a|x'\rangle \sqrt{|g(x')|}\, dx'\, \langle x'|p_b|\psi\rangle \\ &= \int (-i)\,(e_a{}^\mu(x)\partial_\mu + \Gamma_a(x))\, \frac{\delta(x-x')}{\sqrt{|g(x')|}}\, \sqrt{|g(x')|}\, dx' \\ &\quad \times (-i)\,(e_b{}^\nu(x')\partial'_\nu + \Gamma_a(x'))\,\psi(x') \\ &= -(e_a{}^\mu \partial_\mu + \Gamma_a)(e_b{}^\nu \partial_\nu + \Gamma_b)\psi = p_a p_b \psi. \end{aligned} \tag{1.137}$$

This can then be mapped into the coordinate frame:

$$e^a{}_\alpha e^b{}_\beta p_a p_b \psi = -e^a{}_\alpha e^b{}_\beta (e^\mu{}_a \partial_\mu + \Gamma_a)(e^\nu{}_b \partial_\nu + \gamma_b)\psi. \tag{1.138}$$

At this point let us use

$$\partial_\mu e^a{}_\mu - \Gamma^\lambda_{\mu\nu} e^a{}_\lambda + \omega_{ca}{}^\mu e^c{}_\mu = 0, \tag{1.139}$$

from which we have

$$\omega^{ab}{}_\mu = e^{a\nu} e^b{}_{\nu;\mu} \tag{1.140}$$

and

$$\Gamma_a \equiv \tfrac{1}{2}\frac{1}{\sqrt{|g|}} \partial_\mu(\sqrt{|g|}\, e_a{}^\mu) = \tfrac{1}{2} e_a{}^\mu{}_{;\mu} = \tfrac{1}{2}\omega_{ca}{}^\mu e^c{}_\mu. \tag{1.141}$$

Here

$$e^a{}_{\nu;\mu} \equiv \partial_\mu e^a{}_\mu - \Gamma^\lambda_{\mu\nu} e^a{}_\lambda \tag{1.142}$$

is the covariant derivative of the vierbein with respect to the metric.

The relation (1.139) is a consequence of the relation which tells us how the frame field $\gamma^a(x)$ changes with position:

$$\partial_\mu \gamma_a = \omega^a{}_{b\mu} \gamma^b. \tag{1.143}$$

It is illustrative to calculate the product (1.138) first in *flat spacetime*. Then one can always choose a constant frame field, so that $\omega_{ab\mu} = 0$. Then eq. (1.138) becomes

$$e^a{}_\alpha e^b{}_\beta p_a p_b \psi = -e^b{}_\beta \partial_\alpha (e_b{}^\nu \partial_\nu \psi) = -\partial_\alpha \partial_\beta \psi + \Gamma^\nu{}_{\alpha\beta} \partial_\nu \psi = -D_\alpha D_\beta \psi, \tag{1.144}$$

where we have used

$$e^b{}_\beta e_b{}^\nu = \delta_\beta{}^\nu \; , \qquad e^b{}_{\beta,\alpha} e_b{}^\nu = -e^b{}_\beta e_b{}^\nu{}_{,\alpha}, \tag{1.145}$$

and the expression for the affinity

$$\Gamma^\mu{}_{\alpha\beta} = e^{a\mu} e_{a\alpha,\beta} \tag{1.146}$$

which holds in flat spacetime where

$$e^a{}_{\mu,\nu} - e^a{}_{\nu,\mu} = 0. \tag{1.147}$$

We see that the product of operators is just the product of the covariant derivatives.

Let us now return to curved spacetime and consider the expression (1.138). After expansion we obtain

$$-e^a{}_\alpha e^b{}_\beta p_a p_b \psi = D_\alpha D_\beta \psi + \omega^{ab}{}_\alpha e_{a\beta} e_b{}^\nu \partial_\nu \psi + \tfrac{1}{2} e^c{}_\mu \omega_{ca}{}^\mu (e^a{}_\alpha \partial_\beta \psi + e^a{}_\beta \partial_\alpha \psi)$$

$$+ \tfrac{1}{2} e^a{}_\beta e^c{}_\mu \omega_{ca}{}^\mu{}_{;\alpha} \psi - \tfrac{1}{2} e^a{}_\beta e^b{}_\mu \omega_{ca}{}^\mu \omega^c{}_{b\alpha} \psi$$

$$+ \tfrac{1}{4} e^a{}_\alpha e^b{}_\beta e^c{}_\mu e^d{}_\nu \omega_{ca}{}^\mu \omega_{db}{}^\nu \psi. \tag{1.148}$$

This is not a Hermitian operator. In order to obtain a Hermitian operator one has to take a suitable symmetrized combination.

The expression (1.148) simplifies significantly if we contract it by $g^{\alpha\beta}$ and use the relation (see Sec. 6.1) for more details)

$$R = e_a{}^\mu e_b{}^\nu (\omega^{ab}{}_{\nu;\mu} - \omega^{ab}{}_{\mu;\nu} + \omega^{ac}{}_\mu \omega_c{}^b{}_\nu - \omega^{ac}{}_\nu \omega_c{}^b{}_\mu). \tag{1.149}$$

We obtain

$$-g^{\alpha\beta} e^a{}_\alpha e^b{}_\beta p_a p_b \psi = -\eta^{ab} p_a p_b \psi = -p^a p_a \psi$$

The spinless point particle

$$= \left(D_\alpha D^\alpha - \tfrac{1}{4} R - \tfrac{1}{4} e^{a\mu} e_{b\nu} \omega_{ca}{}^\nu \omega^{cb}{}_\mu \right) \quad (1.150)$$

which is a Hermitian operator. If one tries another possible Hermitian combination,

$$\tfrac{1}{4} \left(e^a{}_\mu p_a e^{b\mu} p_b + e^{b\mu} p_a e^a{}_\mu p_b + p_a e^a{}_\mu p_b e^{b\mu} + p_a e^{b\mu} p_b e^a{}_\mu \right), \quad (1.151)$$

one finds, using $[p_a, e^{b\mu}] = -i e_a{}^\nu \partial_\nu e^{b\mu}$, that it is equal to $\eta^{ab} p_a p_b$ since the extra terms cancel out.

1.4. SECOND QUANTIZATION

In the first quantized theory we arrived at the Lorentz invariant Schrödinger equation. Now, following refs. [8]–[17], [19, 20] we shall treat the wave function $\psi(\tau, x^\mu)$ as a field and the Schrödinger equation as a field equation which can be derived from an action. First, we shall consider the classical field theory, and then the quantized field theory of $\psi(\tau, x^\mu)$. We shall construct a Hamiltonian and find out that the equation of motion is just the Heisenberg equation. Commutation relations for our field ψ and its conjugate canonical momentum $i\psi^\dagger$ are defined in a straightforward way and are not quite the same as in the conventional relativistic field theory. The norm of our states is thus preserved, although the states are localized in spacetime. Finally, we point out that for the states with definite masses the expectation value of the energy–momentum and the charge operator coincide with the corresponding expectation values of the conventional field theory. We then compare the new theory with the conventional one.

CLASSICAL FIELD THEORY WITH INVARIANT EVOLUTION PARAMETER

The wave function ψ in the Schrödinger equation (1.67) can be considered as a field derived from the following action

$$I[\psi] = \int d\tau \, d^D x \left(i \psi^* \frac{\partial \psi}{\partial \tau} - \frac{\Lambda}{2} (\partial_\mu \psi^* \partial^\mu \psi - \kappa^2 \psi^* \psi) \right), \quad (1.152)$$

where the Lagrangian density $\mathcal{L}$ depends also on the τ-derivatives. By using the standard techniques of field theory (see, e.g., [34]) we may vary the action (1.152) and the boundaries of the region of integration R:

$$\begin{aligned} \delta I &= \int_R d\tau \, d^D x \, \delta \mathcal{L} + \int_{R-R'} d\tau \, d^D x \, \mathcal{L} \\ &= \int_R d\tau \, d^D x \, \delta \mathcal{L} + \int_B d\tau d\Sigma_\mu \mathcal{L} \delta x^\mu + \int d^D x \, \mathcal{L} \delta \tau \bigg|_{\tau_1}^{\tau_2}. \end{aligned} \quad (1.153)$$

Here $d\Sigma_\mu$ is an element of a $(D-1)$-dimensional closed hypersurface B in spacetime. If we assume that the equations of motion are are satisfied then the change of the action is

$$\delta I = \int d\tau\, d^D x \left[\partial_\mu \left(\frac{\partial \mathcal{L}}{\partial \partial_\mu \psi} \delta\psi + \frac{\partial \mathcal{L}}{\partial \partial_\mu \psi^*} \delta\psi^* + \mathcal{L} \delta x^\mu \right) \right.$$
$$\left. + \frac{\partial}{\partial \tau} \left(\frac{\partial \mathcal{L}}{\partial \partial \psi/\partial \tau} \delta\psi + \frac{\partial \mathcal{L}}{\partial \partial \psi^*/\partial \tau} \delta\psi^* + \mathcal{L} \delta\tau \right) \right], \quad (1.154)$$

where $\delta\psi \equiv \psi'(\tau, x) - \psi(\tau, x)$ is the variation of the field at fixed τ and x^μ. By introducing the total variation

$$\bar\delta\psi \equiv \psi'(\tau', x') - \psi(\tau, x) = \delta\psi + \partial_\mu \psi \delta x^\mu + \frac{\partial \psi}{\partial \tau} \delta\tau \quad (1.155)$$

eq. (1.154) becomes[8] [16]

$$\delta I = \int d\tau\, d^D x \left[\partial_\mu \left(\frac{\partial \mathcal{L}}{\partial \partial_\mu \psi} \bar\delta\psi + \frac{\partial \mathcal{L}}{\partial \partial_\mu \psi^*} \bar\delta\psi^* + T^\mu{}_\nu \delta x^\nu + T^\mu{}_\tau \delta\tau \right) \right.$$
$$\left. + \frac{\partial}{\partial \tau} \left(\frac{\partial \mathcal{L}}{\partial \partial \psi/\partial \tau} \bar\delta\psi + \frac{\partial \mathcal{L}}{\partial \partial \psi^*/\partial \tau} \bar\delta\psi^* + \Theta_\mu \delta x^\mu + \Theta \delta\tau \right) \right],$$
$$(1.156)$$

where

$$\Theta \equiv \mathcal{L} - \frac{\partial \mathcal{L}}{\partial \partial \psi/\tau} \frac{\partial \psi}{\partial \tau} - \frac{\partial \mathcal{L}}{\partial \partial \psi^*/\tau} \frac{\partial \psi^*}{\partial \tau} = \frac{\Lambda}{2}(\partial_\mu \psi^* \partial^\mu \psi - \kappa^2 \psi^* \psi), \quad (1.157)$$

$$\Theta_\mu \equiv -\frac{\partial \mathcal{L}}{\partial \partial \psi/\partial \tau} \partial_\mu \psi - \frac{\partial \mathcal{L}}{\partial \partial \psi^*/\partial \tau} \partial_\mu \psi^* = -i\psi^* \partial_\mu \psi, \quad (1.158)$$

$$T^\mu{}_\tau \equiv -\frac{\partial \mathcal{L}}{\partial \partial_\mu \psi} \frac{\partial \psi}{\partial \tau} - \frac{\partial \mathcal{L}}{\partial \partial_\mu \psi^*} \frac{\partial \psi^*}{\partial \tau} = \frac{\Lambda}{2}\left(\partial^\mu \psi^* \frac{\partial \psi}{\partial \tau} + \partial^\mu \psi \frac{\partial \psi^*}{\partial \tau} \right), \quad (1.159)$$

$$T^\mu{}_\nu \equiv \mathcal{L} \delta^\mu{}_\nu - \frac{\partial \mathcal{L}}{\partial \partial_\mu \psi} \partial_\nu \psi - \frac{\partial \mathcal{L}}{\partial \partial_\mu \psi^*} \partial_\nu \psi^*$$
$$= \left[i\psi^* \frac{\partial \psi}{\partial \tau} - \frac{\Lambda}{2}(\partial_\alpha \psi^* \partial^\alpha \psi - \kappa^2 \psi^* \psi) \right] \delta^\mu{}_\nu$$
$$+ \frac{\Lambda}{2}(\partial^\mu \psi^* \partial_\nu \psi + \partial^\mu \psi \partial_\nu \psi^*). \quad (1.160)$$

---

[8] J. R. Fanchi [10] considered only the terms $T^\mu{}_\nu$ and $T^\mu{}_\tau$, but he omitted the terms Θ and Θ_μ. R. Kubo [13] considered also Θ, but he ommitted Θ_μ.

The spinless point particle 33

Besides the usual stress–energy tensor $T^\mu{}_\nu$ we obtain the additional terms due to the presence of the τ-term in the action (1.152). The quantities Θ, Θ_μ, $T^\mu{}_\tau$ are generalizations of $T^\mu{}_\nu$.

A very important aspect of this parametrized field theory is that *the generator of τ evolution, or the Hamiltonian,* is

$$H = \int d^D x\, \Theta = -\frac{\Lambda}{2}\int d^D x (\partial_\mu \psi^* \partial^\mu \psi - \kappa^2 \psi^* \psi) \tag{1.161}$$

and *the generator of spacetime translations* is

$$H_\mu = \int d^D x\, \Theta_\mu = -i\int d^D x\, \psi^* \partial_\mu \psi. \tag{1.162}$$

This can be straightforwardly verified. Let us take into account that the canonically conjugate variables are

$$\psi,\quad \Pi = \frac{\partial \mathcal{L}}{\partial \partial \psi/\partial \tau} = i\psi^*, \tag{1.163}$$

and they satisfy the following equal τ Poisson bracket relations

$$\{\psi(\tau,x),\Pi(\tau',x')\}|_{\tau'=\tau} = \delta(x-x'), \tag{1.164}$$

or

$$\{\psi(\tau,x),\psi^*\}|_{\tau'=\tau} = -i\,\delta(x-x'), \tag{1.165}$$

and

$$\{\psi(\tau,x),\psi\}|_{\tau'=\tau} = 0,\qquad \{\psi^*(\tau,x),\psi^*\}|_{\tau'=\tau} = 0. \tag{1.166}$$

Let us consider the case $\bar\delta\psi = \bar\delta\psi^* = 0$ and assume that the action does not change under variations δx^ν and $\delta\tau$. Then from (1.156) we obtain the following conservation law:

$$\oint d\Sigma_\mu (T^\mu{}_\nu \delta x^\nu + T^\mu{}_\tau \delta\tau) + \frac{\partial}{\partial \tau}\int d^D x\,(\Theta_\mu \delta x^\nu + \Theta \delta\tau) = 0. \tag{1.167}$$

In general $\psi(\tau,x)$ has indefinite mass and is localized in spacetime. Therefore the boundary term in eq. (1.167) vanishes and we find that the generator

$$G(\tau) = \int d^D x (\Theta_\mu \delta x^\nu + \Theta \delta\tau) \tag{1.168}$$

is conserved at all values of τ (see Fig. 1.3a).

In particular $\psi(\tau,x)$ may have definite mass M, and then from (1.157–1.159) we have

$$\partial\Theta/\partial\tau = 0,\quad i\psi^* \partial\psi/\partial\tau + \kappa^2 \psi^* \psi = M^2 \psi^* \psi,\quad T^\mu{}_\tau = 0,\quad \partial\Theta_\mu/\partial\tau = 0$$

and the conservation law (1.167) becomes in this particular case

$$\oint d\Sigma_\mu T^\mu{}_\nu \delta x^\nu = 0, \qquad (1.169)$$

which means that the generator

$$G(\Sigma) = \int d\Sigma_\mu T^\mu{}_\nu \delta x^\nu \qquad (1.170)$$

is conserved at all space-like hypersurfaces Σ (Fig. 1.3b). The latter case just corresponds to the conventional field theory.

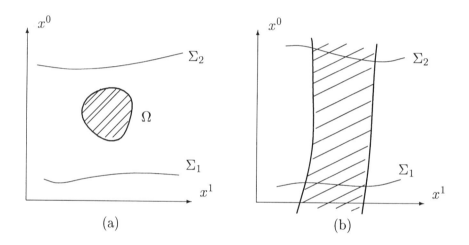

Figure 1.3. A field $\psi(\tau, x)$ is localized within a spacetime region Ω. The field at the boundaries Σ_1 and Σ_2 (which may be at infinity) is zero (a). A field ψ which has definite mass can be localized only in space but not in time (b).

Variation of an arbitrary functional $A[\psi(x), \Pi(x)]$ is given by $G(\tau)$ and can be expressed in terms of the Poisson bracket

$$\delta A = \{A, G(\tau)\}. \qquad (1.171)$$

The Poisson bracket is defined according to

$$\{A[\psi(x'), \Pi(x')], B[\psi(x), \Pi(x)]\}$$

$$= \int d^D x'' \left(\frac{\delta A}{\delta \psi(x'')} \frac{\delta B}{\delta \Pi(x'')} - \frac{\delta B}{\delta \psi(x'')} \frac{\delta A}{\delta \Pi(x'')} \right), \qquad (1.172)$$

The spinless point particle

where $\delta/\delta\psi(x'')$, $\delta/\delta\Pi(x'')$ are functional derivatives.

If in eq. (1.171) we put $A = \psi$ and take into account eq. (1.155) with $\bar{\delta}\psi = 0$, we obtain

$$\frac{\partial \psi}{\partial \tau} = -\{\psi, H\}, \tag{1.173}$$

$$\partial_\mu \psi = -\{\psi, H_\mu\}. \tag{1.174}$$

Eq.(1.173) is equivalent to the Schrödinger equation (1.67).

A generic field ψ which satisfies the Schrödinger equation (1.67) can be written as a superposition of the states with definite p_μ:

$$\psi(\tau, x) = \int d^D p\, c(p) \exp\left[i p_\mu x^\mu\right] \exp\left[-i\frac{\Lambda}{2}(p^2 - \kappa^2)\tau\right]. \tag{1.175}$$

Using (1.175) we have

$$H = -\frac{\Lambda}{2} \int d^D p\, (p^2 - \kappa^2) c^*(p) c(p), \tag{1.176}$$

$$H_\mu = \int d^D p\, p_\mu\, c^*(p) c(p). \tag{1.177}$$

On the other hand, from the spacetime stress–energy tensor (1.160) we find the total energy–momentum

$$P^\mu = \int d\Sigma_\nu T^{\mu\nu}. \tag{1.178}$$

It is not the generator of spacetime translations. Use of P^μ as generator of spacetime translations results in all conceptual and technical complications (including the lack of manifest Lorentz covariance) of the conventional relativistic field theory.

THE CANONICAL QUANTIZATION

Quantization of the parametrized field theory is straightforward. The field $\psi(\tau, x)$ and its canonically conjugate momentum $\Pi = i\psi^\dagger(\tau, x)$ become operators satisfying the following equal τ commutation relations[9]:

$$[\psi(\tau, x), \Pi(\tau', x')]|_{\tau'=\tau} = i\delta(x - x'), \tag{1.179}$$

---

[9]Enatsu [8] has found that the commutation relation (1.180) is equivalent to that of the conventional, on shell, quantum field theory, provided that $\epsilon(x - x')\epsilon(\tau - \tau') = 1$. However, since x^μ and τ are independent, the latter relation is not satisfied in general, and the commutation relation (1.180) is not equivalent to the usual on shell commutation relation. And yet, as far as the expectation values of certain operators are concerned, such as the energy–momentum are concerned, the predictions of both theories are the same; and this is all that matters.

or
$$[\psi(\tau, x), \psi^\dagger(\tau', x')]|_{\tau'=\tau} = \delta(x - x'), \tag{1.180}$$

and
$$[\psi(\tau, x), \psi(\tau', x')]|_{\tau'=\tau} = [\psi^\dagger(\tau, x), \psi^\dagger(\tau', x')]|_{\tau'=\tau} = 0. \tag{1.181}$$

In momentum space the above commutation relations read
$$[c(p), c^\dagger(p')] = \delta(p - p'), \tag{1.182}$$

$$[c(p), c(p')] = [c^\dagger(p), c^\dagger(p')] = 0. \tag{1.183}$$

The general commutator (for $\tau' \neq \tau$) is:
$$[\psi(\tau, x), \psi^\dagger(\tau', x')] = \int d^D p \, e^{ip_\mu(x^\mu - x'^\mu)} e^{-\frac{\Lambda}{2}(p^2 - \kappa^2)(\tau - \tau')}. \tag{1.184}$$

The operators $\psi^\dagger(\tau, x)$ and $\psi(\tau, x)$ are creation and annihilation operators, respectively. The vacuum is defined as a state which satisfies
$$\psi(\tau, x)|0\rangle = 0. \tag{1.185}$$

If we act on the vacuum by $\psi^\dagger(\tau, x)$ we obtain a single particle state with definite position:
$$\psi^\dagger(\tau, x)|0\rangle = |x, \tau\rangle. \tag{1.186}$$

An arbitrary 1-particle state is a superposition
$$|\Psi^{(1)}\rangle = \int dx \, f(\tau, x) \, \psi^\dagger(\tau, x)|0\rangle, \tag{1.187}$$

where $f(\tau, x)$ is the wave function.

A 2-particle state with one particle at x_1 and another at x_2 is obtained by applying the creation operator twice:
$$\psi^\dagger(\tau, x_1)\psi^\dagger(\tau, x_2)|0\rangle = |x_1, x_2, \tau\rangle. \tag{1.188}$$

In general
$$\psi^\dagger(\tau, x_1)...\psi^\dagger(\tau, x_n)|0\rangle = |x_1, ..., x_n, \tau\rangle, \tag{1.189}$$

and an arbitrary n-particle state is the superposition
$$|\Psi^{(n)}\rangle = \int dx_1...dx_n \, f(\tau, x_1, ..., x_n) \psi^\dagger(\tau, x_1)...\psi^\dagger(\tau, x_n)|0\rangle, \tag{1.190}$$

where $f(\tau, x_1, ..., x_n)$ is the wave function the n particles are spread with.

In momentum space the corresponding equations are

$$c(p)|0\rangle = 0, \tag{1.191}$$

$$c^\dagger(p)|0\rangle = |p\rangle, \tag{1.192}$$

$$c^\dagger(p_1)...c^\dagger(p_n)|0\rangle = |p_1, ..., p_n\rangle, \tag{1.193}$$

$$|\Psi^{(n)}\rangle = \int \mathrm{d}p_1...\mathrm{d}p_n \, g(\tau, p_1, ..., p_n) \, c^\dagger(p_1)...c^\dagger(p_n)|0\rangle \tag{1.194}$$

The most general state is a superposition of the states $|\Psi^{(n)}\rangle$ with definite numbers of particles.

According to the commutation relations (1.180)–(1.183) the states $|x_1, ..., x_n\rangle$ and $|p_1, ..., p_n\rangle$ satisfy

$$\langle x_1, ..., x_n | x'_1, ..., x'_n \rangle = \delta(x_1 - x'_1)...\delta(x_n - x'_n), \tag{1.195}$$

$$\langle p_1, ..., p_n | p'_1, ..., p'_n \rangle = \delta(p_1 - p'_1)...\delta(p_n - p'_n), \tag{1.196}$$

which assures that the norm of an arbitrary state $|\Psi\rangle$ is always positive.

We see that the formulation of the unconstrained relativistic quantum field theory goes along the same lines as the well known second quantization of a non-relativistic particle. Instead of the 3-dimensional space we have now a D-dimensional space whose signature is arbitrary. In particular, we may take a 4-dimensional space with Minkowski signature and identify it with *spacetime*.

THE HAMILTONIAN AND THE GENERATOR OF SPACETIME TRANSLATIONS

The generator $G(\tau)$ defined in (1.168) is now an operator. An infinitesimal change of an arbitrary operator A is given by the commutator

$$\delta A = -i[A, G(\tau)], \tag{1.197}$$

which is the quantum analog of eq. (1.171). If we take $A = \psi$, then eq. (1.197) becomes

$$\frac{\partial \psi}{\partial \tau} = i[\psi, H], \tag{1.198}$$

where

$$\frac{\partial \psi}{\partial x^\mu} = i[\psi, H_\mu], \qquad (1.199)$$

$$H = \int d^D x \, \Theta = -\frac{\Lambda}{2}\int d^D x (\partial_\mu \psi^\dagger \partial^\mu \psi - \kappa^2 \psi^\dagger \psi), \qquad (1.200)$$

$$H_\mu = \int d^D x \, \Theta_\mu = -i\int d^D x \, \psi^\dagger \partial_\mu \psi. \qquad (1.201)$$

Using the commutation relations (1.179)–(1.181) we find that eq. (1.198) is equivalent to the field equation (1.67) (the Schrödinger equation). Eq. (1.198) is thus *the Heisenberg equation* for the field operator ψ. We also find that eq. (1.199) gives just the identity $\partial_\mu \psi = \partial_\mu \psi$.

In momentum representation the field operators are expressed in terms of the operators $c(p)$, $c^\dagger(p)$ according to eq. (1.175) and we have

$$H = -\frac{\Lambda}{2}\int d^D p \, (p^2 - \kappa^2) c^\dagger(p) c(p), \qquad (1.202)$$

$$H_\mu = \int d^D p \, p_\mu c^\dagger(p) c(p). \qquad (1.203)$$

The operator H is the Hamiltonian and it generates the τ-evolution, whereas H_μ is the generator of spacetime translations. In particular, H_0 generates translations along the axis x^0 and can be either positive or negative definite.

ENERGY–MOMENTUM OPERATOR

Let us now consider the generator $G(\Sigma)$ defined in eq. (1.170) with $T^\mu{}_\nu$ given in eq. (1.160) in which the classical fields ψ, ψ^* are now replaced by the operators ψ, $\psi^\dagger$. The total energy–momentum P_ν of the field is given by the integration of $T^\mu{}_\nu$ over a space-like hypersurface:

$$P_\nu = \int d\Sigma_\mu \, T^\mu{}_\nu. \qquad (1.204)$$

Instead of P_ν defined in (1.204) it is convenient to introduce

$$\widetilde{P}_\nu = \int ds \, P_\nu, \qquad (1.205)$$

where ds is a distance element along the direction n^μ which is orthogonal to the hypersurface element $d\Sigma_\mu$. The latter can be written as $d\Sigma_\mu = n_\mu d\Sigma$. Using $ds \, d\Sigma = d^D x$ and integrating out x^μ in (1.205) we find that τ-dependence disappears and we obtain (see Box 1.1)

$$\widetilde{P}_\nu = \int d^D p \, \frac{\Lambda}{2}(n_\mu p^\mu) \, p_\nu (c^\dagger(p) c(p) + c(p) c^\dagger(p)). \qquad (1.206)$$

Box 1.1

$$G(\Sigma) = \int d\Sigma_\mu\, d\tau\, T^\mu{}_\nu\, \delta x^\nu$$

$\begin{cases} \delta x^\mu = \xi^\mu \delta s\,, & \xi^\mu \text{ a vector along } \delta x^\mu\text{-direction} \\ d\Sigma_\mu = n_\mu d\Sigma\,, & n_\mu \text{ a vector along } d\Sigma_\mu\text{-direction} \end{cases}$

$$P(n,\xi) = \int d\Sigma\, d\tau\, T^\mu{}_\nu\, n_\mu \xi^\nu$$

integrate over ds and divide by $s_2 - s_1$

$$P(n,\xi) = \lim_{s_{1,2}\to -\infty,\infty} \frac{1}{s_2 - s_1} \int_{s_1}^{s_2} d\Sigma\, ds\, d\tau\, T^\mu{}_\nu\, n_\mu \xi^\nu$$

$d\Sigma\, ds = d^D x$

$$P(n,\xi) = \lim_{s_{1,2}\to -\infty,\infty} \frac{\tau_1 - \tau_2}{s_1 - s_2} \int_{s_1}^{s_2} d^D p\, \frac{\Lambda}{2} (n_\mu p^\mu) p_\nu \xi^\nu \left(c^\dagger(p)c(p) + c(p)c^\dagger(p)\right)$$

$\Lambda = \frac{\Delta s}{\Delta \tau} \frac{1}{\sqrt{p^2}}$ when mass is definite

$$P(n,\xi) = \tfrac{1}{2} \int d^D p\, \epsilon(np)\, p_\nu \xi^\nu \left(c^\dagger(p)c(p) + c(p)c^\dagger(p)\right)$$

When ξ^ν is time-like $P(n,\xi)$ is projection of *energy*, when ξ^ν is space-like $P(n,\xi)$ is projection of *momentum* on ξ^ν

From (1.206) we have that the projection of $\tilde{P}_\nu$ on a time-like vector n^ν is always positive, while the projection on a space-like vector N^ν can be positive or negative, depending on signs of $p_\mu n^\mu$ and $p_\mu N^\mu$. In a special reference frame in which $n^\nu = (1, 0, 0, ..., 0)$ this means that P_0 is always positive, whereas P_r, $r = 1, 2, ..., D-1$, can be positive or negative definite.

What is the effect of the generator $\tilde{P}_\nu \delta x^\nu$ on a field $\psi(\tau, x)$. From (1.197), (1.206) we have

$$\delta\psi = -i[\psi, \tilde{P}_\nu]\delta x^\nu \tag{1.207}$$

$$= -i \int \mathrm{d}^D p\, \Lambda (n_\mu p^\mu) c(p) \exp\left[ip_\mu x^\mu\right] \exp\left[-\frac{i\Lambda}{2}(p^2 - \kappa^2)\tau\right] p_\nu \delta x^\nu$$

Acting on a Fourier component of ψ with definite $p_\mu p^\mu = M^2$ the generator $\tilde{P}_\nu \delta x^\nu$ gives

$$\delta\phi = -i[\phi, \tilde{P}_\nu]\delta x^\nu = -i\int \mathrm{d}^D p\, \Lambda\, M \epsilon(np) c(p) e^{ip_\mu x^\mu} \delta(p^2 - M^2) p_\nu \delta x^\nu$$

$$= -\partial_\nu \phi^{(+)} \delta x^\nu + \partial_\nu \phi^{(-)} \delta x^\nu, \tag{1.208}$$

$$\epsilon(np) \equiv \frac{n^\mu p_\mu}{\sqrt{|p^2|}} = \begin{cases} +1, & n^\mu p_\mu > 0 \\ -1, & n^\mu p_\mu < 0 \end{cases} \tag{1.209}$$

where

$$\phi(x) = \int_{-\infty}^{\infty} \mathrm{d}\tau\, e^{i\mu\tau} \psi(\tau, x)$$

$$= \int_{p_0=-\infty}^{\infty} \mathrm{d}^D p\, \delta(p^2 - M^2) c(p) e^{ip_\mu x^\mu}$$

$$= \int_{p_0=0}^{\infty} \mathrm{d}^D p\, \delta(p^2 - M^2) c(p) e^{ip_\mu x^\mu}$$

$$+ \int_{p_0=-\infty}^{0} \mathrm{d}^D p\, \delta(p^2 - M^2) c(p) e^{ip_\mu x^\mu}$$

$$= \phi^{(+)} + \phi^{(-)} \tag{1.210}$$

As in Sec. 1.3. we identify $\mu^2 + \kappa^2 \equiv M^2$.

We see that the action of $\tilde{P}_\nu \delta x^\nu$ on ϕ differs from the action of $H_\nu \delta x^\nu$. The difference is in the step function $\epsilon(np)$. When acting on $\phi^{(+)}$ the generator $\tilde{P}_\nu \delta x^\nu$ gives the same result as $H_\nu \delta x^\nu$. On the contrary, when acting on $\phi^{(-)}$ it gives the opposite sign. This demonstrates that $\tilde{P}_\nu \delta x^\nu$

The spinless point particle

does not generate translations and Lorentz transformations in Minkowski space M_D. This is a consequence of the fact that P_0 is always positive definite.

PHASE TRANSFORMATIONS AND THE CHARGE OPERATOR

Let us now consider the case when the field is varied at the boundary, while δx^μ and $\delta \tau$ are kept zero. Let the field transform according to

$$\psi' = e^{i\alpha}\psi, \quad \delta\psi = i\alpha\psi, \quad \delta\psi^\dagger = -i\alpha\psi^\dagger. \tag{1.211}$$

Eq. (1.156) then reads

$$\delta I = \oint d\tau \, d\Sigma_\mu \left(-\frac{i\Lambda\alpha}{2}\right)(\partial^\mu\psi^\dagger\psi - \partial^\mu\psi\,\psi^\dagger) + \int d^D x\,(-\alpha)\psi^\dagger\psi \Big|_{\tau_1}^{\tau_2} \tag{1.212}$$

Introducing *the charge density*

$$\rho_c = -\alpha\psi^\dagger\psi \tag{1.213}$$

and *the charge current density*

$$j_c^\mu = -\frac{i\Lambda\alpha}{2}(\partial^\mu\psi^\dagger\psi - \partial^\mu\psi\,\psi^\dagger), \tag{1.214}$$

and assuming that the action is invariant under the phase transformations (1.211) we obtain from (1.212) the following conservation law:

$$\oint d\Sigma_\mu\, j_c^\mu + \frac{d}{d\tau}\int d^D x\, \rho_c = 0, \tag{1.215}$$

or

$$\partial_\mu j_c^\mu + \frac{\partial \rho_c}{\partial \tau} = 0. \tag{1.216}$$

In general, when the field has indefinite mass it is localized in spacetime and the surface term in (1.215) vanishes. So we have that the generator

$$C = -\int d^D x\, \rho_c \tag{1.217}$$

is conserved at all τ-values.

In particular, when ψ has definite mass, then $\psi^\dagger\psi$ is independent of τ and the conservation of C is trivial. Since ψ is not localized along a time-like direction the current density j_c^μ does not vanish at the space-like hypersurface even if taken at infinity. It does vanish, however, at a sufficiently far away time-like hypersurface. Therefore the generator

$$Q = \int d\Sigma_\mu j_c^\mu \tag{1.218}$$

is conserved at all space-like hypersufaces. The quantity Q corresponds to the charge operator of the usual field theory. However, it is not generator of the phase transformations (1.211).

By using the field expansion (1.175) we find

$$C = -\alpha \int \mathrm{d}^D p \, c^\dagger(p) c(p). \tag{1.219}$$

From

$$\delta\psi = -i[\psi, C] \tag{1.220}$$

and the commutation relations (1.179)–(1.183) we have

$$\delta\psi = i\alpha\psi, \quad \delta\psi^\dagger = -i\alpha\psi^\dagger, \tag{1.221}$$

$$\delta c(p) = i\alpha c(p), \quad \delta c^\dagger(p) = -i\alpha c^\dagger(p), \tag{1.222}$$

which are indeed the phase transformations (1.211). Therefore C is *the generator of phase transformations*.

On the other hand, from $\delta\psi = -i[\psi, Q]$ we do not obtain the phase transformations. But we obtain from Q something which is very close to the phase transformations, if we proceed as follows. Instead of Q we introduce

$$\tilde{Q} = \int \mathrm{d}s \, Q, \tag{1.223}$$

where $\mathrm{d}s$ is an infinitesimal interval along a time-like direction n^μ orthogonal to $\mathrm{d}\Sigma_\mu = n_\mu \mathrm{d}\Sigma$. Using $\mathrm{d}s\mathrm{d}\Sigma = \mathrm{d}^D x$ and integrating out x^μ eq. (1.223) becomes

$$\tilde{Q} = \int \left(-\frac{\alpha\Lambda}{2}\right) \mathrm{d}p \, (p^\mu n_\mu)(c^\dagger(p)c(p) + c(p)c^\dagger(p)). \tag{1.224}$$

The generator $\tilde{Q}$ then gives

$$\delta\psi = -i[\psi, \tilde{Q}] = \int i\alpha\Lambda \mathrm{d}^D p \, (p^\nu n_\nu) c(p) \exp[ip_\mu x^\mu] \exp\left[-\frac{i\Lambda}{2}(p^2 - \kappa^2)\tau\right]. \tag{1.225}$$

For a Fourier component with definite $p_\mu p^\mu$ we have

$$\begin{aligned}\delta\phi = -i[\phi, \tilde{Q}] &= \int \mathrm{d}^D p \, i\alpha\Lambda m \, \epsilon(np) c(p) \, e^{ip_\mu x^\mu} \delta(p^2 - M^2) \\ &= i\alpha\phi^{(+)} - i\alpha\phi^{(-)} \end{aligned} \tag{1.226}$$

where $\epsilon(np)$ and $\phi^{(+)}$, $\phi^{(-)}$ are defined in (1.209) and (1.210), respectively.

The spinless point particle 43

Action of operators on states. From the commutation relations (1.180)–(1.180) one finds that when the operators $c^\dagger(p)$ and $c(p)$ act on the eigenstates of the operators H, H_μ, $\widetilde{P}_\mu$, C, $\widetilde{Q}$ they increase and decrease, respectively, the corresponding eigenvalues. For instance, if $H_\mu|\Psi\rangle = p_\mu|\Psi\rangle$ then

$$H_\mu c^\dagger(p')|\Psi\rangle = (p_\mu + p'_\mu)|\Psi\rangle, \qquad (1.227)$$

$$H_\mu c(p')|\Psi\rangle = (p_\mu - p'_\mu)|\Psi\rangle, \qquad (1.228)$$

and similarly for other operators.

THE EXPECTATION VALUES OF THE OPERATORS

Our parametrized field theory is just a straightforward generalization of the non-relativistic field theory in E_3. Instead of the 3-dimensional Euclidean space E_3 we have now the 4-dimensional Minkowski space M_4 (or its D-dimensional generalization). Mathematically the theories are equivalent (apart from the dimensions and signatures of the corresponding spaces). A good exposition of the unconstrained field theory is given in ref. [35]

A state $|\Psi_n\rangle$ of n identical free particles can be given in terms of the generalized wave packet profiles $g^{(n)}(p_1,...,p_n)$ and the generalized state vectors $|p_1,...,p_n\rangle$:

$$|\Psi_n\rangle = \int dp_1...dp_n\, g^{(n)}(p_1,...,p_n)|p_1,...,p_n\rangle. \qquad (1.229)$$

The action of the operators $c(p)$, $c^\dagger(p)$ on the state vector is the following:

$$c(p)|p_1,...,p_n\rangle = \frac{1}{\sqrt{n}}\sum_{i=1}^n \delta(p-p_i)|p_1,..,\check{p}_i,..,p_n\rangle, \qquad (1.230)$$

$$c^\dagger(p)|p_1,...,p_n\rangle = \sqrt{n+1}\,|p,p_1,...,p_n\rangle. \qquad (1.231)$$

The symbol $\check{p}_i$ means that the quantity p_i is not present in the expression. Obviously

$$|p_1,...,p_n\rangle = \frac{1}{\sqrt{n!}}c^\dagger(p_n)...c^\dagger(p_1)|0\rangle. \qquad (1.232)$$

For the product of operators we have

$$c^\dagger(p)c(p)|p_1,...,p_n\rangle = \sum_{i=1}^n \delta(p-p_i)|p_1,...,p_n\rangle \qquad (1.233)$$

The latter operator obviously counts how many there are particles with momentum p. Integrating over p we obtain

$$\int d^D p\, c^\dagger(p)c(p)|p_1,...,p_n\rangle = n|p_1,...,p_n\rangle. \qquad (1.234)$$

44   *THE LANDSCAPE OF THEORETICAL PHYSICS: A GLOBAL VIEW*

The expectation value of the operator $\tilde{P}^\mu$ (see eq. (1.206)) in the state $|\Psi_n\rangle$ is

$$\begin{aligned}\langle|\Psi_n|\tilde{P}^\mu|\Psi\rangle &= \langle\Psi_n|\int d^D p \frac{\Lambda}{2}(n_\nu p^\nu) p^\mu (2c^\dagger(p)c(p) + \delta(0))|\Psi_n\rangle \\ &= \Lambda \sum_{i=1}^n \langle (n_\nu p_i^\nu) p_i^\mu \rangle + \frac{\Lambda}{2}\delta(0)\int d^D p\, (n_\nu p^\nu) p^\mu, \quad (1.235)\end{aligned}$$

where

$$\langle (n_\nu p_i^\nu) p_i^\mu \rangle = \int dp_1...dp_n g^{*(n)}(p_1,...,p_n) g^{(n)}(p_1,...,p_n)(p_i^\nu n_\nu) p_i^\mu.$$

In particular, the wave packet profile $g^{(n)}(p_1,...,p_n)$ can be such that $|\Psi_n\rangle = |p_1,...,p_n\rangle$. Then the expectation value (1.235) becomes

$$\langle p_1,...,p_n|\tilde{P}^\mu|p_1,...,p_n\rangle = \Lambda\sum_{i=1}^n (n_\nu p_i^\nu) p_i^\mu + \frac{\Lambda}{2}\delta(0)\int d^D p\,(n_\nu p^\nu)p^\mu. \quad (1.236)$$

The extra term $\frac{\Lambda}{2}\delta(0)\int d^D p\,(n_\nu p^\nu)p^\mu$ vanishes for a space-like component of p^μ, but it does not vanish for a time-like component.

So far the momenta $p_1,...,p_n$ of n identical particles have not been constrained to a mass shell. Let us now consider the case where all the momenta $p_1,...,p_n$ are constrained to the mass shell, so that $p_i^\mu p_{i\mu} = p_i^2 = p^2 = M^2$, $i=1,2,...,n$. From the classical equation of motion $p_i^\mu = \dot{x}_i^\mu/\Lambda$ we have

$$\Lambda = \frac{ds_i}{d\tau}\frac{1}{\sqrt{p_i^2}}, \quad (1.237)$$

where $ds_i = \sqrt{dx_i^\mu dx_{i\mu}}$. Now, Λ and τ being the same for all particles we see that if p_i^2 is the same for all particles also the 4-dimensional speed $ds_i/d\tau$ is the same. Inserting (1.237) into (1.236) and chosing parametrization such that $ds_i/d\tau = 1$ we obtain the following expectation value of $\tilde{P}^\mu$ in a state on the mass shell:

$$\langle\tilde{P}^\mu\rangle_m = \langle p_1,...,p_n|\tilde{P}^\mu|p_1,...,p_n\rangle_m = \sum_{i=1}^n \epsilon(np_i)p_i^\mu + \frac{\delta(0)}{2}\int d^D p\,\epsilon(np)p^\mu. \quad (1.238)$$

Taking $n_\nu = (1,0,0,....,0)$ we have

$$\langle\tilde{P}^0\rangle_m = \sum_{i=1}^n \epsilon(p_i^0)p_i^0 + \frac{\delta(0)}{2}\int d^D p\,\epsilon(p^0)p^0, \quad (1.239)$$

The spinless point particle   45

$$\langle \tilde{P}^r \rangle_m = \sum_{i=1}^n \epsilon(p_i^0) p_i^r \ , \quad r = 1, 2, ..., D-1. \tag{1.240}$$

So far we have considered momenta as continuous. However, if we imagine a large box and fix the boundary conditions, then the momenta are discrete. Then in eqs. (1.239), (1.240) particles $i = 1, 2, ..., n$ do not necessarily all have different momenta p_i^μ. There can be n_p particles with a given discrete value of momentum p. Eqs. (1.239), (1.240) can then be written in the form

$$\langle \tilde{P}^0 \rangle_m = \sum_{\mathbf{p}} (n_{\mathbf{p}}^+ + n_{\mathbf{p}}^- + \frac{1}{2}) \omega_{\mathbf{p}}, \tag{1.241}$$

$$\langle \tilde{\mathbf{P}} \rangle_m = \sum_{\mathbf{p}} (n_{\mathbf{p}}^+ + n_{\mathbf{p}}^-) \mathbf{p} \tag{1.242}$$

Here

$$\omega_{\mathbf{p}} \equiv |\sqrt{\mathbf{p}^2 + M^2}| = \epsilon(p^0) p^0,$$

whereas $n_{\mathbf{p}}^+ = n_{\omega_{\mathbf{p}}, \mathbf{p}}$ is the number of particles with positive p^0 at a given value of momentum and $n_{\mathbf{p}}^- = n_{-\omega_{\mathbf{p}}, -\mathbf{p}}$. In eq. (1.242) we have used

$$-\sum_{\mathbf{p}=-\infty}^{\infty} n_{-\omega_{\mathbf{p}}, \mathbf{p}} \mathbf{p} = \sum_{\mathbf{p}=-\infty}^{\infty} n_{-\omega_{\mathbf{p}}, -\mathbf{p}} \mathbf{p}.$$

Instead of $\delta(0)$ in eq. (82) we have written 1, since in the case of discrete momenta $\delta(p - p')$ is replaced by $\delta_{pp'}$, and $\delta_{pp} = 1$.

Equations (1.241), (1.242) are just the same expressions as obtained in the conventional, on shell quantized field theory of a non-Hermitian scalar field. In an analogous way we also find that the expectation value of the electric charge operator $\tilde{Q}$, when taken on the mass shell, is identical to that of the conventional field theory.

COMPARISON WITH THE CONVENTIONAL RELATIVISTIC QUANTUM FIELD THEORY

In the unconstrained relativistic quantum field theory mass is indefinite, in general. But in particular it can be definite. The field is then given by the expression (1.72) for a fixed value of μ, say $\mu = 0$, in which momentum is constrained to a mass shell. If we calculate the commutator $[\phi(x), \phi^\dagger(x')]$ using the commutation relations (1.182), (1.183) we obtain

$$[\phi(x), \phi(x')] = \int d^D p \, e^{ip_\mu (x^\mu - x'^\mu)} \delta(p^2 - M^2), \tag{1.243}$$

which differs from zero both for time-like and for space-like separations between x^μ and x'^μ. In this respect the new theory differs significantly from the conventional theory in which the commutator is zero for space-like separations and which assures that the process

$$\langle 0|\phi(x)\phi^\dagger(x')|0\rangle$$

has vanishing amplitude. No faster–than–light propagation is possible in the conventional relativistic field theory.

The commutation relations leading to the conventional field theory are

$$[c(p), c^\dagger(p')] = \epsilon(p^0)\delta^D(p-p'), \qquad (1.244)$$

where $\epsilon(p^0) = +1$ for $p^0 > 0$ and $\epsilon(p^0) = -1$.

By a direct calculation we then find

$$[\phi(x), \phi^\dagger(x')] = \int d^D p \, e^{ip(x-x')} \epsilon(p^0) \delta(p^2 - M^2) \equiv D(x-x'), \qquad (1.245)$$

which is indeed different from zero when $(x-x')^2 > 0$ and zero when $(x-x')^2 < 0$.

The commutation relations (1.244), (1.245) also assure that the Heisenberg equation

$$\frac{\partial \phi}{\partial x^0} = i[\psi, P^0] \qquad (1.246)$$

is equivalent to the field equation

$$(\partial_\mu \partial^\mu + M^2)\phi = 0. \qquad (1.247)$$

Here P^0, which now serves as Hamiltonian, is the 0-th component of the momentum operator defined in terms of the definite mass fields $\phi(x)$, $\phi^\dagger(x)$ (see Box 1.2).

If vacuum is defined according to

$$c(p)|0\rangle = 0, \qquad (1.248)$$

then because of the commutation relation (1.244) the scalar product is

$$\langle p|p'\rangle = \langle 0|c(p)c^\dagger|0\rangle \epsilon(p^0)\delta^D(p-p'), \qquad (1.249)$$

from which it follows that states with negative p^0 have negative norms.

The usual remedy is in the redefinition of vacuum. Writing

$$p^\mu \equiv p = (\mathbf{p}, \omega_\mathbf{p}) \,, \qquad \omega_\mathbf{p} = |\sqrt{M^2 + \mathbf{p}^2}|, \qquad (1.250)$$

The spinless point particle 47

Box 1.2: Momentum on mass shell

$$P^\mu = \int T^{\mu\nu} d\Sigma_\nu$$

$$\begin{cases} T^{\mu\nu} = \dfrac{\Lambda}{2}(\partial^\mu \Phi^\dagger \partial^\nu \Phi + \partial^\mu \Phi \partial^\nu \Phi^\dagger) - \dfrac{\Lambda}{2}(\partial_\alpha \Phi^\dagger \partial^\alpha \Phi - M^2 \Phi^\dagger \Phi)\delta^\mu{}_\nu \\ \Phi = \int dp\, \delta(p^2 - M^2) e^{ipx} c(p) \\ \partial_\mu \Phi = \int dp\, \delta(p^2 - M^2) e^{ipx} c(p) i p_\mu \\ d\Sigma_\nu = n_\nu d\Sigma \end{cases}$$

$$P^\mu = \dfrac{\Lambda}{2} \int dp\, dp'\, \delta(p^2 - M^2)\delta(p'^2 - M^2) p'^\mu p^\nu n_\nu \\ \times \left(c^\dagger(p') c(p) + c(p') c^\dagger(p) \right) e^{i(p-p')x} d\Sigma$$

$$\begin{cases} \delta(p^2 - M^2) = \int ds\, e^{i(p^2 - M^2)s}, \quad ds\, d\Sigma = d^D x \\ \Lambda = \dfrac{1}{\sqrt{p^2}} \qquad \dfrac{ds}{d\tau} = 1 \end{cases}$$

$$P^\mu = \tfrac{1}{2} \int dp\, \delta(p^2 - M^2) \epsilon(pn) p^\mu \left(c^\dagger(p) c(p) + c(p) c^\dagger(p) \right)$$

$$\begin{cases} dp = d\omega\, d\mathbf{p} \quad p = (\omega, \mathbf{p}) \\ \omega_\mathbf{p} = \sqrt{\mathbf{p}^2 + M^2} \end{cases}$$

$$P^\mu = \dfrac{1}{2} \int d\mathbf{p} \begin{pmatrix} \omega_\mathbf{p} \\ \mathbf{p} \end{pmatrix} \dfrac{1}{2\omega_\mathbf{p}} \Big(c^\dagger(\omega_\mathbf{p}, \mathbf{p}) c(\omega_\mathbf{p}, \mathbf{p}) + c(\omega_\mathbf{p}, \mathbf{p}) c^\dagger(\omega_\mathbf{p}, \mathbf{p}) \\ + c^\dagger(-\omega_\mathbf{p}, -\mathbf{p}) c(-\omega_\mathbf{p}, -\mathbf{p}) + c(-\omega_\mathbf{p}, -\mathbf{p}) c^\dagger(-\omega_\mathbf{p}, -\mathbf{p}) \Big)$$

$$\begin{cases} \text{I.} \quad [c(p), c^\dagger(p')] = \delta^D(p - p') \quad c(p)|0\rangle = 0 \\ \dfrac{1}{2\omega_\mathbf{p}} c^\dagger(\omega_\mathbf{p}, \mathbf{p}) = a^\dagger(\mathbf{p}), \quad \dfrac{1}{2\omega_\mathbf{p}} c^\dagger(-\omega_\mathbf{p}, -\mathbf{p}) = b^\dagger(-\mathbf{p}) \\ \dfrac{1}{2\omega_\mathbf{p}} c(\omega_\mathbf{p}, \mathbf{p}) = a(\mathbf{p}), \quad \dfrac{1}{2\omega_\mathbf{p}} c(-\omega_\mathbf{p}, -\mathbf{p}) = b(-\mathbf{p}) \\ a(\mathbf{p})|0\rangle = 0, \quad b(-\mathbf{p})|0\rangle = 0 \end{cases}$$

$$\left\{\begin{array}{l} \text{II.} \quad [c(p),c^\dagger(p')] = \epsilon(p^0)\delta^D(p-p') \\ \frac{1}{2\omega_{\mathbf{p}}}c^\dagger(\omega_{\mathbf{p}},\mathbf{p}) = a^\dagger(\mathbf{p})\,, \quad \frac{1}{2\omega_{\mathbf{p}}}c^\dagger(-\omega_{\mathbf{p}},-\mathbf{p}) = b(\mathbf{p}) \\ \frac{1}{2\omega_{\mathbf{p}}}c(\omega_{\mathbf{p}},\mathbf{p}) = a(\mathbf{p})\,, \quad \frac{1}{2\omega_{\mathbf{p}}}c(-\omega_{\mathbf{p}},-\mathbf{p}) = b^\dagger(\mathbf{p}) \\ a(\mathbf{p})|0\rangle = 0\,, \quad b(\mathbf{p})|0\rangle = 0 \end{array}\right.$$

or

$$P^\mu = \frac{1}{2}\int \mathrm{d}\mathbf{p}\begin{pmatrix}\omega_{\mathbf{p}}\\ \mathbf{p}\end{pmatrix}\left(a^\dagger(\mathbf{p})a(\mathbf{p}) + a(\mathbf{p})a^\dagger(\mathbf{p}) + b^\dagger(\mathbf{p})b(\mathbf{p}) + b(\mathbf{p})b^\dagger(\mathbf{p})\right)$$

For both definitions I and II the expression for P_μ is the same, but the relations of $b(\mathbf{p})$, $b^\dagger(\mathbf{p})$ to $c(p)$ are different.

and denoting

$$\frac{1}{\sqrt{2\omega_{\mathbf{p}}}}c(\mathbf{p},\omega_{\mathbf{p}}) \equiv a(\mathbf{p})\,, \qquad \frac{1}{\sqrt{2\omega_{\mathbf{p}}}}c^\dagger(\mathbf{p},\omega_{\mathbf{p}}) \equiv a^\dagger(\mathbf{p}), \qquad (1.251)$$

$$\frac{1}{\sqrt{2\omega_{\mathbf{p}}}}c^\dagger(-\mathbf{p},-\omega_{\mathbf{p}})\,, \qquad \frac{1}{\sqrt{2\omega_{\mathbf{p}}}}c(-\mathbf{p},-\omega_{\mathbf{p}}) = b^\dagger(\mathbf{p}), \qquad (1.252)$$

let us define the vacuum according to

$$a(\mathbf{p})|0\rangle = 0\,, \qquad b(\mathbf{p})|0\rangle = 0. \qquad (1.253)$$

If we rewrite the commutation relations (1.244) in terms of the operators $a(\mathbf{p})$, $b(\mathbf{p})$, we find (see Box 1.3)

$$[a(\mathbf{p}),a^\dagger(\mathbf{p}')] = \delta(\mathbf{p}-\mathbf{p}'), \qquad (1.254)$$

$$[b(\mathbf{p}),b^\dagger(\mathbf{p}')] = \delta(\mathbf{p}-\mathbf{p}'), \qquad (1.255)$$

$$[a(\mathbf{p}),b(\mathbf{p}')] = [a^\dagger(\mathbf{p}),b^\dagger(\mathbf{p}')] = [a(\mathbf{p}),b^\dagger(\mathbf{p}')] = [a^\dagger(\mathbf{p}),b(\mathbf{p}')] = 0. \qquad (1.256)$$

The spinless point particle

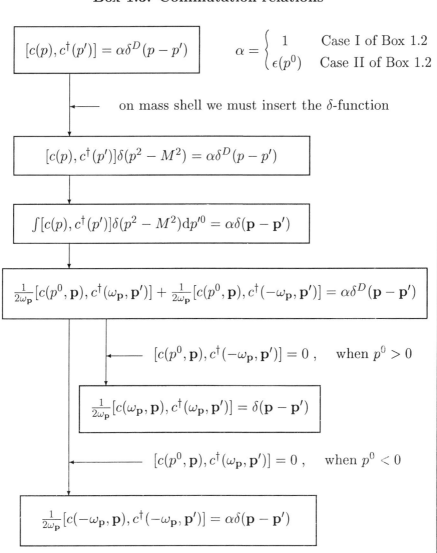

Box 1.3: Commutation relations

$[c(p), c^\dagger(p')] = \alpha \delta^D(p-p')$

$\alpha = \begin{cases} 1 & \text{Case I of Box 1.2} \\ \epsilon(p^0) & \text{Case II of Box 1.2} \end{cases}$

← on mass shell we must insert the δ-function

$[c(p), c^\dagger(p')]\delta(p^2 - M^2) = \alpha \delta^D(p-p')$

$\int [c(p), c^\dagger(p')]\delta(p^2 - M^2) dp'^0 = \alpha \delta(\mathbf{p}-\mathbf{p}')$

$\frac{1}{2\omega_{\mathbf{p}}}[c(p^0, \mathbf{p}), c^\dagger(\omega_{\mathbf{p}}, \mathbf{p}')] + \frac{1}{2\omega_{\mathbf{p}}}[c(p^0, \mathbf{p}), c^\dagger(-\omega_{\mathbf{p}}, \mathbf{p}')] = \alpha \delta^D(\mathbf{p}-\mathbf{p}')$

← $[c(p^0, \mathbf{p}), c^\dagger(-\omega_{\mathbf{p}}, \mathbf{p}')] = 0$, when $p^0 > 0$

$\frac{1}{2\omega_{\mathbf{p}}}[c(\omega_{\mathbf{p}}, \mathbf{p}), c^\dagger(\omega_{\mathbf{p}}, \mathbf{p}')] = \delta(\mathbf{p}-\mathbf{p}')$

← $[c(p^0, \mathbf{p}), c^\dagger(\omega_{\mathbf{p}}, \mathbf{p}')] = 0$, when $p^0 < 0$

$\frac{1}{2\omega_{\mathbf{p}}}[c(-\omega_{\mathbf{p}}, \mathbf{p}), c^\dagger(-\omega_{\mathbf{p}}, \mathbf{p}')] = \alpha \delta(\mathbf{p}-\mathbf{p}')$

$\Rightarrow$
$[a(\mathbf{p}), a^\dagger(\mathbf{p}')] = \delta(\mathbf{p}-\mathbf{p}')$
$[b(\mathbf{p}), b^\dagger(\mathbf{p}')] = \delta(\mathbf{p}-\mathbf{p}')$

True either for Case I or Case II with corresponding definitions of $b(\mathbf{p}), b^\dagger(\mathbf{p})$

Using the operators (1.251), (1.252) and the vacuum definition (1.253) we obtain that the scalar products between the states $|\mathbf{p}\rangle_+ = a^\dagger(\mathbf{p})|0\rangle$, $|\mathbf{p}\rangle_- = b^\dagger(\mathbf{p})|0\rangle$ are non-negative:

$$\langle \mathbf{p}|_+|\mathbf{p'}\rangle_+ = \langle \mathbf{p}|_-|\mathbf{p'}\rangle_- = \delta(\mathbf{p}-\mathbf{p'}),$$
$$\langle \mathbf{p}|_+|\mathbf{p'}\rangle_- = \langle \mathbf{p}|_-|\mathbf{p'}\rangle_+ = 0 \qquad (1.257)$$

If we now calculate, in the presence of the commutation relations (1.244) and the vacuum definition (1.253), the eigenvalues of the operator P^0 we find that they are all positive. Had we used the vacuum definition (1.248) we would have found that P^0 can have negative eigenvalues.

On the other hand, in *the unconstrained theory* the commutation relations (1.182), (1.183) and the vacuum definition (1.248) are valid. Within such a framework the eigenvalues of P^0 are also all positive, if the space-like hypersurface Σ is oriented along a positive time-like direction, otherwise they are all negative. The creation operators are $c^\dagger(p) = c^\dagger(p^0, \mathbf{p})$, and they create states $|p\rangle = |p^0, \mathbf{p}\rangle$ where p^0 can be positive or negative. Not only the states with positive, but also the states with negative 0-th component of *momentum* (equal to frequency if units are such that $\hbar = 1$) have positive energy P^0, regardless of the sign of p^0. One has to be careful not to confuse p^0 with the energy.

That energy is always positive is clear from the following classical example. If we have a matter continuum (fluid or dust) then the energy–momentum is defined as the integral over a hypersurface Σ of the stress–energy tensor:

$$P^\mu = \int T^{\mu\nu} \mathrm{d}\Sigma_\nu. \qquad (1.258)$$

For dust we have $T^{\mu\nu} = \rho u^\mu u^\nu$, where $u^\mu = \mathrm{d}x^\mu/\mathrm{d}s$ and s is the proper time. The dust energy is then $P^0 = \int \rho u^0 u^\nu n_\nu \, \mathrm{d}\Sigma$, where the hypersurface element has been written as $\mathrm{d}\Sigma_\nu = n_\nu \mathrm{d}\Sigma$. Here n^ν is a time-like vector field orthogonal to the hypersurface, pointing along a positive time-like direction (into the "future"), such that there exists a coordinate system in which $n_\nu = (1,0,0,0,...)$. Then $P^0 = \int \mathrm{d}\Sigma \, \rho u^0 u^0$ is obviously positive, even if u^0 is negative. If the dust consists of only one massive particle, then the density is singular on the particle worldline:

$$\rho(x) = m \int \mathrm{d}s \, \delta(x - X(s)), \qquad (1.259)$$

and

$$P^0 = \int \mathrm{d}\Sigma \, \mathrm{d}s \, m u^0 u^\mu n_\mu \, \delta(x - X(s)) = m u^0 \, (u^\mu n_\mu).$$

We see that the point particle energy, defined by means of $T^{\mu\nu}$, differs from $p^0 = m u^0$. The difference is in the factor $u^\mu n_\mu$, which in the chosen coordinate system may be plus or minus one.

Conclusion. Relativistic quantum field theory is one of the most successful physical theories. And yet it is not free from serious conceptual and technical difficulties. This was especially clear to Dirac, who expressed his opinion that quantum field theory, because it gives infinite results, should be taken only as a provisional theory waiting to be replaced by a better theory.

In this section I have challenged some of the cherished basic assumptions which, in my opinion, were among the main stumbling blocks preventing a further real progress in quantum field theory and its relation to gravity. These assumptions are:

1) *Identification of negative frequency with negative energy.* When a field is expanded both positive and negative $p^0 = \hbar\omega = \omega$, $\hbar = 1$, occur. It is then taken for granted that, by the correspondence principle, negative frequencies mean negative energies. *A tacit assumption is that the quantity p^0 is energy, while actually it is not energy*, as it is shown here.

2) *Identification of causality violation with propagation along space-like separations.* It is widely believed that faster–than–light propagation violates causality. A series of thought experiments with faster–than–light particles has been described [18] and concluded that they all lead to causal loops which are paradoxical. All those experiments are classical and tell us nothing about how the situation would change if quantum mechanics were taken into account[10]. In Sec. 13.1 I show that quantum mechanics, properly interpreted, eliminates the causality paradoxes of tachyons (and also of worm holes, as already stated by Deutsch). In this section, therefore, I have assumed that amplitudes do not need to vanish at space-like separations. What I gain is a very elegant, manifestly covariant quantum field theory based on the straightforward commutation relations (1.181)–(1.183) in which fields depend not only on spacetime coordinates x^μ, but also on the Poincaré invariant parameter τ (see Sec. 1.2). Evolution and causality are related to τ. The coordinate x^0 has nothing to do with evolution and causality considerations.

To sum up, we have here a consistent classical and quantum unconstrained (or "parametrized") field theory which is manifestly Lorentz covariant and in which the fields depend on an invariant parameter τ. Although the quantum states are localized in spacetime there is no problem with negative norm states and unitarity. This is a result of the fact that the commutation relations between the field operators are not quite the

---

[10] Usually it is argued that tachyons are even worse in quantum field theory, because their negative energies would have caused vacuum instability. Such an argument is valid in the conventional quantum field theory, but not in its generalization based on the invariant evolution parameter τ.

same as in the conventional relativistic quantum field theory. While in the conventional theory evolution of a state goes along the coordinate x^0, and is governed by the components P^0 of the momentum operator, in the unconstrained theory evolutions goes along τ and is governed by the covariant Hamiltonian H (eq. (1.161). The commutation relations between the field operators are such that the Heisenberg equation of motion determined by H is equivalent to the field equation which is just the Lorentz covariant Schrödinger equation (1.67). Comparison of the parametrized quantum field theory with the conventional relativistic quantum field theory reveals that the expectation values of energy–momentum and charge operator in the states with definite masses are the same for both theories. Only the free field case has been considered here. I expect that inclusion of interactions will be straightforward (as in the non-relativistic quantum field theory), very instructive and will lead to new, experimentally testable predictions. Such a development waits to be fully worked out, but many partial results have been reported in the literature [19, 20]. Although very interesting, it is beyond the scope of this book.

We have seen that the conventional field theory is obtained from the unconstrained theory if in the latter we take the definite mass fields $\phi(x)$ and treat negative frequencies differently from positive one. Therefore, strictly speaking, the conventional theory is not a special case of the unconstrained theory. When the latter theory is taken on mass shell then the commutator $[\phi(x), \phi(x')]$ assumes the form (1.243) and thus differs from the conventional commutator (1.245) which has the function $\epsilon(p^0) = (1/2)(\theta(p^0) - \theta(-p^0))$ under the integral.

Later we shall argue that the conventional relativistic quantum field theory is a special case, not of the unconstrained point particle, but of the unconstrained string theory, where the considered objects are the time-like strings (i.e. worldlines) moving in spacetime.

Chapter 2

POINT PARTICLES AND CLIFFORD ALGEBRA

Until 1992 I considered the tensor calculus of general relativity to be the most useful language by which to express physical theories. Although I was aware that differential forms were widely considered to be superior to tensor calculus I had the impression that they did not provide a sufficiently general tool for all the cases. Moreover, whenever an actual calculation had to be performed, one was often somehow forced to turn back to tensor calculus. A turning point in my endeavors to understand better physics and its mathematical formalization was in May 1992 when I met professor Waldyr Rodrigues, Jr.[1], who introduced me into the subject of *Clifford algebra*. After the one or two week discussion I became a real enthusiast of the geometric calculus based on Clifford algebra. This was the tool I always missed and that enabled me to grasp geometry from a wider perspective.

In this chapter I would like to forward my enthusiasm to those readers who are not yet enthusiasts themselves. I will describe the subject from a physicist's point of view and concentrate on the usefulness of Clifford algebra as a language for doing physics and geometry. I am more and more inclined towards the view that Clifford algebra is not only *a* language but but *the* language of a "unified theory" which will encompass all our current knowledge in fundamental theoretical physics.

I will attempt to introduce the ideas and the concepts in such a way as to satisfy one's intuition and facilitate an easy understanding. For more rigorous mathematical treatments the reader is advised to look at the existing literature [22, 23, 36].

---

[1] We were both guests of Erasmo Recami at The Institute of Theoretical Physics, Catania, Italia.

2.1. INTRODUCTION TO GEOMETRIC CALCULUS BASED ON CLIFFORD ALGEBRA

We have seen that point particles move in some kind of space. In non relativistic physics the space is 3-dimensional and Euclidean, while in the theory of relativity space has 4-dimensions and pseudo-Euclidean signature, and is called spacetime. Moreover, in general relativity spacetime is curved, which provides gravitation. If spacetime has even more dimensions —as in Kaluza–Klein theories— then such a higher-dimensional gravitation contains 4-dimensional gravity and Yang–Mills fields (including the fields associated with electromagnetic, weak, and strong forces). Since physics happens to take place in a certain space which has the role of a stage or arena, it is desirable to understand its geometric properties as deeply as possible.

Let V_n be a continuous space of arbitrary dimension n. To every point of V_n we can ascribe n *parameters* x^μ, $\mu = 1, 2, ..., n$, which are also called *coordinates*. Like house numbers they can be freely chosen, and once being fixed they specify points of the space[2].

When considering points of a space we ask ourselves what are the distances between the points. *The distance* between two infinitesimally separated points is given by

$$\mathrm{d}s^2 = g_{\mu\nu}\, \mathrm{d}x^\mu\, \mathrm{d}x^\nu. \tag{2.1}$$

Actually, this is the square of the distance, and $g_{\mu\nu}(x)$ is the metric tensor. The quantity $\mathrm{d}s^2$ is *invariant* with respect to general coordinate transformations $x^\mu \to x'^\mu = f^\mu(x)$.

Let us now consider *the square root* of the distance. Obviously it is $\sqrt{g_{\mu\nu}\mathrm{d}x^\mu\, \mathrm{d}x^\nu}$. But the latter expression is not linear in $\mathrm{d}x^\mu$. We would like to define an object which is *linear* in $\mathrm{d}x^\mu$ and whose square is eq. (2.1). Let such object be given by the expression

$$\mathrm{d}x = \mathrm{d}x^\mu\, e_\mu \tag{2.2}$$

It must satisfy

$$\mathrm{d}x^2 = e_\mu e_\nu\, \mathrm{d}x^\mu \mathrm{d}x^\nu = \tfrac{1}{2}(e_\mu e_\nu + e_\nu e_\mu)\, \mathrm{d}x^\mu \mathrm{d}x^\nu = g_{\mu\nu}\, \mathrm{d}x^\mu\, \mathrm{d}x^\nu = \mathrm{d}s^2, \tag{2.3}$$

from which it follows that

$$\tfrac{1}{2}(e_\mu e_\nu + e_\nu e_\mu) = g_{\mu\nu} \tag{2.4}$$

---

[2] See Sec. 6.2, in which the subtleties related to specification of spacetime points are discussed.

Point particles and Clifford algebra

The quantities e^μ so introduced are a new kind of number, called *Clifford numbers*. They do not commute, but satisfy eq. (2.4) which is a characteristic of *Clifford algebra*.

In order to understand what is the meaning of the object dx introduced in (2.2) let us study some of its properties. For the sake of avoiding use of differentials let us write (2.2) in the form

$$\frac{dx}{d\tau} = \frac{dx^\mu}{d\tau} e_\mu, \qquad (2.5)$$

where τ is an arbitrary parameter invariant under general coordinate transformations. Denoting $dx/d\tau \equiv a$, $dx^\mu/d\tau = a^\mu$, eq. (2.5) becomes

$$a = a^\mu e_\mu. \qquad (2.6)$$

Suppose we have two such objects a and b. Then

$$(a+b)^2 = a^2 + ab + ba + b^2 \qquad (2.7)$$

and

$$\tfrac{1}{2}(ab + ba) = \tfrac{1}{2}(e_\mu e_\nu + e_\nu e_\mu) a^\mu b^\nu = g_{\mu\nu} a^\mu a^\nu. \qquad (2.8)$$

The last equation algebraically corresponds to *the inner product* of two *vectors* with components a^μ and b^ν. Therefore we denote

$$a \cdot b \equiv \tfrac{1}{2}(ab + ba). \qquad (2.9)$$

From (2.7)–(2.8) we have that the sum $a + b$ is an object whose square is also a scalar.

What about the antisymmetric combinations? We have

$$\frac{1}{2}(ab - ba) = \tfrac{1}{2}(a^\mu b^\nu - a^\nu b^\mu) e_\mu e_\nu \qquad (2.10)$$

This is nothing but *the outer product* of the vectors. Therefore we denote it as

$$a \wedge b \equiv \tfrac{1}{2}(ab - ba) \qquad (2.11)$$

In 3-space this is related to the familiar vector product $a \times b$ which is the dual of $a \wedge b$.

The object $a = a^\mu e_\mu$ is thus nothing but a *vector*: a^μ are its components and e^μ are n linearly independent basic vectors of V_n. Obviously, if one changes parametrization, a or dx remains the same. Since under a general coordinate transformation the components a^μ and dx^μ do change, e_μ should also change in such a way that the vectors a and dx remain invariant.

An important lesson we have learnt so far is that

- the "square root" of the distance is a vector;
- vectors are Clifford numbers;
- vectors are objects which, like distance, are *invariant* under general coordinate transformations.

Box 2.1: Can we add apples and oranges?

When I asked my daughter, then ten years old, how much is 3 apples and 2 oranges plus 1 apple and 1 orange, she immediately replied "4 apples and 3 oranges". If a child has no problems with adding apples and oranges, it might indicate that contrary to the common wisdom, often taught at school, such an addition has mathematical sense after all. The best example that this is indeed the case is complex numbers. Here instead of 'apples' we have real and, instead of 'oranges', imaginary numbers. The sum of a real and imaginary number is a complex number, and summation of complex numbers is a mathematically well defined operation. Analogously, in Clifford algebra we can sum Clifford numbers of different degrees. In other words, summation of scalar, vectors, bivectors, etc., is a well defined operation.

The basic operation in Clifford algebra is *the Clifford product ab*. It can be decomposed into the symmetric part $a \cdot b$ (defined in (2.9)) and the antisymmetric part $a \wedge b$ (defined in (2.11)):

$$ab = a \cdot b + a \wedge b \qquad (2.12)$$

We have seen that $a \cdot b$ is a scalar. On the contrary, eq. (2.10) shows that $a \wedge b$ is not a scalar. Decomposing the product $e_\mu e_\nu$ according to (2.12),

$$e_\mu e_\nu = e_\mu \cdot e_\nu + e_\mu \wedge e_\nu = g_{\mu\nu} + e_\mu \wedge e_\nu,$$

we can rewrite (2.10) as

$$a \wedge b = \frac{1}{2}(a^\mu b^\nu - a^\nu b^\mu) e_\mu \wedge e_\nu, \qquad (2.13)$$

which shows that $a \wedge b$ is a new type of geometric object, called *bivector*, which is neither a scalar nor a vector.

Point particles and Clifford algebra 57

The geometric product (2.12) is thus the sum of a scalar and a bivector. The reader who has problems with such a sum is advised to read Box 2.1.

A *vector* is an algebraic representation of *direction* in a space V_n; it is associated with an oriented line.

A *bivector* is an algebraic representation of an oriented plane.

This suggests a generalization to trivectors, quadrivectors, etc. It is convenient to introduce the name *r-vector* and call r its *degree*:

0-vector	s	scalar
1-vector	a	vector
2-vector	$a \wedge b$	bivector
3-vector	$a \wedge b \wedge c$	trivector
.	.	.
.	.	.
.	.	.
r-vector	$A_r = a_1 \wedge a_2 \wedge ... \wedge a_r$	multivector

In a space of finite dimension this cannot continue indefinitely: an n-vector is the highest r-vector in V_n and an $(n+1)$-*vector* is identically zero. An r-vector A_r represents an oriented r-volume (or r-direction) in V_n.

Multivectors A_r are elements of the *Clifford algebra* C_n of V_n. An element of C_n will be called a *Clifford number*. Clifford numbers can be multiplied amongst themselves and the results are Clifford numbers of mixed degrees, as indicated in the basic equation (2.12). The theory of multivectors, based on Clifford algebra, was developed by Hestenes [22]. In Box 2.2 some useful formulas are displayed without proofs.

Let $e_1, e_2, ..., e_n$ be linearly independent vectors, and $\alpha, \alpha^i, \alpha^{i_1 i_2}, ...$ scalar coefficients. A generic Clifford number can then be written as

$$A = \alpha + \alpha^i e_i + \frac{1}{2!} \alpha^{i_1 i_2} e_{i_1} \wedge e_{i_2} + ... \frac{1}{n!} \alpha^{i_1 ... i_n} e_{i_1} \wedge ... \wedge e_{i_n}. \qquad (2.14)$$

Since it is a superposition of multivectors of all possible grades it will be called *polyvector*.[3] Another name, also often used in the literature, is *Clifford aggregate*. These mathematical objects have far reaching geometrical and physical implications which will be discussed and explored to some extent in the rest of the book.

---

[3]Following a suggestion by Pezzaglia [23] I call a generic Clifford number *polyvector* and reserve the name *multivector* for an r-vector, since the latter name is already widely used for the corresponding object in the calculus of differential forms.

Box 2.2: Some useful basic equations

For a vector a and an r-vector A_r the inner and the outer product are defined according to

$$a \cdot A_r \equiv \tfrac{1}{2}(aA_r - (-1)^r A_r a) = -(-1)^r A_r \cdot a, \qquad (2.15)$$

$$a \wedge A_r = \tfrac{1}{2}(aA_r + (-1)^r A_r a) = (-1)^r A_r \wedge a. \qquad (2.16)$$

The inner product has symmetry opposite to that of the outer product, therefore the signs in front of the second terms in the above equations are different.
Combining (2.15) and (2.16) we find

$$aA_r = a \cdot A_r + a \wedge A_r. \qquad (2.17)$$

For $A_r = a_1 \wedge a_2 \wedge ... \wedge a_r$ eq. (2.15) can be evaluated to give the useful expansion

$$a \cdot (a_1 \wedge ... \wedge a_r) = \sum_{k=1}^{r} (-1)^{k+1}(a \cdot a_k) a_1 \wedge ... a_{k-1} \wedge a_{k+1} \wedge ... a_r. \qquad (2.18)$$

In particular,

$$a \cdot (b \wedge c) = (a \cdot b)c - (a \cdot c)b. \qquad (2.19)$$

It is very convenient to introduce, besides the basic vectors e_μ, another set of basic vectors e^ν by the condition

$$e_\mu \cdot e^\nu = \delta_\mu{}^\nu. \qquad (2.20)$$

Each e^μ is a linear combination of e_ν:

$$e^\mu = g^{\mu\nu} e_\nu, \qquad (2.21)$$

from which we have

$$g^{\mu\alpha} g_{\alpha\nu} = \delta_\mu{}^\nu \qquad (2.22)$$

and

$$g^{\mu\nu} = e^\mu \cdot e^\nu = \tfrac{1}{2}(e^{\mu\nu} + e^\nu e^\mu). \qquad (2.23)$$

Point particles and Clifford algebra 59

2.2. ALGEBRA OF SPACETIME

In spacetime we have 4 linearly independent vectors e_μ, $\mu = 0,1,2,3$. Let us consider *flat* spacetime. It is then convenient to take orthonormal basis vectors γ_μ

$$\gamma_\mu \cdot \gamma_\nu = \eta_{\mu\nu}, \qquad (2.24)$$

where $\eta_{\mu\nu}$ is the diagonal metric tensor with signature $(+ - - -)$.

The Clifford algebra in V_4 is called the *Dirac algebra*. Writing $\gamma_{\mu\nu} \equiv \gamma_\mu \wedge \gamma_\nu$ for a basis bivector, $\gamma_{\mu\nu\rho} \equiv \gamma_\mu \wedge \gamma_\nu \wedge \gamma_\rho$ for a basis trivector, and $\gamma_{\mu\nu\rho\sigma} \equiv \gamma_\mu \wedge \gamma_\nu \wedge \gamma_\rho \wedge \gamma_\sigma$ for a basis quadrivector we can express an arbitrary number of the Dirac algebra as

$$D = \sum_r D_r = d + d^\mu \gamma_\mu + \frac{1}{2!} d^{\mu\nu} \gamma_{\mu\nu} + \frac{1}{3!} d^{\mu\nu\rho} \gamma_{\mu\nu\rho} + \frac{1}{4!} d^{\mu\nu\rho\sigma} \gamma_{\mu\nu\rho\sigma}, \quad (2.25)$$

where $d, d^\mu, d^{\mu\nu}, \ldots$ are scalar coefficients.

Let us introduce

$$\gamma_5 \equiv \gamma_0 \wedge \gamma_1 \wedge \gamma_2 \wedge \gamma_3 = \gamma_0 \gamma_1 \gamma_2 \gamma_3, \qquad \gamma_5^2 = -1, \qquad (2.26)$$

which is the unit element of 4-dimensional volume and is called a *pseudoscalar*. Using the relations

$$\gamma_{\mu\nu\rho\sigma} = \gamma_5 \epsilon_{\mu\nu\rho\sigma}, \qquad (2.27)$$

$$\gamma_{\mu\nu\rho} = \gamma_{\mu\nu\rho\sigma} \gamma^\sigma, \qquad (2.28)$$

where $\epsilon_{\mu\nu\rho\sigma}$ is the totally antisymmetric tensor and introducing the new coefficients

$$S \equiv d, \quad V^\mu \equiv d^\mu, \quad T^{\mu\nu} \equiv \tfrac{1}{2} d^{\mu\nu},$$
$$C_\sigma \equiv \frac{1}{3!} d^{\mu\nu\rho} \epsilon_{\mu\nu\rho\sigma}, \quad P \equiv \frac{1}{4!} d^{\mu\nu\rho\sigma} \epsilon_{\mu\nu\rho\sigma}, \qquad (2.29)$$

we can rewrite D of eq. (2.25) as the sum of scalar, vector, bivector, pseudovector and pseudoscalar parts:

$$D = S + V^\mu \gamma_\mu + T^{\mu\nu} \gamma_{\mu\nu} + C^\mu \gamma_5 \gamma_\mu + P \gamma_5. \qquad (2.30)$$

POLYVECTOR FIELDS

A polyvector may depend on spacetime points. Let $A = A(x)$ be an r-vector field. Then one can define the *gradient operator* according to

$$\partial = \gamma^\mu \partial_\mu, \qquad (2.31)$$

where ∂_μ is the usual partial derivative. The gradient operator ∂ can act on any r-vector field. Using (2.17) we have

$$\partial A = \partial \cdot A + \partial \wedge A. \qquad (2.32)$$

Example. Let $A = a = a_\nu \gamma^\nu$ be a 1-vector field. Then

$$\begin{aligned}\partial a &= \gamma^\mu \partial_\mu (a_\nu \gamma^\nu) = \gamma^\mu \cdot \gamma^\nu \, \partial_\mu a^\nu + \gamma^\mu \wedge \gamma^\nu \partial_\mu a_\nu \\ &= \partial_\mu a^\mu + \tfrac{1}{2}(\partial_\mu a_\nu - \partial_\nu a_\mu)\gamma^\mu \wedge \gamma^\nu.\end{aligned} \qquad (2.33)$$

The simple expression ∂a thus contains a scalar and a bivector part, the former being the usual divergence and the latter the usual curl of a vector field.

Maxwell equations. We shall now demonstrate by a concrete physical example the usefulness of Clifford algebra. Let us consider the electromagnetic field which, in the language of Clifford algebra, is a bivector field F. The source of the field is the electromagnetic current j which is a 1-vector field. Maxwell's equations read

$$\partial F = 4\pi j. \qquad (2.34)$$

The grade of the gradient operator ∂ is 1. Therefore we can use the relation (2.32) and we find that eq. (2.34) becomes

$$\partial \cdot F + \partial \wedge F = 4\pi j, \qquad (2.35)$$

which is equivalent to

$$\partial \cdot F = -4\pi j, \qquad (2.36)$$

$$\partial \wedge F = 0, \qquad (2.37)$$

since the first term on the left of eq. (2.35) is a vector and the second term is a bivector. This results from the general relation (2.35). It can also be explicitly demonstrated. Expanding

$$F = \tfrac{1}{2} F^{\mu\nu} \gamma_\mu \wedge \gamma_\nu, \qquad (2.38)$$

Point particles and Clifford algebra

$$j = j^\mu \gamma_\mu, \qquad (2.39)$$

we have

$$\partial \cdot F = \gamma^\alpha \partial_\alpha \cdot (\tfrac{1}{2} F^{\mu\nu} \gamma_\mu \wedge \gamma_\nu) = \tfrac{1}{2} \gamma^\alpha \cdot (\gamma_\mu \wedge \gamma_\nu) \partial_\alpha F^{\mu\nu}$$

$$= \tfrac{1}{2} ((\gamma^\alpha \cdot \gamma_\mu) \gamma_\nu - (\gamma^\alpha \cdot \gamma_\nu) \gamma_\mu) \partial_\alpha F^{\mu\nu} = \partial_\mu F^{\mu\nu} \gamma_\nu, \qquad (2.40)$$

$$\partial \wedge F = \tfrac{1}{2} \gamma^\alpha \wedge \gamma_\mu \wedge \gamma_\nu \partial_\alpha F^{\mu\nu} = \tfrac{1}{2} \epsilon^\alpha{}_{\mu\nu\rho} \partial_\alpha F^{\mu\nu} \gamma_5 \gamma^\rho, \qquad (2.41)$$

where we have used (2.19) and eqs.(2.27), (2.28). From the above considerations it then follows that the compact equation (2.34) is equivalent to the usual tensor form of Maxwell equations

$$\partial_\nu F^{\mu\nu} = -4\pi j^\mu, \qquad (2.42)$$

$$\epsilon^\alpha{}_{\mu\nu\rho} \partial_\alpha F^{\mu\nu} = 0. \qquad (2.43)$$

Applying the gradient operator ∂ to the left and to the right side of eq. (2.34) we have

$$\partial^2 F = \partial j. \qquad (2.44)$$

Since $\partial^2 = \partial \cdot \partial + \partial \wedge \partial = \partial \cdot \partial$ is a scalar operator, $\partial^2 F$ is a bivector. The right hand side of eq. (2.44) gives

$$\partial j = \partial \cdot j + \partial \wedge j. \qquad (2.45)$$

Equating the terms of the same grade on the left and the right hand side of eq. (2.44) we obtain

$$\partial^2 F = \partial \wedge j, \qquad (2.46)$$

$$\partial \cdot j = 0. \qquad (2.47)$$

The last equation expresses the conservation of the electromagnetic current.

Motion of a charged particle. In this example we wish to go a step forward. Our aim is not only to describe how a charged particle moves in an electromagnetic field, but also include a particle's(classical) spin. Therefore, following Pezzaglia [23], we define the *momentum polyvector P* as the *vector momentum p* plus the bivector *spin angular momentum S*,

$$P = p + S, \qquad (2.48)$$

or in components

$$P = p^\mu \gamma_\mu + \tfrac{1}{2} S^{\mu\nu} \gamma_\mu \wedge \gamma_\nu. \qquad (2.49)$$

We also assume that the condition $p_\mu S^{\mu\nu} = 0$ is satisfied. The latter condition ensures the spin to be a simple bivector, which is purely space-like in the rest frame of the particle. The polyvector equation of motion is

$$\dot P \equiv \frac{\mathrm{d}P}{\mathrm{d}\tau} = \frac{e}{2m}[P,F], \qquad (2.50)$$

where $[P,F] \equiv PF - FP$. The vector and bivector parts of eq. (2.50) are

$$\dot p^\mu = \frac{e}{m} F^\mu{}_\nu p^\nu, \qquad (2.51)$$

$$\dot S^{\mu\nu} = \frac{e}{2m}(F^\mu{}_\alpha S^{\alpha\nu} - F^\nu{}_\alpha S^{\alpha\mu}). \qquad (2.52)$$

These are just the equations of motion for linear momentum and spin, respectively.

2.3. PHYSICAL QUANTITIES AS POLYVECTORS

The compact equations at the end of the last section suggest a generalization that every physical quantity is a polyvector. We shall explore such an assumption and see how far we can come.

In 4-dimensional spacetime *the momentum polyvector* is

$$P = \mu + p^\mu e_\mu + S^{\mu\nu} e_\mu e_\nu + \pi^\mu e_5 e_\mu + m e_5, \qquad (2.53)$$

and *the velocity polyvector* is

$$\dot X = \dot\sigma + \dot x^\mu e_\mu + \dot\alpha^{\mu\nu} e_\mu e_\nu + \dot\xi^\mu e_5 e_\mu + \dot s e_5, \qquad (2.54)$$

where e_μ are four basis vectors satisfying

$$e_\mu \cdot e_\nu = \eta_{\mu\nu}, \qquad (2.55)$$

and $e_5 \equiv e_0 e_1 e_2 e_3$ is the pseudoscalar. For the purposes which will become clear later we now use the symbols e_μ, e_5 instead of γ_μ and γ_5.

We associate with each particle the velocity polyvector $\dot X$ and its conjugate momentum polyvector P. These quantities are generalizations of the point particle 4-velocity $\dot x$ and its conjugate momentum p. Besides a vector part we now include the scalar part $\dot\sigma$, the bivector part $\dot\alpha^{\mu\nu} e_\mu e_\nu$, the pseudovector part $\dot\xi^\mu e_5 e_\mu$ and the pseudoscalar part $\dot s e_5$ into the definition

Point particles and Clifford algebra

of the particle's velocity, and analogously for the particle's momentum. We would now like to derive the equations of motion which will tell us how those quantities depend on the evolution parameter τ. For simplicity we consider a free particle.

Let the action be a straightforward generalization of the first order or phase space action (1.11) of the usual constrained point particle relativistic theory:

$$I[X, P, \lambda] = \frac{1}{2} \int d\tau \left(P\dot{X} + \dot{X}P - \lambda(P^2 - K^2) \right), \qquad (2.56)$$

where λ is a scalar Lagrange multiplier and K a polyvector constant[4]:

$$K^2 = \kappa^2 + k_\mu e_\mu + K^{\mu\nu} e_\mu e_\nu + K^\mu e_5 e_\mu + k^2 e_5. \qquad (2.57)$$

It is a generalization of particle's mass squared. In the usual, unconstrained, theory, mass squared was a scalar constant, but here we admit that, in principle, mass squared is a polyvector. Let us now insert the explicit expressions (2.53),(2.54) and (2.57) into the Lagrangian

$$L = \tfrac{1}{2} \left(P\dot{X} + \dot{X}P - \lambda(P^2 - K^2) \right) = \sum_{r=0}^{4} \langle L \rangle_r, \qquad (2.58)$$

and evaluate the corresponding multivector parts $\langle L \rangle_r$. Using

$$e_\mu \wedge e_\nu \wedge e_\rho \wedge e_\sigma = e_5 \, \epsilon_{\mu\nu\rho\sigma}, \qquad (2.59)$$

$$e_\mu \wedge e_\nu \wedge e_\rho = (e_\mu \wedge e_\nu \wedge e_\rho \wedge e_\sigma) e^\sigma = e_5 \, \epsilon_{\mu\nu\rho\sigma} e^\sigma, \qquad (2.60)$$

$$e_\mu \wedge e_\nu = -\tfrac{1}{2}(e_\mu \wedge e_\nu \wedge e_\rho \wedge e_\sigma)(e^\rho \wedge e^\sigma) = -\tfrac{1}{2} e_5 \, \epsilon_{\mu\nu\rho\sigma} e^\rho \wedge e^\sigma, \qquad (2.61)$$

we obtain

$$\langle L \rangle_0 = \mu\dot{\sigma} - m\dot{s} + p_\mu \dot{x}^\mu + \pi_\mu \dot{\xi}^\mu + S^{\mu\nu} \dot{\alpha}^{\rho\sigma} \eta_{\mu\sigma} \eta_{\nu\rho}$$
$$- \frac{\lambda}{2}(\mu^2 + p^\mu p_\mu + \pi^\mu \pi_\mu - m^2 - 2S^{\mu\nu} S_{\mu\nu} - \kappa^2), \qquad (2.62)$$

$$\langle L \rangle_1 = \left[\dot{\sigma} p_\sigma + \mu \dot{x}_\sigma - (\dot{\xi}^\rho S^{\mu\nu} + \pi^\rho \dot{\alpha}^{\mu\nu}) \epsilon_{\mu\nu\rho\sigma} \right] e^\sigma$$
$$- \lambda(\mu p_\sigma - S^{\mu\nu} \pi^\rho \epsilon_{\mu\nu\rho\sigma} - \tfrac{1}{2} k_\sigma) e^\sigma, \qquad (2.63)$$

---

[4]The scalar part is not restricted to positive values, but for later convenience we write it as κ^2, on the understanding that κ^2 can be positive, negative or zero.

$$\langle L \rangle_2 =$$
$$\left[\tfrac{1}{2}(\pi^\mu \dot{x}^\nu - p^\mu \dot{\xi}^\nu + \dot{s} S^{\mu\nu} + m\dot{\alpha}^{\mu\nu})\epsilon_{\mu\nu\rho\sigma} + \dot{\sigma} S_{\rho\sigma} + \mu\dot{\alpha}_{\rho\sigma} + 2 S_{\rho\nu}\dot{\alpha}^\nu_\sigma \right] e^\rho \wedge e^\sigma$$
$$- \frac{\lambda}{2} \left[(\pi^\mu p^\nu + m S^{\mu\nu})\epsilon_{\mu\nu\rho\sigma} + 2\mu S_{\rho\sigma} - K_{\rho\sigma} \right] e^\rho \wedge e^\sigma , \qquad (2.64)$$

$$\langle L \rangle_3 = \left[\dot{\sigma}\pi^\sigma + \mu\dot{\xi}^\sigma + (S^{\mu\nu}\dot{x}^\rho + \dot{\alpha}^{\mu\nu}p^\rho)\epsilon_{\mu\nu\rho}{}^\sigma - \lambda(\mu\pi^\sigma + S^{\mu\nu}p^\rho \epsilon_{\mu\nu\rho}{}^\sigma - \tfrac{1}{2}\kappa_\sigma) \right] e_5 e^\sigma , \qquad (2.65)$$

$$\langle L \rangle_4 = \left[m\dot{\sigma} + \mu\dot{s} - \tfrac{1}{2} S^{\mu\nu} \dot{\alpha}^{\rho\sigma} \epsilon_{\mu\nu\rho\sigma} - \frac{\lambda}{2}(2\mu m + S^{\mu\nu} S^{\rho\sigma} \epsilon_{\mu\nu\rho\sigma} - k^2) \right] e_5. \qquad (2.66)$$

The equations of motion are obtained for each pure grade multivector $\langle L \rangle_r$ separately. That is, when varying the polyvector action I, we vary each of its r-vector parts separately. From the scalar part $\langle L \rangle_0$ we obtain

$$\delta\mu \quad : \quad \dot{\sigma} - \lambda\mu = 0, \qquad (2.67)$$
$$\delta m \quad : \quad -\dot{s} + \lambda m = 0, \qquad (2.68)$$
$$\delta s \quad : \quad \dot{m} = 0 \qquad (2.69)$$
$$\delta\sigma \quad : \quad \dot{\mu} = 0, \qquad (2.70)$$
$$\delta p_\mu \quad : \quad \dot{x}^\mu - \lambda p^\mu = 0, \qquad (2.71)$$
$$\delta\pi_\mu \quad : \quad \dot{\xi}^\mu - \lambda\pi^\mu = 0, \qquad (2.72)$$
$$\delta x^\mu \quad : \quad \dot{p}^\mu = 0 \qquad (2.73)$$
$$\delta\xi^\mu \quad : \quad \dot{\pi}_\mu = 0, \qquad (2.74)$$
$$\delta\alpha^{\mu\nu} \quad : \quad \dot{S}_{\mu\nu} = 0, \qquad (2.75)$$
$$\delta S_{\mu\nu} \quad : \quad \dot{\alpha}^{\mu\nu} - \lambda S^{\mu\nu} = 0. \qquad (2.76)$$

From the r-vector parts $\langle L \rangle_r$ for $r = 1, 2, 3, 4$ we obtain the same set of equations (2.67)–(2.76). Each individual equation results from varying a different variable in $\langle L \rangle_0$, $\langle L \rangle_1$, etc.. Thus, for instance, the μ-equation of motion (2.67) from $\langle L \rangle_0$ is the same as the p_μ equation from $\langle L \rangle_1$ and the same as the m-equation from $\langle L \rangle_4$, and similarly for all the other equations (2.67)–(2.76). Thus, as far as the variables μ, m, s, σ, p_μ, π_μ, $S_{\mu\nu}$, ξ^μ and $\alpha^{\mu\nu}$ are considered, the higher grade parts $\langle L \rangle_r$ of the Lagrangian L contains the same information about the equations of motion. The difference occurs if we consider the *Lagrange multiplier* λ. Then every r-vector part of L gives a different equation of motion:

$$\frac{\partial \langle L \rangle_0}{\partial \lambda} = 0 : \quad \mu^2 + p^\mu p_\mu + \pi^\mu \pi_\mu - m^2 - 2 S^{\mu\nu} S_{\mu\nu} - \kappa^2 = 0, (2.77)$$

$$\frac{\partial \langle L \rangle_1}{\partial \lambda} = 0 : \quad \mu \pi_\sigma - S^{\mu\nu} \pi^\rho \epsilon_{\mu\nu\rho\sigma} - \tfrac{1}{2} k_\sigma = 0, \tag{2.78}$$

$$\frac{\partial \langle L \rangle_2}{\partial \lambda} = 0 : \quad (\pi^\mu \pi^\nu + m S^{\mu\nu}) \epsilon_{\mu\nu\rho\sigma} + 2\mu S_{\rho\sigma} - K_{\rho\sigma} = 0, \tag{2.79}$$

$$\frac{\partial \langle L \rangle_3}{\partial \lambda} = 0 : \quad \mu \pi_\sigma + S^{\mu\nu} p^\rho \epsilon_{\mu\nu\rho\sigma} + \tfrac{1}{2} \kappa_\sigma = 0, \tag{2.80}$$

$$\frac{\partial \langle L \rangle_4}{\partial \lambda} = 0 : \quad 2\mu m + S^{\mu\nu} S^{\rho\sigma} \epsilon_{\mu\nu\rho\sigma} - k^2 = 0. \tag{2.81}$$

The above equations represent constraints among the dynamical variables. Since our Lagrangian is not a scalar but a polyvector, we obtain more than one constraint.

Let us rewrite eqs.(2.80),(2.78) in the forms

$$\pi_\sigma - \frac{\kappa_\sigma}{2\mu} = -\frac{1}{\mu} S^{\mu\nu} p^\rho \epsilon_{\mu\nu\rho\sigma}, \tag{2.82}$$

$$p_\sigma - \frac{k_\sigma}{2\mu} = \frac{1}{\mu} S^{\mu\nu} \pi^\rho \epsilon_{\mu\nu\rho\sigma}. \tag{2.83}$$

We see from (2.82) that the vector momentum p_μ and its pseudovector partner π_μ are related in such a way that $\pi_\mu - \kappa_\mu/2\mu$ behaves as the well known *Pauli-Lubanski spin pseudo vector*. A similar relation (2.83) holds if we interchange p_μ and π_μ.

Squaring relations (2.82), (2.83) we find

$$(\pi_\sigma - \frac{\kappa_\sigma}{2\mu})(\pi^\sigma - \frac{\kappa^\sigma}{2\mu}) = -\frac{2}{\mu^2} p_\sigma p^\sigma S_{\mu\nu} S^{\mu\nu} + \frac{4}{\mu^2} p_\mu p^\nu S^{\mu\sigma} S_{\nu\sigma}, \tag{2.84}$$

$$(p_\sigma - \frac{k_\sigma}{2\mu})(p^\sigma - \frac{k^\sigma}{2\mu}) = -\frac{2}{\mu^2} \pi_\sigma \pi^\sigma S_{\mu\nu} S^{\mu\nu} + \frac{4}{\mu^2} \pi_\mu \pi^\nu S^{\mu\sigma} S_{\nu\sigma}. \tag{2.85}$$

From (2.82), (2.83) we also have

$$\left(\pi_\sigma - \frac{\kappa_\sigma}{2\mu}\right) p^\sigma = 0, \quad \left(p_\sigma - \frac{k_\sigma}{2\mu}\right) \pi^\sigma = 0. \tag{2.86}$$

Additional interesting equations which follow from (2.82), (2.83) are

$$p^\rho p_\rho \left(1 + \frac{2}{\mu^2} S_{\mu\nu} S^{\mu\nu}\right) - \frac{4}{\mu^2} p_\mu p^\nu S^{\mu\sigma} S_{\nu\sigma} = -\frac{\kappa_\rho \kappa^\rho}{4\mu^2} + \frac{1}{2\mu}(p_\rho k^\rho + \pi_\rho \kappa^\rho), \tag{2.87}$$

$$\pi^\rho \pi_\rho \left(1 + \frac{2}{\mu^2} S_{\mu\nu} S^{\mu\nu}\right) - \frac{4}{\mu^2} \pi_\mu \pi^\nu S^{\mu\sigma} S_{\nu\sigma} = -\frac{k_\rho k^\rho}{4\mu^2} + \frac{1}{2\mu}(p_\rho k^\rho + \pi_\rho \kappa^\rho). \tag{2.88}$$

Contracting (2.82), (2.83) by $\epsilon^{\alpha_1\alpha_2\alpha_3\sigma}$ we can express $S^{\mu\nu}$ in terms of p^ρ and π^σ

$$S^{\mu\nu} = \frac{\mu}{2p^\alpha p_\alpha} \epsilon^{\mu\nu\rho\sigma} p_\rho \left(\pi_\sigma - \frac{\kappa_\sigma}{2\mu}\right) = -\frac{\mu}{2\pi^\alpha \pi_\alpha} \epsilon^{\mu\nu\rho\sigma} \pi_\rho \left(p_\sigma - \frac{k_\sigma}{2\mu}\right), \quad (2.89)$$

provided that we assume the following extra condition:

$$S^{\mu\nu} p_\nu = 0, \qquad S^{\mu\nu} \pi_\nu = 0. \quad (2.90)$$

Then for positive $p^\sigma p_\sigma$ it follows from (2.84) that $(\pi_\sigma - \kappa_\sigma/2\mu)^2$ is negative, i.e., $\pi_\sigma - \kappa_\sigma/2\mu$ are components of a space-like (pseudo-) vector. Similarly, it follows from (2.85) that when $\pi^\sigma \pi_\sigma$ is negative, $(p_\sigma - k_\sigma/2\mu)^2$ is positive, so that $p_\sigma - k_\sigma/2\mu$ is a time-like vector. Altogether we thus have that p_σ, k_σ are time-like and π_σ, κ_σ are space-like. Inserting (2.89) into the remaining constraint (2.81) and taking into account the condition (2.90) we obtain

$$2m\mu - k^2 = 0. \quad (2.91)$$

The polyvector action (2.56) is thus shown to represent a very interesting classical dynamical system with spin. The interactions could be included by generalizing the minimal coupling prescription. *Gravitational interaction* is included by generalizing (2.55) to

$$e_\mu \cdot e_\nu = g_{\mu\nu}, \quad (2.92)$$

where $g_{\mu\nu}(x)$ is the spacetime metric tensor. A *gauge interaction* is included by introducing a polyvector gauge field A, a polyvector coupling constant G, and assume an action of the kind

$$I[X, P, \lambda] = \tfrac{1}{2} \int d\tau \left[P\dot{X} + \dot{X}P - \lambda\left((P - G \star A)^2 - K^2\right)\right], \quad (2.93)$$

where '$\star$' means the scalar product between Clifford numbers, so that $G \star A \equiv \langle GA \rangle_0$. The polyvector equations of motion can be elegantly obtained by using the Hestenes formalism for multivector derivatives. We shall not go into details here, but merely sketch a plausible result,

$$\dot{\Pi} = \lambda[G \star \partial_X A, P], \qquad \Pi \equiv P - G \star A, \quad (2.94)$$

which is a generalized Lorentz force equation of motion, a more particular case of which is given in (2.50).

After this short digression let us return to our free particle case. One question immediately arises, namely, what is the physical meaning of the polyvector mass squared K^2. Literally this means that a particle is characterized not only by a scalar and/or a pseudoscalar mass squared, but also by a vector, bivector and pseudovector mass squared. To a particle are thus associated a constant vector, 2-vector, and 3-vector which point into fixed directions in spacetime, regardless of the direction of particle's motion. For a given particle the Lorentz symmetry is thus broken, since there exists a preferred direction in spacetime. This cannot be true at the fundamental level. Therefore the occurrence of the polyvector K^2 in the action must be a result of a more fundamental dynamical principle, presumably an action in a higher-dimensional spacetime without such a fixed term K^2. It is well known that the scalar mass term in 4-dimensions can be considered as coming from a massless action in 5 or more dimensions. Similarly, also the 1-vector, 2-vector, and 3-vector terms of K^2 can come from a higher-dimensional action without a K^2-term. Thus *in 5-dimensions*:

(i) *the scalar constraint* will contain the term $p^A p_A = p^\mu p_\mu + p^5 p_5$, and the constant $-p^5 p_5$ takes the role of the scalar mass term in 4-dimensions;

(ii) *the vector constraint* will contain a term like $P_{ABC} S^{AB} e^C$, $A, B = 0, 1, 2, 3, 5$, containing the term $P_{\mu\nu\alpha} S^{\mu\nu} e^\alpha$ (which, since $P_{\mu\nu\alpha} = \epsilon_{\mu\nu\alpha\beta}\pi^\beta$, corresponds to the term $S^{\mu\nu}\pi^\rho \epsilon_{\mu\nu\rho\sigma} e^\sigma$) plus an extra term $P_{5\nu\alpha} S^{5\alpha} e^\alpha$ which corresponds to the term $k^\alpha e_\alpha$.

In a similar manner we can generate the 2-vector term $K_{\mu\nu}$ and the 3-vector term κ_σ from 5-dimensions.

The polyvector mass term K^2 in our 4-dimensional action (2.93) is arbitrary in principle. Let us find out what happens if we set $K^2 = 0$. Then, in the presence of the condition (2.90), eqs. (2.87) or (2.88) imply

$$S_{\mu\nu} S^{\mu\nu} = -\frac{\mu^2}{2}, \tag{2.95}$$

that is $S_{\mu\nu} S^{\mu\nu} < 0$. On the other hand $S_{\mu\nu} S^{\mu\nu}$ in the presence of the condition (2.90) can only be positive (or zero), as can be straightforwardly verified. In 4-dimensional spacetime $S_{\mu\nu} S^{\mu\nu}$ were to be negative only if in the particle's rest frame the spin components S^{0r} were different from zero which would be the case if (2.90) would not hold.

Let us assume that $K^2 = 0$ and that condition (2.90) does hold. Then the constraints (2.78)–(2.81) have a solution[5]

$$S^{\mu\nu} = 0, \quad \pi^\mu = 0, \quad \mu = 0. \tag{2.96}$$

---

[5]This holds even if we keep κ^2 different from zero, but take vanishing values for k^2, κ_μ, k_μ and $K_{\mu\nu}$.

The only remaining constraint is thus

$$p^\mu p_\mu - m^2 = 0, \tag{2.97}$$

and the polyvector action (2.56) is simply

$$\begin{aligned} I[X, P, \lambda] &= I[s, m, x^\mu, p_\mu, \lambda] \\ &= \int d\tau \left[-m\dot{s} + p_\mu \dot{x}^\mu - \frac{\lambda}{2}(p^\mu p_\mu - m^2) \right], \end{aligned} \tag{2.98}$$

in which the mass m is a dynamical variable conjugate to s. In the action (2.98) mass is thus just a pseudoscalar component of the polymomentum

$$P = p^\mu e_\mu + m e_5, \tag{2.99}$$

and $\dot{s}$ is a pseudoscalar component of the velocity polyvector

$$\dot{X} = \dot{x}^\mu e_\mu + \dot{s} e_5. \tag{2.100}$$

Other components of the polyvectors $\dot{X}$ and P (such as $S^{\mu\nu}$, π^μ, μ), when $K^2 = 0$ (or more weakly, when $K^2 = \kappa^2$), are automatically eliminated by the constraints (2.77)–(2.81).

From a certain point of view this is very good, since our analysis of the polyvector action (2.56) has driven us close to the conventional point particle theory, with the exception that mass is now a dynamical variable. This reminds us of the Stueckelberg point particle theory [2]–[15] in which mass is a constant of motion. This will be discussed in the next section. We have here demonstrated in a very elegant and natural way that the Clifford algebra generalization of the classical point particle in four dimensions tells us that a fixed mass term in the action cannot be considered as fundamental. This is not so obvious for the scalar (or pseudoscalar) part of the polyvector mass squared term K^2, but becomes dramatically obvious for the 1-vector, 2-vector and 4-vector parts, because they imply a preferred direction in spacetime, and such a preferred direction cannot be fundamental if the theory is to be Lorentz covariant.

This is a very important point and I would like to rephrase it. We start with the well known relativistic constrained action

$$I[x^\mu, p_\mu, \lambda] = \int d\tau \left(p_\mu \dot{x}^\mu - \frac{\lambda}{2}(p^2 - \kappa^2) \right). \tag{2.101}$$

Faced with the existence of the geometric calculus based on Clifford algebra, it is natural to generalize this action to polyvectors. Concerning the fixed mass constant κ^2 it is natural to replace it by a fixed polyvector or to discard it. If we discard it we find that mass is nevertheless present, because now momentum is a polyvector and as such it contains a pseudoscalar part me_5. If we keep the fixed mass term then we must also keep, in principle, its higher grade parts, but this is in conflict with Lorentz covariance. Therefore the fixed mass term in the action is not fundamental but comes, for instance, from higher dimensions. Since, without the K^2 term, in the presence of the condition $S^{\mu\nu}p_\nu = 0$ we cannot have classical spin in four dimensions (eq. (2.95) is inconsistent), this points to the existence of higher dimensions. Spacetime must have more than four dimensions, where we expect that the constraint $P^2 = 0$ (without a fixed polyvector mass squared term K) allows for nonvanishing classical spin.

The "fundamental" classical action is thus a polyvector action in higher dimensions without a fixed mass term. Interactions are associated with the metric of V_N. Reduction to four dimensions gives us gravity plus gauge interactions, such as the electromagnetic and Yang–Mills interactions, and also the classical spin which is associated with the bivector dynamical degrees of freedom sitting on the particle, for instance the particle's finite extension, magnetic moment, and similar.

There is a very well known problem with Kaluza–Klein theory, since in four dimensions a charged particle's mass cannot be smaller that the Planck mass. Namely, when reducing from five to four dimensions mass is given by $p^\mu p_\mu = \hat{m}^2 + \hat{p}_5^2$, where $\hat{m}$ is the 5-dimensional mass. Since $\hat{p}_5$ has the role of electric charge e, the latter relation is problematic for the electron: in the units in which $\hbar = c = G = 1$ the charge e is of the order of the Planck mass, so $p^\mu p_\mu$ is also of the same order of magnitude. There is no generally accepted mechanism for solving such a problem. In the polyvector generalization of the theory, the scalar constraint is (2.77) and in five or more dimensions it assumes an even more complicated form. The terms in the constraint have different signs, and the 4-dimensional mass $p^\mu p_\mu$ is not necessarily of the order of the Planck mass: there is a lot of room to "make" it small.

All those considerations clearly illustrate why the polyvector generalization of the point particle theory is of great physical interest.

2.4. THE UNCONSTRAINED ACTION FROM THE POLYVECTOR ACTION

FREE PARTICLE

In the previous section we have found that when the polyvector fixed mass squared K^2 is zero then a possible solution of the equations of motion satisfies (2.96) and the generic action (2.56) simplifies to

$$I[s,m,x^\mu,p_\mu,\lambda] = \int \mathrm{d}\tau \left[-m\dot{s} + p_\mu \dot{x}^\mu - \frac{\lambda}{2}(p^\mu p_\mu - m^2)\right]. \qquad (2.102)$$

At this point let us observe that a similar action, with a scalar variable s, has been considered by DeWitt [25] and Rovelli [26]. They associate the variable s with the clock carried by the particle. We shall say more about that in Sec. 6.2.

We are now going to show that the latter action is equivalent to the Stueckelberg action discussed in Chapter 1.

The equations of motion resulting from (2.102) are

$$\delta s \; : \quad \dot{m} = 0, \qquad (2.103)$$
$$\delta m \; : \quad \dot{s} - \lambda m = 0, \qquad (2.104)$$
$$\delta x^\mu \; : \quad \dot{p}_\mu = 0, \qquad (2.105)$$
$$\delta p_\mu \; : \quad \dot{x}^\mu - \lambda p^\mu =, 0 \qquad (2.106)$$
$$\delta \lambda \; : \quad p^\mu p_\mu - m^2 = 0. \qquad (2.107)$$
$$\qquad (2.108)$$

We see that in this dynamical system mass m is one of the dynamical variables; it is canonically conjugate to the variable s. From the equations of motion we easily read out that s is the proper time. Namely, from (2.104), (2.106) and (2.107) we have

$$p^\mu = \frac{\dot{x}^\mu}{\lambda} = m \frac{\mathrm{d}x^\mu}{\mathrm{d}s}, \qquad (2.109)$$

$$\dot{s}^2 = \lambda^2 m^2 = \dot{x}^2, \quad \text{i.e} \quad \mathrm{d}s^2 = \mathrm{d}x^\mu \mathrm{d}x_\mu. \qquad (2.110)$$

Using eq. (2.104) we find that

$$-m\dot{s} + \frac{\lambda}{2}\kappa^2 = -\frac{m\dot{s}}{2} = -\frac{1}{2}\frac{\mathrm{d}(ms)}{\mathrm{d}\tau}. \qquad (2.111)$$

The action (2.102) then becomes

$$I = \int \mathrm{d}\tau \left(\frac{1}{2}\frac{\mathrm{d}(ms)}{\mathrm{d}\tau} + p_\mu \dot{x}^\mu - \frac{\lambda}{2}p^\mu p_\mu\right), \qquad (2.112)$$

where λ should be no more considered as a quantity to be varied, but it is now fixed: $\lambda = \Lambda(\tau)$. The total derivative in (2.112) can be omitted, and the action is simply

$$I[x^\mu,p_\mu] = \int \mathrm{d}\tau (p_\mu \dot{x}^\mu - \frac{\Lambda}{2}p^\mu p_\mu). \qquad (2.113)$$

Point particles and Clifford algebra

This is just *the Stueckelberg action* (1.36) with $\kappa^2 = 0$. The equations of motion derived from (2.113) are

$$\dot{x}^\mu - \Lambda p^\mu = 0, \qquad (2.114)$$

$$\dot{p}_\mu = 0. \qquad (2.115)$$

From (2.115) it follows that $p_\mu p^\mu$ is a constant of motion. Denoting the latter constant of motion as m and using (2.114) we obtain that momentum can be written as

$$p^\mu = m \frac{\dot{x}^\mu}{\sqrt{\dot{x}^\nu \dot{x}_\nu}} = m \frac{dx^\mu}{ds}, \quad ds = dx^\mu dx_\mu, \qquad (2.116)$$

which is the same as in eq. (2.109). The equations of motion for x^μ and p_μ derived from the Stueckelberg action (2.99) are the same as the equations of motion derived from the action (2.102). *A generic Clifford algebra action (2.56) thus leads directly to the Stueckleberg action.*

The above analysis can be easily repeated for a more general case where the scalar constant κ^2 is different from zero, so that instead of (2.98) or (2.102) we have

$$I[s, m, x^\mu, p_\mu, \lambda] = \int d\tau \left[-m\dot{s} + p_\mu \dot{x}^\mu - \frac{\lambda}{2}(p^\mu p_\mu - m^2 - \kappa^2) \right]. \qquad (2.117)$$

Then instead of (2.113) we obtain

$$I[x^\mu, p_\mu] = \int d\tau \left(p_\mu \dot{x}^\mu - \frac{\Lambda}{2}(p^\mu p_\mu - \kappa^2) \right). \qquad (2.118)$$

The corresponding Hamiltonian is

$$H = \frac{\Lambda}{2}(p^\mu p_\mu - \kappa^2), \qquad (2.119)$$

and in the quantized theory the Schrödinger equation reads

$$i \frac{\partial \psi}{\partial \tau} = \frac{\Lambda}{2}(p^\mu p_\mu - \kappa^2)\psi. \qquad (2.120)$$

Alternatively, in the action (2.102) or (2.117) we can first eliminate λ by the equation of motion (2.104). So we obtain

$$I[s, m, x^\mu, p_\mu] = \int d\tau \left[-\frac{m\dot{s}}{2} + p_\mu \dot{x}^\mu - \frac{\dot{s}}{2m}(p^\mu p_\mu - \kappa^2) \right]. \qquad (2.121)$$

The equations of motion are

$$\delta s : \quad -\frac{\dot{m}}{2} - \frac{\mathrm{d}}{\mathrm{d}\tau}\left(\frac{p^\mu p_\mu - \kappa^2}{2m}\right) = 0 \Rightarrow \dot{m} = 0, \quad (2.122)$$

$$\delta m : \quad -\frac{1}{2} + \frac{1}{2m^2}(p^\mu p_\mu - \kappa^2) = 0 \Rightarrow p^\mu p_\mu - m^2 - \kappa^2 = 0, \quad (2.123)$$

$$\delta x^\mu : \quad \dot{p}_\mu = 0, \quad (2.124)$$

$$\delta p_\mu : \quad \dot{x}^\mu - \frac{\dot{s}}{m} p^\mu = 0 \Rightarrow p^\mu = \frac{m\dot{x}^\mu}{\dot{s}} = m\frac{\mathrm{d}x^\mu}{\mathrm{d}s}. \quad (2.125)$$

Then we can choose a "solution" for $s(\tau)$, write $\dot{s}/m = \Lambda$, and omit the first term, since in view of (2.122) it is a total derivative. So again we obtain the Stueckelberg action (2.118).

The action that we started from, e.g., (2.121) or (2.102) has a constraint on the variables x^μ, s or on the p_μ, m, but the action (2.118) which we arrived at contains only the variables x^μ, p_μ and has no constraint.

In the action (2.121) we can use the relation $\dot{s} = \mathrm{d}s/\mathrm{d}\tau$ and write it as

$$I[m, p_\mu, x^\mu] = \int \mathrm{d}s \left[-\frac{m}{2} + p_\mu \frac{\mathrm{d}x^\mu}{\mathrm{d}s} - \frac{1}{2m}(p^\mu p_\mu - \kappa^2)\right]. \quad (2.126)$$

The evolution parameter is now s, and again variation with respect to m gives the constraint $p^\mu p_\mu - m^2 - \kappa^2 = 0$. Eliminating m from the action (2.126) by the the latter constraint, written in the form

$$m = \sqrt{p^\mu p_\mu - \kappa^2} \quad (2.127)$$

we obtain *the unconstrained action*

$$I[x^\mu, p_\mu] = \int \mathrm{d}s \left(p_\mu \frac{\mathrm{d}x^\mu}{\mathrm{d}s} - \sqrt{p^\mu p_\mu - \kappa^2}\right), \quad (2.128)$$

which is also equivalent to the original action (2.117). The Hamiltonian corresponding to (2.128) is

$$H = p_\mu \frac{\mathrm{d}x^\mu}{\mathrm{d}s} - L = \sqrt{p^\mu p_\mu - \kappa^2}. \quad (2.129)$$

Such a Hamiltonian is not very practical for quantization, since the Schrödinger equation contains the square root of operators

$$i\frac{\partial \psi}{\partial s} = \sqrt{p^\mu p_\mu - \kappa^2}\, \psi. \quad (2.130)$$

In order to perform the quantization properly one has to start directly from the original polyvector action (2.56). This will be discussed in Sec. 2.5.

However, in the approximation[6] $p_\mu p^\mu \ll -\kappa^2$ eq. (2.128) becomes

$$I[x^\mu, p_\mu] \approx \int ds \left(p_\mu \frac{dx^\mu}{ds} - \frac{1}{2\sqrt{-\kappa^2}} p^\mu p_\mu - \sqrt{-\kappa^2} \right) \tag{2.131}$$

which is again the Stueckelberg action, but with $1/\sqrt{-\kappa^2} = \Lambda$. It is very interesting that on the one hand the Stueckelberg action arises exactly from the polyvector action, and on the other hand it arises as an approximation.

PARTICLE IN A FIXED BACKGROUND FIELD

Let us now consider the action (2.117) and modify it so that it will remain covariant under the transformation

$$L \to L' = L + \frac{d\phi}{d\tau}, \tag{2.132}$$

where

$$\phi = \phi(s, x^\mu) \tag{2.133}$$

For this purpose we have to introduce the gauge fields A_μ and V which transform according to

$$eA'_\mu = eA_\mu + \partial_\mu \phi, \tag{2.134}$$

$$eV' = eV + \frac{\partial \phi}{\partial s}. \tag{2.135}$$

The covariant action is then

$$I = \int d\tau \left[-m\dot{s} + p_\mu \dot{x}^\mu - \frac{\lambda}{2}(\pi_\mu \pi^\mu - \mu^2 - \kappa^2) \right], \tag{2.136}$$

where we have introduced the kinetic momentum

$$\pi_\mu = p_\mu - eA_\mu \tag{2.137}$$

and its pseudoscalar counterpart

$$\mu = m + eV. \tag{2.138}$$

The symbol 'μ' here should not be confused with the same symbol used in Sec. 2.3 for a completely different quantity.

---

[6] Remember that κ^2 comes from the scalar part of the polyvector mass squared term (2.57) and that it was a matter of our convention of writing it in the form κ^2. We could have used another symbol without square, e.g., α, and then it would be manifestly clear that α can be negative.

From (2.136) we derive the following equations of motion:

$$\delta x^\mu \; : \quad \dot{\pi}_\mu = e F_{\mu\nu} \dot{x}^\nu - \dot{s} e \left(\frac{\partial A_\mu}{\partial s} - \partial_\mu V \right), \qquad (2.139)$$

$$\delta s \; : \quad \dot{\mu} = -\dot{x}^\nu e \left(\frac{\partial A_\nu}{\partial s} - \partial_\nu V \right), \qquad (2.140)$$

$$\delta p_\mu \; : \quad \lambda \pi_\mu = \dot{x}_\mu, \qquad (2.141)$$

$$\qquad (2.142)$$

$$\delta m \; : \quad \dot{s} = \lambda \mu. \qquad (2.143)$$

These equations of motion are the same as those from the Stueckelberg action (2.147).

From (2.140) and (2.143) we have

$$-m\dot{s} + \frac{\lambda}{2}\mu^2 = -\tfrac{1}{2}\frac{\mathrm{d}}{\mathrm{d}\tau}(\mu s) + e\dot{s}V - \frac{1}{2}e\dot{x}^\nu \left(\frac{\partial A_\nu}{\partial s} - \partial_\nu V \right) s. \qquad (2.144)$$

Inserting the latter expression into the action (2.136) we obtain

$$I = \int \mathrm{d}\tau \bigg[-\tfrac{1}{2}\frac{\mathrm{d}(\mu s)}{\mathrm{d}\tau} + p_\mu \dot{x}^\mu - \frac{\lambda}{2}(\pi^\mu \pi_\mu - \kappa^2) + e\dot{s}V$$

$$- \frac{1}{2}e\dot{x}^\nu \left(\frac{\partial A_\nu}{\partial s} - \partial_\nu V \right) s \bigg], \qquad (2.145)$$

which is analogous to eq. (2.112). However, in general $\dot{\mu}$ is now not zero, and as a result we cannot separate the variables m, s into a total derivative term as we did in (2.117).

Let us consider a particular case when the background fields A_μ, V satisfy

$$\frac{\partial A_\mu}{\partial s} = 0, \qquad \partial_\mu V = 0. \qquad (2.146)$$

Then the last term in (2.145) vanishes; in addition we may set $V = 0$. Omitting the total derivative term, eq. (2.145) becomes

$$I[x^\mu, p_\mu] = \int \mathrm{d}s \left[p_\mu \frac{\mathrm{d}x^\mu}{\mathrm{d}s} - \frac{\Lambda}{2}(\pi^\mu \pi_\mu - \kappa^2) \right], \qquad (2.147)$$

where $\Lambda = \lambda/\dot{s}$ is now fixed. This is precisely the Stueckelberg action in the presence of a fixed electromagnetic field, and s corresponds to the Stueckelberg Lorentz invariant parameter τ.

However, when we gauged the free particle Stueckelberg action we obtained in general a τ-dependent gauge field A_μ and also a scalar field V.

We shall now see that such a general gauged Stueckelberg action is an approximation to the action (2.136). For this purpose we shall repeat the procedure of eqs. (2.121)–(2.128). Eliminating λ from the action (2.136) by using the equation of motion (2.143) we obtain an equivalent action

$$I[x^\mu, p_\mu, s, m] = \int d\tau \left[-m\dot{s} + p_\mu \dot{x}^\mu - \frac{\dot{s}}{2\mu}(\pi^\mu \pi_\mu - \mu^2 - \kappa^2) \right] \quad (2.148)$$

whose variation with respect to m again gives the constraint $\pi^\mu \pi_\mu - \mu^2 - \kappa^2 = 0$. From (2.148), using (2.138) we have

$$\begin{aligned} I &= \int d\tau \left[p_\mu \dot{x}^\mu - \frac{\dot{s}}{2\mu}(\pi^\mu \pi_\mu - \kappa^2) + e\dot{s}V - \frac{\mu \dot{s}}{2} \right] \\ &= \int d\tau \left[p_\mu \dot{x}^\mu - \dot{s}(\pi^\mu \pi_\mu - \kappa^2)^{1/2} + e\dot{s}V \right] \\ &\approx \int d\tau \left[p_\mu \dot{x}^\mu - \frac{\dot{s}}{2\sqrt{-\kappa^2}} \pi^\mu \pi_\mu - \dot{s}\sqrt{-\kappa^2} + e\dot{s}V \right]. \quad (2.149) \end{aligned}$$

Thus

$$I[x^\mu, p_\mu] = \int ds \left[p_\mu \frac{dx^\mu}{ds} - \frac{1}{2\sqrt{-\kappa^2}} \pi^\mu \pi_\mu - \sqrt{-\kappa^2} + eV \right]. \quad (2.150)$$

The last step in eq. (2.149) is valid under the approximation $\pi^\mu \pi_\mu \ll -\kappa^2$, where we assume $-\kappa^2 > 0$. In (2.150) we indeed obtain an action which is equivalent to the gauged Stueckelberg action (2.147) if we make the correspondence $1\sqrt{-\kappa^2} \to \Lambda$. The constant terms $-\sqrt{-\kappa^2}$ in (2.150) and $\Lambda \kappa^2/2$ in eq. (2.147) have no influence on the equations of motion.

We have thus found a very interesting relation between the Clifford algebra polyvector action and the Stueckelberg action in the presence of an electromagnetic and pseudoscalar field. If the electromagnetic field A_μ does not depend on the pseudoscalar parameter s and if there is no force owed to the pseudoscalar field V, then the kinetic momentum squared $\pi^\mu \pi_\mu$ is a *constant of motion*, and the gauged Clifford algebra action (2.136) is exactly equivalent to the Stueckelberg action. In the presence of a pseudoscalar force, i.e., when $\partial_\mu V \neq 0$ and/or when $\partial A_\mu/\partial s \neq 0$, the action (2.136) is approximately equivalent to the gauged Stueckelberg action (2.147) if the kinetic momentum squared $\pi^\mu \pi_\mu$ is much smaller than the scalar mass constant squared $-\kappa^2$.

2.5. QUANTIZATION OF THE POLYVECTOR ACTION

We have assumed that a point particle's classical motion is governed by the polyvector action (2.56). Variation of this action with respect to λ gives

the polyvector constraint
$$P^2 - K^2 = 0. \tag{2.151}$$

In the quantized theory the position and momentum polyvectors $X = X^J e_J$ and $P = P^J e_J$, where $e_J = (1, e_\mu, e_\mu e_\nu, e_5 e_\mu, e_5)$, $\mu < \nu$, become the operators
$$\widehat{X} = \widehat{X}^J e_J, \qquad \widehat{P} = \widehat{P}^J e_J;, \tag{2.152}$$
satisfying
$$[\widehat{X}^J, \widehat{P}_K] = i \delta^J{}_K. \tag{2.153}$$
Using the explicit expressions like (2.53),(2.54) the above equations imply
$$[\hat{\sigma}, \hat{\mu}] = i, \quad [\hat{x}^\mu, \hat{p}_\nu] = i\delta^\mu{}_\nu, \quad [\hat{\alpha}^{\mu\nu}, \widehat{S}_{\alpha,\beta}] = i\delta^{\mu\nu}{}_{\alpha\beta}, \tag{2.154}$$

$$[\hat{\xi}^\mu, \hat{\pi}_\nu] = i\delta^\mu{}_\nu, \quad [\hat{s}, \hat{m}] = i. \tag{2.155}$$
In a particular representation in which $\widehat{X}^J$ are diagonal, the momentum polyvector operator is represented by the multivector derivative (see Sec. 6.1).
$$\widehat{P}_J = -i \frac{\partial}{\partial X^J} \tag{2.156}$$
Explicitly, the later relation means
$$\hat{m} = -i\frac{\partial}{\partial \sigma}, \ \hat{p}_\mu = -i\frac{\partial}{\partial x^\mu}, \ \widehat{S}_{\mu\nu} = -i\frac{\partial}{\partial \alpha^{\mu\nu}}, \ \hat{\pi}_\mu = -i\frac{\partial}{\partial \xi^\mu}, \ \hat{m} = -i\frac{\partial}{\partial s}. \tag{2.157}$$

Let us assume that a quantum state can be represented by a polyvector-valued wave function $\Phi(X)$ of the position polyvector X. A possible physical state is a solution to the equation
$$(\widehat{P}^2 - K^2)\Phi = 0, \tag{2.158}$$
which replaces the classical constraint (2.151).

When $K^2 = \kappa^2 = 0$ eq. (2.158) becomes
$$\widehat{P}^2 \Phi = 0. \tag{2.159}$$

Amongst the set of functions $\Phi(X)$ there are some such that satisfy
$$\widehat{P}\Phi = 0. \tag{2.160}$$

Let us now consider a special case where Φ has definite values of the operators $\hat{\mu}$, $\hat{S}_{\mu\nu}$, $\hat{\pi}_\mu$:

$$\hat{\mu}\Phi = 0, \quad \hat{S}_{\mu\nu}\Phi = 0, \quad \hat{\pi}_\mu\Phi = 0 \tag{2.161}$$

Then

$$\hat{P}\Phi = (\hat{p}^\mu e_\mu + \hat{m}e_5)\Phi = 0. \tag{2.162}$$

or

$$(\hat{p}^\mu \gamma_\mu - \hat{m})\Phi = 0, \tag{2.163}$$

where

$$\gamma_\mu \equiv e_5 e_\mu, \quad \gamma_5 = \gamma_0\gamma_1\gamma_2\gamma_3 = e_0 e_1 e_2 e_3 = e_5. \tag{2.164}$$

When Φ is an eigenstate of $\hat{m}$ with definite value m, i.e., when $\hat{m}\phi = m\Phi$, then eq. (2.163) becomes the familiar *Dirac equation*

$$(\hat{p}_\mu \gamma^\mu - m)\Phi = 0. \tag{2.165}$$

A polyvector wave function which satisfies eq. (2.165) is a *spinor*. We have arrived at the very interesting result that *spinors can be represented by particular polyvector wave functions*.

3-dimensional case

To illustrate this let us consider the 3-dimensional space V_3. Basis vectors are σ_1, σ_2, σ_3 and they satisfy the Pauli algebra

$$\sigma_i \cdot \sigma_j \equiv \tfrac{1}{2}(\sigma_i\sigma_j + \sigma_j\sigma_i) = \delta_{ij}, \quad i,j = 1,2,3. \tag{2.166}$$

The unit pseudoscalar

$$\sigma_1\sigma_2\sigma_3 \equiv I \tag{2.167}$$

commutes with all elements of the Pauli algebra and its square is $I^2 = -1$. It behaves as the ordinary imaginary unit i. Therefore, in 3-space, we may identify the imaginary unit i with the unit pseudoscalar I.

An arbitrary polyvector in V_3 can be written in the form

$$\Phi = \alpha^0 + \alpha^i\sigma_i + i\beta^i\sigma_i + i\beta = \Phi^0 + \Phi^i\sigma_i, \tag{2.168}$$

where Φ^0, Φ^i are formally complex numbers.

We can decompose [22]:

$$\Phi = \Phi\tfrac{1}{2}(1+\sigma_3) + \Phi\tfrac{1}{2}(1-\sigma_3) = \Phi_+ + \Phi_-, \tag{2.169}$$

where $\Phi_+ \in \mathcal{I}_+$ and $\Phi_- \in \mathcal{I}_-$ are independent minimal *left* ideals (see Box 3.2).

Box 3.2: Definition of ideal

A left ideal $\mathcal{I}_L$ in an algebra C is a set of elements such that if $a \in \mathcal{I}_L$ and $c \in C$, then $ca \in \mathcal{I}_L$. If $a \in \mathcal{I}_L$, $b \in \mathcal{I}_L$, then $(a + b) \in \mathcal{I}_L$. A right ideal $\mathcal{I}_R$ is defined similarly except that $ac \in \mathcal{I}_R$. A left (right) minimal ideal is a left (right) ideal which contains no other ideals but itself and the null ideal.

A basis in $\mathcal{I}_+$ is given by two polyvectors

$$u_1 = \tfrac{1}{2}(1 + \sigma_3) , \quad u_2 = (1 - \sigma_3)\sigma_1 , \tag{2.170}$$

which satisfy

$$\begin{array}{lll} \sigma_3 u_1 = u_1, & \sigma_1 u_1 = u_2, & \sigma_2 u_1 = iu_2, \\ \sigma_3 u_2 = -u_2, & \sigma_1 u_2 = u_1, & \sigma_2 u_2 = -iu_1. \end{array} \tag{2.171}$$

These are precisely the well known relations for basis spinors. Thus we have arrived at the very profound result that the polyvectors u_1, u_2 behave as basis spinors.

Similarly, a basis in $\mathcal{I}_+$ is given by

$$v_1 = \tfrac{1}{2}(1 + \sigma_3)\sigma_1 , \quad v_2 = \tfrac{1}{2}(1 - \sigma_3) \tag{2.172}$$

and satisfies

$$\begin{array}{lll} \sigma_3 v_1 = v_1, & \sigma_1 v_1 = v_2, & \sigma_2 v_1 = iv_2, \\ \sigma_3 v_2 = -v_2, & \sigma_1 v_2 = v_1, & \sigma_2 v_2 = -iv_1. \end{array} \tag{2.173}$$

A polyvector Φ can be written in *spinor basis* as

$$\Phi = \Phi_+^1 u_1 + \Phi_+^2 u_2 + \Phi_-^1 v_1 + \Phi_-^2 v_2 , \tag{2.174}$$

where

$$\begin{array}{ll} \Phi_+^1 = \Phi^0 + \Phi^3 , & \Phi_-^1 = \Phi^1 - i\Phi^2 \\ \Phi_+^2 = \Phi^1 + i\Phi^2 , & \Phi_-^2 = \Phi^0 - \Phi^3 \end{array} \tag{2.175}$$

Eq. (2.174) is an alternative expansion of a polyvector. We can expand the same polyvector Φ either according to (2.168) or according to (2.174).

Introducing the matrices

$$\xi_{ab} = \begin{pmatrix} u_1 & v_1 \\ u_2 & v_2 \end{pmatrix}, \qquad \Phi^{ab} = \begin{pmatrix} \Phi_+^1 & \Phi_-^1 \\ \Phi_+^2 & \Phi_-^2 \end{pmatrix} \qquad (2.176)$$

we can write (2.174) as

$$\Phi = \Phi^{ab}\xi_{ab}. \qquad (2.177)$$

Thus a polyvector can be represented as a *matrix* Φ^{ab}. The decomposition (2.169) then reads

$$\Phi = \Phi_+ + \Phi_- = (\Phi_+^{ab} + \Phi_-^{ab})\xi_{ab}, \qquad (2.178)$$

where

$$\Phi_+^{ab} = \begin{pmatrix} \Phi_+^1 & 0 \\ \Phi_+^2 & 0 \end{pmatrix}, \qquad (2.179)$$

$$\Phi_-^{ab} = \begin{pmatrix} 0 & \Phi_-^1 \\ 0 & \Phi_-^2 \end{pmatrix}. \qquad (2.180)$$

From (2.177) we can directly calculate the matrix elements Φ^{ab}. We only need to introduce the new elements $\xi^{\dagger ab}$ which satisfy

$$(\xi^{\dagger ab}\xi_{cd})_S = \delta^a{}_c \delta^b{}_d. \qquad (2.181)$$

The superscript $\dagger$ means Hermitian conjugation [22]. If

$$A = A_S + A_V + A_B + A_P \qquad (2.182)$$

is a Pauli number, then

$$A^\dagger = A_S + A_V - A_B - A_P. \qquad (2.183)$$

This means that the order of basis vectors σ_i in the expansion of $A^\dagger$ is reversed. Thus $u_1^\dagger = u_1$, but $u_2^\dagger = \frac{1}{2}(1 + \sigma_3)\sigma_1$. Since $(u_1^\dagger u_1)_S = \frac{1}{2}$, $(u_2^\dagger u_2)_S = \frac{1}{2}$, it is convenient to introduce $u^{\dagger^1} = 2u_1$ and $u^{\dagger^2} = 2u_2$ so that $(u^{\dagger^1}u_1)_S = 1$, $(u^{\dagger^2}u_2)_S = 1$. If we define similar relations for v_1, v_2 then we obtain (2.181).

From (2.177) and (2.181) we have

$$\Phi^{ab} = (\xi^{\dagger ab}\Phi)_I. \qquad (2.184)$$

Here the subscript I means *invariant part*, i.e., scalar plus pseudoscalar part (remember that pseudoscalar unit has here the role of imaginary unit and that Φ^{ab} are thus complex numbers).

The relation (2.184) tells us how from an arbitrary polyvector Φ (i.e., a Clifford number) can we obtain its *matrix representation* Φ^{ab}.

Φ in (2.184) is an arbitrary Clifford number. In particular, Φ may be any of the basis vectors σ_i.

Example $\Phi = \sigma_1$:

$$\begin{aligned}
\Phi^{11} &= (\xi^{\dagger^{11}} \sigma_1)_I = (u^{\dagger^1} \sigma_1)_I = ((1+\sigma_3)\sigma_1)_I = 0, \\
\Phi^{12} &= (\xi^{\dagger^{12}} \sigma_1)_I = (v^{\dagger^1} \sigma_1)_I = ((1-\sigma_3)\sigma_1\sigma_1)_I = 1, \\
\Phi^{21} &= (\xi^{\dagger^{21}} \sigma_1)_I = (u^{\dagger^2} \sigma_1)_I = ((1+\sigma_3)\sigma_1\sigma_1)_I = 1, \\
\Phi^{22} &= (\xi^{\dagger^{22}} \sigma_1)_I = (v^{\dagger^2} \sigma_1)_I = ((1-\sigma_3)\sigma_1)_I = 0.
\end{aligned} \quad (2.185)$$

Therefore

$$(\sigma_1)^{ab} = \begin{pmatrix} 0 & 1 \\ 1 & 0 \end{pmatrix}. \qquad (2.186)$$

Similarly we obtain from (2.184) when $\Phi = \sigma_2$ and $\Phi = \sigma_3$, respectively, that

$$(\sigma_2)^{ab} = \begin{pmatrix} 0 & -i \\ i & 0 \end{pmatrix}, \quad (\sigma_3)^{ab} = \begin{pmatrix} 1 & 0 \\ 0 & -1 \end{pmatrix}. \qquad (2.187)$$

So we have obtained the matrix representation of the basis vectors σ_i. Actually (2.186), (2.187) are the well known *Pauli matrices*.

When $\Phi = u_1$ and $\Phi = u_2$, respectively, we obtain

$$(u_1)^{ab} = \begin{pmatrix} 1 & 0 \\ 0 & 0 \end{pmatrix}, \quad (u_2)^{ab} = \begin{pmatrix} 0 & 0 \\ 1 & 0 \end{pmatrix} \qquad (2.188)$$

which are a matrix representation of the *basis spinors* u_1 and u_2.

Similarly we find

$$(v_1)^{ab} = \begin{pmatrix} 0 & 1 \\ 0 & 0 \end{pmatrix}, \quad (v_2)^{ab} = \begin{pmatrix} 0 & 0 \\ 0 & 1 \end{pmatrix} \qquad (2.189)$$

In general a *spinor* is a superposition

$$\psi = \psi^1 u_1 + \psi^2 u_2, \qquad (2.190)$$

and its matrix representation is

$$\psi \to \begin{pmatrix} \psi^1 & 0 \\ \psi^2 & 0 \end{pmatrix}. \qquad (2.191)$$

Point particles and Clifford algebra 81

Another independent spinor is

$$\chi = \chi^1 v_1 + \chi^2 v_2, \qquad (2.192)$$

with matrix representation

$$\chi \to \begin{pmatrix} 0 & \chi^1 \\ 0 & \chi^2 \end{pmatrix}. \qquad (2.193)$$

If we multiply a spinor ψ from the left by any element R of the Pauli algebra we obtain another spinor

$$\psi' = R\psi \to \begin{pmatrix} \psi'^1 & 0 \\ \psi'^2 & 0 \end{pmatrix} \qquad (2.194)$$

which is an element of the same minimal left ideal. Therefore, if only multiplication from the left is considered, a spinor can be considered as a column matrix

$$\psi \to \begin{pmatrix} \psi^1 \\ \psi^2 \end{pmatrix}. \qquad (2.195)$$

This is just the common representation of spinors. But it is not general enough to be valid for all the interesting situations which occur in the Clifford algebra.

We have thus arrived at a very important finding. *Spinors are just particular Clifford numbers*: they belong to a left or right minimal ideal. For instance, a generic *spinor* is

$$\psi = \psi^1 u_1 + \psi^2 u_2 \quad \text{with} \quad \Phi^{ab} = \begin{pmatrix} \psi^1 & 0 \\ \psi^2 & 0 \end{pmatrix}. \qquad (2.196)$$

A *conjugate spinor* is

$$\psi^\dagger = \psi^{1*} u_1^\dagger + \psi^{2*} u_2^\dagger \quad \text{with} \quad (\Phi^{ab})^* = \begin{pmatrix} \psi^{1*} & \psi^{2*} \\ 0 & 0 \end{pmatrix} \qquad (2.197)$$

and it is an element of a minimal *right ideal*.

4-dimensional case

The above considerations can be generalized to 4 or more dimensions. Thus

$$\psi = \psi^0 u_0 + \psi^1 u_1 + \psi^2 u_2 + \psi^3 u_3 \to \begin{pmatrix} \psi^0 & 0 & 0 & 0 \\ \psi^1 & 0 & 0 & 0 \\ \psi^2 & 0 & 0 & 0 \\ \psi^3 & 0 & 0 & 0 \end{pmatrix} \qquad (2.198)$$

and
$$\psi^\dagger = \psi^{*0} u_0^\dagger + \psi^{*1} u_1^\dagger + \psi^{*2} u_2^\dagger + \psi^{*3} u_3^\dagger \rightarrow \begin{pmatrix} \psi^{*0} & \psi^{*1} & \psi^{*2} & \psi^{*3} \\ 0 & 0 & 0 & 0 \\ 0 & 0 & 0 & 0 \\ 0 & 0 & 0 & 0 \end{pmatrix}, \quad (2.199)$$

where u_0, u_1, u_2, u_3 are four basis spinors in spacetime, and $\psi^0, \psi^1, \psi^2, \psi^3$ are complex scalar coefficients.

In 3-space the pseudoscalar unit can play the role of the imaginary unit i. This is not the case of the 4-space V_4, since $e_5 = e_0 e_1 e_2 e_3$ does not commute with all elements of the Clifford algebra in V_4. Here the approaches taken by different authors differ. A straightforward possibility [37] is just to use the complex Clifford algebra with complex coefficients of expansion in terms of multivectors. Other authors prefer to consider real Clifford algebra $\mathcal{C}$ and ascribe the role of the imaginary unit i to an element of $\mathcal{C}$ which commutes with all other elements of $\mathcal{C}$ and whose square is -1. Others [22, 36] explore the possibility of using a non-commuting element as a substitute for the imaginary unit. I am not going to review all those various approaches, but I shall simply assume that the expansion coefficients are in general complex numbers. In Sec. 7.2 I explore the possibility that such complex numbers which occur in the quantized theory originate from the Clifford algebra description of the $(2 \times n)$-dimensional phase space (x^μ, p_μ). In such a way we still conform to the idea that complex numbers are nothing but special Clifford numbers.

A Clifford number ψ expanded according to (2.198) is an element of a left minimal ideal if the four elements u_0, u_1, u_2, u_3 satisfy

$$C u_\lambda = C_{0\lambda} u_0 + C_{1\lambda} u_1 + C_{2\lambda} u_2 + C_{3\lambda} u_3 \qquad (2.200)$$

for an arbitrary Clifford number C. General properties of u_λ were investigated by Teitler [37]. In particular, he found the following representation for u_λ:

$$\begin{aligned}
u_0 &= \tfrac{1}{4}(1 - e^0 + ie^{12} - ie^{012}), \\
u_1 &= -e^{13} u_0 = \tfrac{1}{4}(-e^{13} + e^{013} + ie^{23} - ie^{023}), \\
u_2 &= -ie^3 u_0 = \tfrac{1}{4}(-ie^3 - ie^{03} + e^{123} + e^{0123}), \\
u_3 &= -ie^1 u_0 = \tfrac{1}{4}(-ie^1 - ie^{01} - e^2 - e^{02}),
\end{aligned} \qquad (2.201)$$

from which we have

$$\begin{aligned}
e^0 u_0 &= -u_0, \\
e^1 u_0 &= iu_3, \\
e^2 u_0 &= -u_3, \\
e^3 u_0 &= iu_2.
\end{aligned} \qquad (2.202)$$

Using the representation (2.201) we can calculate from (2.200) the matrix elements $C_{\rho\lambda}$ of any Clifford number. For the spacetime basis vectors $e^\mu \equiv (e^0, e^i)$, $i = 1, 2, 3$, we obtain

$$e^0 = \begin{pmatrix} -1 & 0 \\ 0 & 1 \end{pmatrix}, \quad e^i = \begin{pmatrix} 0 & i\sigma^i \\ i\sigma^i & 0 \end{pmatrix}, \quad (2.203)$$

which is one of the standard matrix representations of e^μ (the Dirac matrices).

If a spinor is multiplied from the left by an arbitrary Clifford number, it remains a spinor. But if is multiplied from the right, it in general transforms into another Clifford number which is no more a spinor. *Scalars, vectors, bivectors, etc., and spinors can be reshuffled by the elements of Clifford algebra: scalars, vectors, etc., can be transformed into spinors, and vice versa.*

Quantum states are assumed to be represented by polyvector wave functions (i.e., Clifford numbers). If the latter are pure scalars, vectors, bivectors, pseudovectors, pseudovectors, and pseudoscalars they describe *bosons*. If, on the contrary, wave functions are spinors, then they describe *fermions*. Within Clifford algebra we have thus transformations which change bosons into fermions! It remains to be investigated whether this kind of "supersymmetry" is related to the well known supersymmetry.

2.6. ON THE SECOND QUANTIZATION OF THE POLYVECTOR ACTION

If we first quantize the polyvector action (2.117) we obtain the wave equation

$$(\hat{p}^\mu \hat{p}_\mu - \hat{m}^2 - \kappa^2)\phi = 0, \quad (2.204)$$

where

$$\hat{p}_\mu = -\partial/\partial x^\mu \equiv \partial_\mu \, , \, \hat{m} = -i\partial/\partial s \, ,$$

and κ is a fixed constant. The latter wave equation can be derived from the action

$$\begin{aligned} I[\phi] &= \tfrac{1}{2} \int ds\, d^d x\, \phi(-\partial^\mu \partial_\mu + \frac{\partial^2}{\partial s^2} - \kappa^2)\phi \\ &= \frac{1}{2} \int ds\, d^d x \left(\partial_\mu \phi \partial^\mu \phi - \left(\frac{\partial \phi}{\partial s}\right)^2 - \kappa^2 \phi^2 \right), \quad (2.205) \end{aligned}$$

where in the last step we have omitted the surface and the total derivative terms.

The canonical momentum is

$$\pi(s,x) = \frac{\partial \mathcal{L}}{\partial \partial \phi/s} = -\frac{\partial \phi}{\partial s} \qquad (2.206)$$

and the Hamiltonian is

$$H[\phi,\pi] = \int d^d x \left(\pi \frac{\partial \phi}{\partial s} - \mathcal{L}\right) = \tfrac{1}{2}\int d^d x\,(-\pi^2 - \partial_\mu \phi \partial^\mu \phi + \kappa^2 \phi^2). \qquad (2.207)$$

If instead of one field ϕ there are two fields ϕ_1, ϕ_2 we have

$$I[\phi_1,\phi_2] = \int ds\,d^d x \left[\partial_\mu \phi_1 \partial^\mu \phi_1 - \left(\frac{\partial \phi_1}{\partial s}\right)^2 - \kappa^2 \phi_1^2 \right.$$

$$\left. + \partial_\mu \phi_2 \partial^\mu \phi_2 - \left(\frac{\partial \phi_2}{\partial s}\right)^2 - \kappa^2 \phi_2^2 \right]. \qquad (2.208)$$

The canonical momenta are

$$\pi_1 = \frac{\partial \mathcal{L}}{\partial \partial \phi_1/s} = -\frac{\partial \phi_1}{\partial s}, \qquad \pi_2 = \frac{\partial \mathcal{L}}{\partial \partial \phi_2/s} = -\frac{\partial \phi_2}{\partial s} \qquad (2.209)$$

and the Hamiltonian is

$$H[\phi_1,\phi_2,\pi_1,\pi_2] = \int d^d x \left(\pi_1 \frac{\partial \phi_1}{\partial s} + \pi_2 \frac{\partial \phi_2}{\partial s} - \mathcal{L}\right)$$

$$= \tfrac{1}{2}\int d^d x\,(-\pi_1^2 - \partial_\mu \phi_1 \partial^\mu \phi_1 + \kappa^2 \phi_1^2$$

$$- \pi_2^2 - \partial_\mu \phi_2 \partial^\mu \phi_2 + \kappa^2 \phi_2^2). \qquad (2.210)$$

Introducing the complex fields

$$\phi = \phi_1 + i\phi_2, \qquad \pi = \pi_1 + i\pi_2$$
$$\phi^* = \phi_1 - i\phi_2, \qquad \pi^* = \pi_1 - i\pi_2 \qquad (2.211)$$

we have

$$I[\phi,\phi^*] = \tfrac{1}{2}\int ds\,d^d x \left(\frac{\partial \phi^*}{\partial s}\frac{\partial \phi}{\partial s} - \partial_\mu \phi^* \partial^\mu \phi - \kappa^2 \phi^* \phi\right) \qquad (2.212)$$

and

$$H[\phi,\phi^*,\pi,\pi^*] = \tfrac{1}{2}\int d^d x(-\pi^* \pi - \partial_\mu \phi^* \partial^\mu \phi + \kappa^2 \phi^* \phi). \qquad (2.213)$$

Comparing the latter Hamiltonian with the one of the Stueckelberg field theory (1.161), we see that it is the same, except for the additional term

$-\pi^* \pi$ which is absent in the Stueckelberg field theory. We see also that the theory described by (2.212) and (2.213) has the same structure as the conventional field theory, except for the number of dimensions. In the conventional theory we have time t and three space-like coordinates x^i, $i = 1,2,3$, while here we have s and four or more coordinates x^μ, one of them being time-like.

As the non-relativistic field theory is an approximation to the relativistic field theory, so the field theory derived from the Stueckelberg action is an approximation to the field theory derived from the polyvector action.

On the other hand, at the classical level (as we have seen in Sec. 2.4) Stueckelberg action, in the absence of interaction, arises exactly from the polyvector action. Even in the presence of the electromagnetic interaction both actions are equivalent, since they give the same equations of motion. However, the field theory based on the latter action differs from the field theory based on the former action by the term $\pi^* \pi$ in the Hamiltonian (2.213). While at the classical level Stueckelberg and the polyvector action are equivalent, at the first and the second quantized level differences arise which need further investigation.

Second quantization then goes along the usual lines: ϕ_i and π_i becomes operators satisfying the equal s commutation relations:

$$[\phi_i(s,x), \phi_j(s,x')] = 0, \qquad [\pi_i(s,x), \pi_j(s,x')] = 0,$$
$$[\phi_i(s,x), \pi_j(s,x')] = i\delta_{ij}\delta(x-x'). \qquad (2.214)$$

The field equations are then just the Heisenberg equations

$$\dot{\pi}_i = i[\pi_i, H]. \qquad (2.215)$$

We shall not proceed here with formal development, since it is in many respects just a repetition of the procedure expounded in Sec. 1.4. But we shall make some remarks. First of all it is important to bear in mind that the usual arguments about causality, unitarity, negative energy, etc., do not apply anymore, and must all be worked out again. Second, whilst in the conventional quantum field theory the evolution parameter is a time-like coordinate $x^0 \equiv t$, in the field theory based on (2.212), (2.213) the evolution parameter is the pseudoscalar variable s. In even-dimensions it is invariant with respect to the Poincaré and the general coordinate transformations of x^μ, including the inversions. And what is very nice here is that the (pseudo)scalar parameter s naturally arises from the straightforward polyvector extension of the conventional reparametrization invariant theory.

Instead of the *Heisenberg picture* we can use the *Schrödinger picture* and the coordinate representation in which the operators $\phi_i(0,x) \equiv \phi_i(x)$ are diagonal, i.e., they are just ordinary functions. The momentum operator is represented by functional derivative[7]

$$\pi_j = -i \frac{\delta}{\delta \phi_j(x)}. \quad (2.216)$$

A state $|\Psi\rangle$ is represented by a *wave functional* $\Psi[\phi(x)] = \langle \phi(x)|\Psi\rangle$ and satisfies the Schrödinger equation

$$i\frac{\partial \Psi}{\partial s} = H\Psi \quad (2.217)$$

in which the evolution parameter is s. Of course s is invariant under the Lorentz transformations, and H contains all four components of the 4-momentum. Equation (2.217) is just like the Stueckelberg equation, the difference being that Ψ is now not a wave function, but a wave functional.

We started from the *constrained* polyvector action (2.117). Performing the *first quantization* we obtained the wave equation (2.204) which follows from the action (2.205) for the field $\phi(s, x^\mu)$. The latter action is *unconstrained*. Therefore we can straightforwardly quantize it, and thus perform the *second quantization*. The state vector $|\Psi\rangle$ in the Schrödinger picture evolves in s which is a *Lorentz invariant evolution parameter*. $|\Psi\rangle$ can be represented by a wave functional $\Psi[s, \phi(x)]$ which satisfied the functional Schrödinger equation. Whilst upon the first quantization the equation of motion for the field $\phi(s, x^\mu)$ contains the second order derivative with respect to s, upon the second quantization only the first order s-derivative remains in the equation of motion for the state functional $\Psi[s, \phi(x)]$.

An analogous procedure is undertaken in the usual approach to quantum field theory (see, e.g., [38]), with the difference that the evolution parameter becomes one of the space time coordinates, namely $x^0 \equiv t$. When trying to quantize the gravitational field it turns out that the evolution parameter t does not occur at all in the Wheeler–DeWitt equation! This is the well known problem of time in quantum gravity. We anticipate that a sort of polyvector generalization of the Einstein–Hilbert action should be taken, which would contain the scalar or pseudoscalar parameter s, and retain it in the generalized Wheeler–DeWitt equation. Some important research in that direction has been pioneered by Greensite and Carlini [39]

---

[7] A detailed discussion of the Schrödinger representation in field theory is to be found in ref. [38].

Point particles and Clifford algebra 87

2.7. SOME FURTHER IMPORTANT CONSEQUENCES OF CLIFFORD ALGEBRA

RELATIVITY OF SIGNATURE

In previous sections we have seen how Clifford algebra can be used in the formulation of the point particle classical and quantum theory. The metric of spacetime was assumed, as usually, to have the Minkowski signature, and we have used the choice $(+---)$. We are now going to find out that within Clifford algebra the signature is a matter of choice of basis vectors amongst available Clifford numbers.

Suppose we have a 4-dimensional space V_4 with signature $(++++)$. Let e_μ, $\mu = 0, 1, 2, 3$, be basis vectors satisfying

$$e_\mu \cdot e_\nu \equiv \tfrac{1}{2}(e_\mu e_\nu + e_\nu e_\mu) = \delta_{\mu\nu}, \qquad (2.218)$$

where $\delta_{\mu\nu}$ is the *Euclidean signature* of V_4. The vectors e_μ can be used as generators of Clifford algebra $\mathcal{C}$ over V_4 with a generic Clifford number (also called polyvector or Clifford aggregate) expanded in term of $e_J = (1, e_\mu, e_{\mu\nu}, e_{\mu\nu\alpha}, e_{\mu\nu\alpha\beta})$, $\mu < \nu < \alpha < \beta$,

$$A = a^J e_J = a + a^\mu e_\mu + a^{\mu\nu} e_\mu e_\nu + a^{\mu\nu\alpha} e_\mu e_\nu e_\alpha + a^{\mu\nu\alpha\beta} e_\mu e_\nu e_\alpha e_\beta. \qquad (2.219)$$

Let us consider the set of four Clifford numbers $(e_0, e_i e_0)$, $i = 1, 2, 3$, and denote them as

$$\begin{aligned} e_0 &\equiv \gamma_0, \\ e_i e_0 &\equiv \gamma_i. \end{aligned} \qquad (2.220)$$

The Clifford numbers γ_μ, $\mu = 0, 1, 2, 3$, satisfy

$$\tfrac{1}{2}(\gamma_\mu \gamma_\nu + \gamma_\nu \gamma_\mu) = \eta_{\mu\nu}, \qquad (2.221)$$

where $\eta_{\mu\nu} = \text{diag}(1, -1, -1, -1)$ is the *Minkowski tensor*. We see that the γ_μ behave as basis vectors in a 4-dimensional space $V_{1,3}$ with signature $(+---)$. We can form a Clifford aggregate

$$\alpha = \alpha^\mu \gamma_\mu \qquad (2.222)$$

which has the properties of a *vector* in $V_{1,3}$. From the point of view of the space V_4 the same object α is a linear combination of a vector and bivector:

$$\alpha = \alpha^0 e_0 + \alpha^i e_i e_0. \qquad (2.223)$$

We may use γ_μ as generators of the Clifford algebra $\mathcal{C}_{1,3}$ defined over the pseudo-Euclidean space $V_{1,3}$. The basis elements of $\mathcal{C}_{1,3}$ are $\gamma_J =$

$(1, \gamma_\mu, \gamma_{\mu\nu}, \gamma_{\mu\nu\alpha}, \gamma_{\mu\nu\alpha\beta})$, with $\mu < \nu < \alpha < \beta$. A generic Clifford aggregate in $\mathcal{C}_{1,3}$ is given by

$$B = b^J \gamma_J = b + b^\mu \gamma_\mu + b^{\mu\nu} \gamma_\mu \gamma_\nu + b^{\mu\nu\alpha} \gamma_\mu \gamma_\nu \gamma_\alpha + b^{\mu\nu\alpha\beta} \gamma_\mu \gamma_\nu \gamma_\alpha \gamma_\beta. \quad (2.224)$$

With suitable choice of the coefficients $b^J = (b, b^\mu, b^{\mu\nu}, b^{\mu\nu\alpha}, b^{\mu\nu\alpha\beta})$ we have that B of eq. (2.224) is equal to A of eq.(2.219). Thus the same number A can be described either within $\mathcal{C}_4$ or within $\mathcal{C}_{1,3}$. The expansions (2.224) and (2.219) exhaust all possible numbers of the Clifford algebras $\mathcal{C}_{1,3}$ and $\mathcal{C}_4$. The algebra $\mathcal{C}_{1,3}$ is isomorphic to the algebra $\mathcal{C}_4$, and actually they are just two different representations of the same set of Clifford numbers (also being called polyvectors or Clifford aggregates).

As an alternative to (2.220) we can choose

$$e_0 e_3 \equiv \tilde{\gamma}_0,$$
$$e_i \equiv \tilde{\gamma}_i, \quad (2.225)$$

from which we have

$$\tfrac{1}{2}(\tilde{\gamma}_\mu \tilde{\gamma}_\nu + \tilde{\gamma}_\nu \tilde{\gamma}_\mu) = \tilde{\eta}_{\mu\nu} \quad (2.226)$$

with $\tilde{\eta}_{\mu\nu} = \mathrm{diag}(-1,1,1,1)$. Obviously $\tilde{\gamma}_\mu$ are basis vectors of a pseudo-Euclidean space $\tilde{V}_{1,3}$ and they generate the Clifford algebra over $\tilde{V}_{1,3}$ which is yet another representation of the same set of objects (i.e., polyvectors). But the spaces V_4, $V_{1,3}$ and $\tilde{V}_{1,3}$ are not the same and they span different subsets of polyvectors. In a similar way we can obtain spaces with signatures $(+-++)$, $(++-+)$, $(+++-)$, $(-+--)$, $(--+-)$, $(---+)$ and corresponding higher dimensional analogs. But we cannot obtain signatures of the type $(++--)$, $(+-+-)$, etc. In order to obtain such signatures we proceed as follows.

4-space. First we observe that the bivector $\bar{I} = e_3 e_4$ satisfies $\bar{I}^2 = -1$, commutes with e_1, e_2 and anticommutes with e_3, e_4. So we obtain that the set of Clifford numbers $\gamma_\mu = (e_1 \bar{I}, e_2 \bar{I}, e_3, e_3)$ satisfies

$$\gamma_\mu \cdot \gamma_\nu = \bar{\eta}_{\mu\nu}, \quad (2.227)$$

where $\bar{\eta} = \mathrm{diag}(-1,-1,1,1)$.

8-space. Let e_A be basis vectors of 8-dimensional vector space with signature $(+++++++)$. Let us decompose

$$e_A = (e_\mu, e_{\bar{\mu}}), \quad \mu = 0,1,2,3,$$
$$\bar{\mu} = \bar{0}, \bar{1}, \bar{2}, \bar{3}. \quad (2.228)$$

The inner product of two basis vectors

$$e_A \cdot e_B = \delta_{AB}, \quad (2.229)$$

then splits into the following set of equations:

$$\begin{aligned} e_\mu \cdot e_\nu &= \delta_{\mu\nu}, \\ e_{\bar\mu} \cdot e_{\bar\nu} &= \delta_{\bar\mu\bar\nu}, \\ e_\mu \cdot e_{\bar\nu} &= 0. \end{aligned} \quad (2.230)$$

The number $\bar I = e_{\bar 0} e_{\bar 1} e_{\bar 2} e_{\bar 3}$ has the properties

$$\begin{aligned} \bar I^2 &= 1, \\ \bar I e_\mu &= e_\mu \bar I, \\ \bar I e_{\bar\mu} &= -e_{\bar\mu} \bar I. \end{aligned} \quad (2.231)$$

The set of numbers

$$\begin{aligned} \gamma_\mu &= e_\mu, \\ \gamma_{\bar\mu} &= e_{\bar\mu} \bar I \end{aligned} \quad (2.232)$$

satisfies

$$\begin{aligned} \gamma_\mu \cdot \gamma_\nu &= \delta_{\mu\nu}, \\ \gamma_{\bar\mu} \cdot \gamma_{\bar\nu} &= -\delta_{\mu\nu}, \\ \gamma_\mu \cdot \gamma_{\bar\mu} &= 0. \end{aligned} \quad (2.233)$$

The numbers $(\gamma_\mu, \gamma_{\bar\mu})$ thus form a set of basis vectors of a vector space $V_{4,4}$ with signature $(+ + + + - - - -)$.

10-space. Let $e_A = (e_\mu, e_{\bar\mu})$, $\mu = 1,2,3,4,5$; $\bar\mu = \bar 1, \bar 2, \bar 3, \bar 4, \bar 5$ be basis vectors of a 10-dimensional Euclidean space V_{10} with signature $(+ + + ...)$. We introduce $\bar I = e_{\bar 1} e_{\bar 2} e_{\bar 3} e_{\bar 4} e_{\bar 5}$ which satisfies

$$\begin{aligned} \bar I^2 &= 1, \\ e_\mu \bar I &= -\bar I e_\mu, \\ e_{\bar\mu} \bar I &= \bar I e_{\bar\mu}. \end{aligned} \quad (2.234)$$

Then the Clifford numbers

$$\begin{aligned} \gamma_\mu &= e_\mu \bar I, \\ \gamma_{\bar\mu} &= e_{\bar\mu} \end{aligned} \quad (2.235)$$

satisfy

$$\begin{aligned} \gamma_\mu \cdot \gamma_\nu &= -\delta_{\mu\nu}, \\ \gamma_{\bar\mu} \cdot \gamma_{\bar\nu} &= \delta_{\bar\mu\bar\nu}, \\ \gamma_\mu \cdot \gamma_{\bar\mu} &= 0. \end{aligned} \quad (2.236)$$

The set $\gamma_A = (\gamma_\mu, \gamma_{\bar\mu})$ therefore spans the vector space of signature $(-----++++)$.

The examples above demonstrate how vector spaces of various signatures are obtained within a given set of polyvectors. Namely, vector spaces of different signature are different subsets of polyvectors within the same Clifford algebra.

This has important physical implications. We have argued that physical quantities are polyvectors (Clifford numbers or Clifford aggregates). Physical space is then not simply a vector space (e.g., Minkowski space), but a space of polyvectors. The latter is a pandimensional continuum $\mathcal{P}$ [23] of points, lines, planes, volumes, etc., altogether. Minkowski space is then just a subspace with pseudo-Euclidean signature. Other subspaces with other signatures also exist within the pandimensional continuum $\mathcal{P}$ and they all have physical significance. If we describe a particle as moving in Minkowski spacetime $V_{1,3}$ we consider only certain physical aspects of the object considered. We have omitted its other physical properties like spin, charge, magnetic moment, etc.. We can as well describe the same object as moving in an Euclidean space V_4. Again such a description would reflect only a part of the underlying physical situation described by Clifford algebra.

GRASSMAN NUMBERS FROM CLIFFORD NUMBERS

In Sec. 2.5 we have seen that certain Clifford aggregates are spinors. Now we shall find out that also Grassmann (anticommuting) numbers are Clifford aggregates. As an example let us consider 8-dimensional space $V_{4,4}$ with signature $(+----+++)$ spanned by basis vectors $\gamma_A = (\gamma_\mu, \gamma_{\bar\mu})$. The numbers

$$\begin{aligned}\theta_\mu &= \tfrac{1}{2}(\gamma_\mu + \gamma_{\bar\mu}), \\ \theta_\mu^\dagger &= \tfrac{1}{2}(\gamma_\mu - \gamma_{\bar\mu})\end{aligned} \quad (2.237)$$

satisfy

$$\{\theta_\mu, \theta_\nu\} = \{\theta_\mu^\dagger, \theta_\nu^\dagger\} = 0 , \quad (2.238)$$

$$\{\theta_\mu, \theta_\nu^\dagger\} = \eta_{\mu\nu} , \quad (2.239)$$

where $\{A, B\} \equiv AB + BA$. From (2.238) we read out that θ_μ anticommute among themselves and are thus Grassmann numbers. Similarly $\theta_\mu^\dagger$ form a set of Grassmann numbers. But, because of (2.239), θ_μ and $\theta_\mu^\dagger$ altogether do not form a set of Grassmann numbers. They form yet another set of basis elements which generate Clifford algebra.

Point particles and Clifford algebra 91

A Clifford number in $V_{4,4}$ can be expanded as

$$C = c + c^{A_1}\gamma_{A_1} + c^{A_1 A_2}\gamma_{A_1}\gamma_{A_2} + ... + c^{A_1 A_2...A_8}\gamma_{A_1}\gamma_{A_2}...\gamma_{A_8}. \quad (2.240)$$

Using (2.237), the same Clifford number C can be expanded in terms of θ_μ, $\theta_\mu^\dagger$:

$$\begin{aligned}
C = c &+ a^\mu \theta_\mu + a^{\mu\nu}\theta_\mu\theta_\nu + a^{\mu\nu\alpha}\theta_\mu\theta_\nu\theta_\alpha + a^{\mu\nu\alpha\beta}\theta_\mu\theta_\nu\theta_\alpha\theta_\beta \\
&+ \bar{a}^\mu \theta_\mu^\dagger + \bar{a}^{\mu\nu}\theta_\mu^\dagger\theta_\nu^\dagger + \bar{a}^{\mu\nu\alpha}\theta_\mu^\dagger\theta_\nu^\dagger\theta_\alpha^\dagger + \bar{a}^{\mu\nu\alpha\beta}\theta_\mu^\dagger\theta_\nu^\dagger\theta_\alpha^\dagger\theta_\beta^\dagger \\
&+ \text{(mixed terms like } \theta_\mu^\dagger \theta_\nu, \text{ etc.)}
\end{aligned} \quad (2.241)$$

where the coefficients a^μ, $a^{\mu\nu}$, ..., $\bar{a}^\mu$, $\bar{a}^{\mu\nu}$... are linear combinations of coefficients c^{A_i}, $c^{A_i A_j}$,...

In a particular case, coefficients c, $\bar{a}^\mu$, $\bar{a}^{\mu\nu}$, etc., can be zero and our Clifford number is then a *Grassmann number* in 4-space:

$$\xi = a^\mu \theta_\mu + a^{\mu\nu}\theta_\mu\theta_\nu + a^{\mu\nu\alpha}\theta_\mu\theta_\nu\theta_\alpha + a^{\mu\nu\alpha\beta}\theta_\mu\theta_\nu\theta_\alpha\theta_\beta. \quad (2.242)$$

Grassmann numbers expanded according to (2.242), or analogous expressions in dimensions other than 4, are much used in contemporary theoretical physics. Recognition that Grassmann numbers can be considered as particular numbers within a more general set of numbers, namely Clifford numbers (or polyvectors), leads in my opinion to further progress in understanding and development of the currently fashionable supersymmetric theories, including superstrings, D-branes and M-theory.

We have seen that a Clifford number C in 8-dimensional space can be expanded in terms of the basis vectors $(\gamma_\mu, \gamma_{\bar{\mu}})$ or $(\theta_\mu, \theta_\mu^\dagger)$. Besides that, one can expand C also in terms of (γ_μ, θ_μ):

$$\begin{aligned}
C = c &+ b^\mu \gamma_\mu + b^{\mu\nu}\gamma_\mu\gamma_\nu + b^{\mu\nu\alpha}\gamma_\mu\gamma_\nu\gamma_\alpha + b^{\mu\nu\alpha\beta}\gamma_\mu\gamma_\nu\gamma_\alpha\gamma_\beta \\
&+ \beta^\mu \theta_\mu + \beta^{\mu\nu}\theta_\mu\theta_\nu + \beta^{\mu\nu\alpha}\theta_\mu\theta_\nu\theta_\alpha + \beta^{\mu\nu\alpha\beta}\theta_\mu\theta_\nu\theta_\alpha\theta_\beta \\
&+ \text{(mixed terms such as } \theta_\mu\gamma_\nu, \text{ etc.)}.
\end{aligned} \quad (2.243)$$

The basic vectors γ_μ span the familiar 4-dimensional spacetime, while θ_μ span an extra space, often called Grassmann space. Usually it is stated that besides four spacetime coordinates x^μ there are also four extra Grassmann coordinates θ_μ and their conjugates $\theta_\mu^\dagger$ or $\bar{\theta}_\mu = \gamma_0 \theta_\mu^\dagger$. This should be contrasted with the picture above in which θ_μ are basis *vectors* of an extra space, and not *coordinates*.

2.8. THE POLYVECTOR ACTION AND DE WITT–ROVELLI MATERIAL REFERENCE SYSTEM

Following an argument by Einstein [42], that points of spacetime are not *a priori* distinguishable, DeWitt [25] introduced a concept of reference fluid. Spacetime points are then defined with respect to the reference fluid. The idea that we can localize points by means of some matter has been further elaborated by Rovelli [26]. As a starting model he considers a reference system consisting of a single particle and a clock attached to it. Besides the particle coordinate variables $X^\mu(\tau)$ there is also the clock variable $T(\tau)$, attached to the particle, which grows monotonically along the particle trajectory. Rovelli then assumes the following action for the variables $X^\mu(\tau)$, $T(\tau)$:

$$I[X^\mu, T] = m \int \mathrm{d}\tau \left(\frac{\mathrm{d}X^\mu}{\mathrm{d}\tau} \frac{\mathrm{d}X_\mu}{\mathrm{d}\tau} - \frac{1}{\omega^2} \left(\frac{\mathrm{d}T}{\mathrm{d}\tau} \right)^2 \right)^{1/2}. \qquad (2.244)$$

If we make replacement $m \to \kappa$ and $T/\omega \to s$ the latter action reads

$$I[X^\mu, s] = \int \mathrm{d}\tau \kappa \left(\dot{X}^\mu \dot{X}_\mu - \dot{s}^2 \right)^{1/2}. \qquad (2.245)$$

If, on the other hand, we start from the polyvector action (2.117) and eliminate m, p_μ, λ by using the equations of motion, we again obtain the action (2.245). *Thus the pseudoscalar variable $s(\tau)$ entering the polyvector action may be identified with Rovelli's clock variable.* Although Rovelli starts with a single particle and clock, he later fills space with these objects. We shall return to Rovelli's reference systems when we discuss extended objects.

Chapter 3

HARMONIC OSCILLATOR IN PSEUDO-EUCLIDEAN SPACE

One of the major obstacles to future progress in our understanding of the relation between quantum field theory and gravity is the problem of the cosmological constant [43]. Since a quantum field is an infinite set of harmonic oscillators, each having a non-vanishing zero point energy, a field as a whole has infinite vacuum energy density. The latter can be considered, when neglecting the gravitational field, just as an additive constant with no influence on dynamics. However, the situation changes dramatically if one takes into account the gravitational field which feels the absolute energy density. As a consequence the infinite (or, more realistically, the Planck scale cutoff) cosmological constant is predicted which is in drastic contradiction with its observed small value. No generally accepted solution of this problem has been found so far.

In this chapter I study the system of uncoupled harmonic oscillators in a space with arbitrary metric signature. One interesting consequence of such a model is vanishing zero point energy of the system when the number of positive and negative signature coordinates is the same, and so there is no cosmological constant problem. As an example I first solve a system of two oscillators in the space with signature $(+-)$ by using a straightforward, though surprisingly unexploited, approach based on the energy being here just a quadratic form consisting of positive and negative terms [44, 45]. Both positive and negative energy components are on the same footing, and the difference in sign does not show up until we consider gravitation. In the quantized theory the vacuum state can be defined straightforwardly and the zero point energies cancel out. If the action is written in a covariant notation one has to be careful of how to define the vacuum state. Usually it is required that energy be positive, and then, in the absence of a cutoff, the formalism contains an infinite vacuum energy and negative norm states

[46]. In the formalism I am going to adopt energy is not necessarily positive, negative energy states are also stable, and there are no negative norm states [3].

The formalism will be then applied to the case of scalar fields. The extension of the spinor–Maxwell field is also discussed. Vacuum energy density is zero and the cosmological constant vanishes. However, the theory contains the negative energy fields which couple to the gravitational field in a different way than the usual, positive energy, fields: the sign of coupling is reversed. This is the price to be paid if one wants to obtain a small cosmological constant in a straightforward way. One can consider this as a prediction of the theory to be tested by suitably designed experiments. Usually, however, the argument is just the opposite of the one proposed here and classical matter is required to satisfy certain (essentially positive) energy conditions [47] which can only be violated by quantum field theory.

3.1. THE 2-DIMENSIONAL PSEUDO-EUCLIDEAN HARMONIC OSCILLATOR

Instead of the usual harmonic oscillator in 2-dimensional space let us consider that given by the Lagrangian

$$L = \tfrac{1}{2}(\dot{x}^2 - \dot{y}^2) - \tfrac{1}{2}\omega^2(x^2 - y^2). \tag{3.1}$$

The corresponding equations of motion are

$$\ddot{x} + \omega^2 x = 0, \qquad \ddot{y} + \omega^2 y = 0. \tag{3.2}$$

Note that in spite of the minus sign in front of the y-terms in the Lagrangian (3.1), the x and y components satisfy the same type equations of motion.

The canonical momenta conjugate to x,y are

$$p_x = \frac{\partial L}{\partial \dot{x}} = \dot{x}, \qquad p_y = \frac{\partial L}{\partial \dot{y}} = -\dot{y}. \tag{3.3}$$

The Hamiltonian is

$$H = p_x \dot{x} + p_y \dot{y} - L = \tfrac{1}{2}(p_x^2 - p_y^2) + \tfrac{1}{2}\omega^2(x^2 - y^2) \tag{3.4}$$

We see immediately that the energy so defined may have positive or negative values, depending on the initial conditions. Even if the system happens to have negative energy it is stable, since the particle moves in a closed curve around the point (0,0). The motion of the harmonic oscillator based on the

Harmonic oscillator in pseudo-Euclidean space

Lagrangian (3.1) does not differ from that of the usual harmonic oscillator. The difference occurs when one considers the gravitational fields around the two systems.

The Hamiltonian equations of motion are

$$\dot{x} = \frac{\partial H}{\partial p_x} = \{x, H\} = p_x, \quad \dot{y} = \frac{\partial H}{\partial p_y} = \{y, H\} = -p_y, \qquad (3.5)$$

$$\dot{p}_x = -\frac{\partial H}{\partial x} = \{p_x, H\} = -\omega^2 x, \quad \dot{p}_y = -\frac{\partial H}{\partial y} = \{p_y, H\} = \omega^2 y, \qquad (3.6)$$

where the basic Poisson brackets are $\{x, p_x\} = 1$ and $\{y, p_y\} = 1$.

Quantizing our system we have

$$[x, p_x] = i, \qquad [y, p_y] = i \qquad (3.7)$$

Introducing the non-Hermitian operators according to

$$c_x = \frac{1}{\sqrt{2}}\left(\sqrt{\omega}x + \frac{i}{\sqrt{\omega}}p_x\right), \quad c_x^\dagger = \frac{1}{\sqrt{2}}\left(\sqrt{\omega}x - \frac{i}{\sqrt{\omega}}p_x\right), \qquad (3.8)$$

$$c_y = \frac{1}{\sqrt{2}}\left(\sqrt{\omega}x + \frac{i}{\sqrt{\omega}}p_y\right), \quad c_y^\dagger = \frac{1}{\sqrt{2}}\left(\sqrt{\omega}x - \frac{i}{\sqrt{\omega}}p_y\right), \qquad (3.9)$$

we have

$$H = \frac{\omega}{2}(c_x^\dagger c_x + c_x c_x^\dagger - c_y^\dagger c_y - c_y c_y^\dagger). \qquad (3.10)$$

From the commutation relations (3.7) we obtain

$$[c_x, c_x^\dagger] = 1, \qquad [c_y, c_y^\dagger] = 1, \qquad (3.11)$$

and the normal ordered Hamiltonian then becomes

$$H = \omega(c_x^\dagger c_x - c_y^\dagger c_y). \qquad (3.12)$$

The vacuum state is defined as

$$c_x|0\rangle = 0, \qquad c_y|0\rangle = 0. \qquad (3.13)$$

The eigenvalues of H are

$$E = \omega(n_x - n_y), \qquad (3.14)$$

where n_x and n_y are eigenvalues of the operators $c_x^\dagger c_x$ and $c_y^\dagger c_y$, respectively.

The zero point energies belonging to the x and y components cancel. *Our 2-dimensional pseudo harmonic oscillator has vanishing zero point energy!*. This is a result we obtain when applying the standard Hamilton procedure to the Lagrangian (3.1).

In the (x,y) representation the vacuum state $\langle x,y|0\rangle \equiv \psi_0(x,y)$ satisfies

$$\frac{1}{\sqrt{2}}\left(\sqrt{\omega}\,x + \frac{1}{\sqrt{\omega}}\frac{\partial}{\partial x}\psi_0(x,y)\right) = 0\,,\quad \frac{1}{\sqrt{2}}\left(\sqrt{\omega}\,y + \frac{1}{\sqrt{\omega}}\frac{\partial}{\partial y}\psi_0(x,y)\right) = 0\,, \tag{3.15}$$

which comes straightforwardly from (3.13). A solution which is in agreement with the probability interpretation,

$$\psi_0 = \frac{2\pi}{\omega}\exp[-\tfrac{1}{2}\omega(x^2+y^2)] \tag{3.16}$$

is normalized according to $\int \psi_0^2\, dx\, dy = 1$.

We see that our particle is localized around the origin. The excited states obtained by applying $c_x^\dagger$, $c_y^\dagger$ to the vacuum state are also localized. This is in agreement with the property that also according to the classical equations of motion (3.2), the particle is localized in the vicinity of the origin. All states $|\psi\rangle$ have positive norm. For instance,

$$\langle 0|cc^\dagger|0\rangle = \langle 0|[c,c^\dagger]|0\rangle = \langle 0|0\rangle = \int \psi_0^2\, dx\, dy = 1.$$

3.2. HARMONIC OSCILLATOR IN d-DIMENSIONAL PSEUDO-EUCLIDEAN SPACE

Extending (3.1) to arbitrary dimension it is convenient to use the compact (covariant) index notation

$$L = \frac{1}{2}\dot{x}^\mu \dot{x}_\mu - \tfrac{1}{2}\omega^2 x^\mu x_\mu\,, \tag{3.17}$$

where for arbitrary vector A^μ the quadratic form is $A^\mu A_\mu \equiv \eta_{\mu\nu}A^\mu A^\nu$. The metric tensor $\eta_{\mu\nu}$ has signature $(+++\ldots ---\ldots)$. The Hamiltonian is

$$H = \tfrac{1}{2}p^\mu p_\mu + \tfrac{1}{2}\omega^2 x^\mu x_\mu \tag{3.18}$$

Conventionally one introduces

$$a^\mu = \frac{1}{\sqrt{2}}\left(\sqrt{\omega}\,x^\mu + \frac{i}{\sqrt{\omega}}p^\mu\right),\quad a^{\mu\dagger} = \frac{1}{\sqrt{2}}\left(\sqrt{\omega}\,x^\mu - \frac{i}{\sqrt{\omega}}p^\mu\right). \tag{3.19}$$

In terms of a^μ, $a^{\mu\dagger}$ the Hamiltonian reads

$$H = \frac{\omega}{2}(a^{\mu\dagger}a_\mu + a_\mu a^{\mu\dagger}). \tag{3.20}$$

Harmonic oscillator in pseudo-Euclidean space 97

Upon quantization we have

$$[x^\mu, p_\nu] = i\delta^\mu{}_\nu \quad \text{or} \quad [x^\mu, p^\nu] = i\eta^{\mu\nu} \qquad (3.21)$$

and

$$[a^\mu, a^\dagger_\nu] = \delta^\mu{}_\nu \quad \text{or} \quad [a^\mu, a^{\nu\dagger}] = \eta^{\mu\nu}. \qquad (3.22)$$

We shall now discuss two possible definitions of the vacuum state. The first possibility is the one usually assumed, whilst the second possibility [44, 45] is the one I am going to adopt.

Possibility I. The vacuum state can be defined according to

$$a^\mu|0\rangle = 0 \qquad (3.23)$$

and the Hamiltonian, normal ordered with respect to the vacuum definition (3.23), becomes, after using (3.22),

$$H = \omega\left(a^{\mu\dagger}a_\mu + \frac{d}{2}\right), \quad d = \eta^{\mu\nu}\eta_{\mu\nu}. \qquad (3.24)$$

Its eigenvalues are all positive and there is the non-vanishing zero point energy $\omega d/2$. In the x representation the vacuum state is

$$\psi_0 = \left(\frac{2\pi}{\omega}\right)^{d/2} \exp[-\tfrac{1}{2}\omega x^\mu x_\mu] \qquad (3.25)$$

It is a solution of the Schrödinger equation $-\tfrac{1}{2}\partial^\mu\partial_\mu\psi_0 + (\omega^2/2)x^\mu x_\mu\psi_0 = E_0\psi_0$ with positive $E_0 = \omega(\tfrac{1}{2} + \tfrac{1}{2} + ...)$. The state ψ_0 as well as excited states can not be normalized to 1. Actually, there exist negative norm states. For instance, if $\eta^{33} = -1$, then

$$\langle 0|a^3 a^{3\dagger}|0\rangle = \langle 0|[a^3, a^{3\dagger}]|0\rangle = -\langle 0|0\rangle.$$

Possibility II. Let us split $a^\mu = (a^\alpha, a^{\bar\alpha})$, where the indices α, $\bar\alpha$ refer to the components with positive and negative signature, respectively, and define the vacuum according to[1]

$$a^\alpha|0\rangle = 0, \qquad a^{\bar\alpha\dagger}|0\rangle = 0. \qquad (3.26)$$

---

[1]Equivalently, one can define annihilation and creation operators in terms of x^μ and the canonically conjugate momentum $p_\mu = \eta_{\mu\nu}p^\nu$ according to $c^\mu = (1/\sqrt{2})(\sqrt{\omega}x^\mu + (i/\sqrt{\omega})p_\mu)$ and $c^{\mu\dagger} = (1/\sqrt{2})(\sqrt{\omega}x^\mu - (i/\sqrt{\omega})p_\mu)$, satisfying $[c^\mu, c^{\nu\dagger}] = \delta^{\mu\nu}$. The vacuum is then defined as $c^\mu|0\rangle = 0$. This is just the higher-dimensional generalization of c_x, c_y (eq. (3.8),(3.9) and the vacuum definition (3.13).

Using (3.22) we obtain the normal ordered Hamiltonian with respect to the vacuum definition (3.26)

$$H = \omega \left(a^{\alpha \dagger} a_\alpha + \frac{r}{2} + a_{\bar{\alpha}} a^{\bar{\alpha}\dagger} - \frac{s}{2} \right), \qquad (3.27)$$

where $\delta_\alpha{}^\alpha = r$ and $\delta_{\bar\alpha}{}^{\bar\alpha} = s$. If the number of positive and negative signature components is the same, i.e., $r = s$, then the Hamiltonian (3.27) has vanishing zero point energy:

$$H = \omega(a^{\alpha\dagger} a_\alpha + a_{\bar\alpha} a^{\bar\alpha\dagger}). \qquad (3.28)$$

Its eigenvalues are positive or negative, depending on which components (positive or negative signature) are excited. In the x-representation the vacuum state (3.26) is

$$\psi_0 = \left(\frac{2\pi}{\omega}\right)^{d/2} \exp[-\tfrac{1}{2}\omega \delta_{\mu\nu} x^\mu x^\nu], \qquad (3.29)$$

where the Kronecker symbol $\delta_{\mu\nu}$ has the values +1 or 0. It is a solution of the Schrödinger equation $-\tfrac{1}{2}\partial^\mu \partial_\mu \psi_0 + (\omega^2/2) x^\mu x_\mu \psi_0 = E_0 \psi_0$ with $E_0 = \omega(\tfrac{1}{2} + \tfrac{1}{2} + ... - \tfrac{1}{2} - \tfrac{1}{2} - ...)$. One can also easily verify that there are no negative norm states.

Comparing *Possibility I* with *Possibility II* we observe that the former has positive energy vacuum invariant under pseudo-Euclidean rotations, whilst the latter has the vacuum invariant under Euclidean rotations and having vanishing energy (when $r = s$). In other words, we have: either (i) non-vanishing energy and pseudo-Euclidean invariance or (ii) vanishing energy and Euclidean invariance of the vacuum state. In the case (ii) the vacuum state ψ_0 changes under the pseudo-Euclidean rotations, but its energy remains zero.

The invariance group of our Hamiltonian (3.18) and the corresponding Schrödinger equation consists of pseudo-rotations. Though a solution of the Schrödinger equation changes under a pseudo-rotation, the theory is covariant under the pseudo-rotations, in the sense that the set of all possible solutions does not change under the pseudo-rotations. Namely, the solution $\psi_0(x')$ of the Schrödinger equation

$$-\tfrac{1}{2}\partial'^\mu \partial'_\mu \psi_0(x') + (\omega^2/2) x'^\mu x'_\mu \psi_0(x') = 0 \qquad (3.30)$$

in a pseudo-rotated frame S' is

$$\psi_0(x') = \frac{2\pi}{\omega}^{d/2} \exp[-\tfrac{1}{2}\omega \delta_{\mu\nu} x'^\mu x'^\nu]. \qquad (3.31)$$

If observed from the frame S the latter solution reads

$$\psi'_0(x) = \frac{2\pi}{\omega}^{d/2} \times \exp[-\tfrac{1}{2}\omega\delta_{\mu\nu}L^\mu{}_\rho x^\rho L^\nu{}_\sigma x^\sigma], \qquad (3.32)$$

where $x'^\mu = L^\mu{}_\rho x^\rho$. One finds that $\psi'_0(x)$ as well as $\psi_0(x)$ (eq. (3.29)) are solutions of the Schrödinger equation in S and they both have the same vanishing energy. In general, in a given reference frame we have thus a degeneracy of solutions with the same energy [44]. This is so also in the case of excited states.

In principle it seem more natural to adopt *Possibility II*, because the classically energy of our harmonic oscillator is nothing but a quadratic form $E = \tfrac{1}{2}(p^\mu p_\mu + \omega^2 x^\mu x_\mu)$, which in the case of a metric of pseudo-Euclidean signature can be positive, negative, or zero.

3.3. A SYSTEM OF SCALAR FIELDS

Suppose we have a system of two scalar fields described by the action[2]

$$I = \tfrac{1}{2}\int d^4x \,(\partial_\mu\phi_1\,\partial^\mu\phi_1 - m^2\phi_1^2 - \partial_\mu\phi_2\,\partial^\mu\phi_2 + m^2\phi_2^2). \qquad (3.33)$$

This action differs from the usual action for a charged field in the sign of the ϕ_2 term. It is a field generalization of our action for the point particle harmonic oscillator in 2-dimensional pseudo-Euclidean space.

The canonical momenta are

$$\pi_1 = \dot{\phi}_1, \qquad \pi_2 = -\dot{\phi}_2 \qquad (3.34)$$

satisfying

$$[\phi_1(\mathbf{x}),\pi_1(\mathbf{x}')] = i\delta^3(\mathbf{x}-\mathbf{x}'), \qquad [\phi_2(\mathbf{x}),\pi_2(\mathbf{x}')] = i\delta^3(\mathbf{x}-\mathbf{x}'). \qquad (3.35)$$

The Hamiltonian is

$$H = \tfrac{1}{2}\int d^3\mathbf{x}\,(\pi_1^2 + m^2\phi_1^2 - \partial_i\phi_1\,\partial^i\phi_1 - \pi_2^2 - m^2\phi_2^2 + \partial_i\phi_2\,\partial^i\phi_2). \qquad (3.36)$$

We use the spacetime metric with signature $(+---)$ so that $-\partial_i\phi_1\,\partial^i\phi_1 = (\nabla\phi)^2$, $i=1,2,3$. Using the expansion ($\omega_\mathbf{k} = (m^2+\mathbf{k}^2)^{1/2}$)

$$\phi_1 = \int \frac{d^3\mathbf{k}}{(2\pi)^3}\frac{1}{2\omega_\mathbf{k}}\left(c_1(\mathbf{k})e^{-ikx} + c_1^\dagger(\mathbf{k})e^{ikx}\right), \qquad (3.37)$$

---

[2] Here, for the sake of demonstration, I am using the formalism of the conventional field theory, though in my opinion a better formalism involves an invariant evolution parameter, as discussed in Sec. 1.4

$$\phi_2 = \int \frac{d^3\mathbf{k}}{(2\pi)^3} \frac{1}{2\omega_\mathbf{k}} \left(c_2(\mathbf{k}) e^{-ikx} + c_2^\dagger(\mathbf{k}) e^{ikx} \right) \tag{3.38}$$

we obtain

$$H = \frac{1}{2} \int \frac{d^3\mathbf{k}}{(2\pi)^3} \frac{\omega_\mathbf{k}}{2\omega_\mathbf{k}} \left(c_1^\dagger(\mathbf{k}) c_1(\mathbf{k}) + c_1(\mathbf{k}) c_1^\dagger(\mathbf{k}) - c_2^\dagger(\mathbf{k}) c_2(\mathbf{k}) - c_2(\mathbf{k}) c_2^\dagger(\mathbf{k}) \right). \tag{3.39}$$

The commutation relations are

$$[c_1(\mathbf{k}), c_1^\dagger(\mathbf{k}')] = (2\pi)^3 2\omega_\mathbf{k} \delta^3(\mathbf{k} - \mathbf{k}'), \tag{3.40}$$

$$[c_2(\mathbf{k}), c_2^\dagger(\mathbf{k}')] = (2\pi)^3 2\omega_\mathbf{k} \delta^3(\mathbf{k} - \mathbf{k}'). \tag{3.41}$$

The Hamiltonian can be written in the form

$$H = \frac{1}{2} \int \frac{d^3\mathbf{k}}{(2\pi)^3} \frac{\omega_\mathbf{k}}{2\omega_\mathbf{k}} \left(c_1^\dagger(\mathbf{k}) c_1(\mathbf{k}) - c_2^\dagger(\mathbf{k}) c_2(\mathbf{k}) \right) \tag{3.42}$$

If we define the vacuum according to

$$c_1(\mathbf{k})|0\rangle = 0, \qquad c_2(\mathbf{k})|0\rangle = 0, \tag{3.43}$$

then the Hamiltonian (3.42) contains the creation operators on the left and has no zero point energy. However, it is not positive definite: it may have positive or negative eigenvalues. But, as is obvious from our analysis of the harmonic oscillator (3.1), negative energy states in our formalism are not automatically unstable; they can be as stable as positive energy states.

Extension of the action (3.33) to arbitrary number os fields ϕ^a is straightforward. Let us now include also the gravitational field $g_{\mu\nu}$. The action is then

$$I = \frac{1}{2} \int d^4x \sqrt{-g} \left(g^{\mu\nu} \partial_\mu \phi^a \partial_\nu \phi^b \gamma_{ab} - m^2 \phi^a \phi^b \gamma_{ab} + \frac{1}{16\pi G} R \right), \tag{3.44}$$

where γ_{ab} is the metric tensor in the space of ϕ^a. Variation of (3.44) with respect to $g^{\mu\nu}$ gives the Einstein equations

$$R_{\mu\nu} - \tfrac{1}{2} g_{\mu\nu} R = -8\pi G T_{\mu\nu}, \tag{3.45}$$

where the stress–energy tensor is

$$T_{\mu\nu} = \frac{2}{\sqrt{-g}} \frac{\partial \mathcal{L}}{\partial g^{\mu\nu}} = \left[\partial_\mu \phi^a \partial_\nu \phi^b - \tfrac{1}{2} g_{\mu\nu} \left(g^{\rho\sigma} \partial_\rho \phi^a \partial_\sigma \phi^b - m^2 \phi^a \phi^b \right) \right] \gamma_{ab}. \tag{3.46}$$

If γ_{ab} has signature $(+++...---)$, with the same number of plus and minus signs, then the vacuum contributions to $T_{\mu\nu}$ cancel, so that the expectation value $\langle T_{\mu\nu} \rangle$ remains finite. In particular, we have

$$T_{00} = \tfrac{1}{2}(\dot\phi^a \dot\phi^b - \partial_i \phi^a \, \partial^i \phi^b + m^2 \phi^a \phi^b)\gamma_{ab}, \qquad (3.47)$$

which is just the Hamiltonian H of eq. (3.36) generalized to an arbitrary number of fields ϕ^a.

A procedure analogous to that before could be carried out for other types of fields such as charged scalar, spinor and gauge fields. The notorious cosmological constant problem does not arise in our model, since the vacuum expectation value $\langle 0|T_{\mu\nu}|0\rangle = 0$. We could reason the other way around: since experiments clearly show that the cosmological constant is small, this indicates (especially in the absence of any other acceptable explanation) that to every field there corresponds a companion field with the signature opposite of the metric eigenvalue in the space of fields. The companion field need not be excited —and thus observed— at all. Its mere existence is sufficient to be manifested in the vacuum energy.

However, there is a price to be paid. If negative signature fields are excited then $\langle T_{00}\rangle$ can be negative, which implies a repulsive gravitational field around such a source. Such a prediction of the theory could be considered as an annoyance, on the one hand, or a virtue, on the other. If the latter point of view is taken then we have here so called exotic matter with negative energy density, which is necessary for the construction of the stable wormholes with time machine properties [48]. Also the Alcubierre warp drive [49] which enables faster than light motion with respect to a distant observer requires matter of negative energy density.

As an example let me show how the above procedure works for spinor and gauge fields. Neglecting gravitation the action is

$$I = \int \mathrm{d}^4 x \left[i\bar\psi^a \gamma^\mu (\partial_\mu \psi^b + ie A_\mu{}^b{}_c \psi^c) - m\bar\psi^a \psi^b + \frac{1}{16\pi} F_{\mu\nu}{}^{ac} F^{\mu\nu}{}_c{}^b \right] \gamma_{ab}, \qquad (3.48)$$

where

$$F_{\mu\nu}{}^{ab} = \partial_\mu A_\nu{}^{ab} - \partial_\nu A_\mu{}^{ab} - (A_\mu{}^{ac} A_\nu{}^{db} - A_\nu{}^{ac} A_\mu{}^{db})\gamma_{cd}. \qquad (3.49)$$

This action is invariant under local rotations

$$\psi'^a = U^a{}_b \phi^b, \quad \bar\psi'^a = U^a{}_b \psi^b, \quad A'_\mu{}^a{}_b = U^{*c}{}_b A_\mu{}^d{}_c U^a{}_d + iU^{*c}{}_b \partial_\mu U^a{}_c, \qquad (3.50)$$

which are generalization of the usual $SU(N)$ transformations to the case of the metric $\gamma_{ab} = \mathrm{diag}(1,1,...,-1,-1)$.

In the case in which there are two spinor fields ψ_1, ψ_2, and $\gamma_{ab} = \mathrm{diag}(1,-1)$, the equations of motion derived from (3.48) admit a solution

$\psi_2 = 0$, $A_\mu^{12} = A_\mu^{21} = A_\mu^{22} = 0$. In the quantum field theory such a solution can be interpreted as that when the fermions of type ψ_2 (the companion or negative signature fields) are not excited (not present) the gauge fields A_μ^{12}, A_μ^{21}, A_μ^{22} are also not excited (not present). What remains are just the ordinary $\psi_1 \equiv \psi$ fermion quanta and $U(1)$ gauge field $A_\mu^{11} \equiv A^\mu$ quanta. The usual spinor–Maxwell electrodynamics is just a special solution of the more general system given by (3.48).

Although having vanishing vacuum energy, such a model is consistent with the well known experimentally observed effects which are manifestations of vacuum energy. Namely, the companion particles ψ_2 are expected to be present in the material of the Earth in small amounts at most, because otherwise the gravitational field around the Earth would be repulsive. So, when considering vacuum effects, there remain only (or predominantly) the interactions between the fermions ψ_1 and the virtual photons A_μ^{11}. For instance, in the case of the Casimir effect [50] the fermions ψ_1 in the two conducting plates interact with the virtual photons A_μ^{11} in the vacuum, and hence impose the boundary conditions on the vacuum modes of A_μ^{11} in the presence of the plates. As a result we have a net force between the plates, just as in the usual theory.

When gravitation is not taken into account, the fields within the doublet (ψ_1, ψ_2) as described by the action (3.48) are not easily distinguishable, since they have the same mass, charge, and spin. They mutually interact only through the mixed coupling terms in the action (3.48), and unless the effects of this mixed coupling are specifically measured the two fields can be mis-identified as a single field. Its double character could manifest itself straightforwardly in the presence of a gravitational field to which the members of a doublet couple with the opposite sign. In order to detect such doublets (or perhaps multiplets) of fields one has to perform suitable experiments. The description of such experiments is beyond the scope of this book, which only aims to bring attention to such a possibility. Here I only mention that difficulties and discrepancies in measuring the precise value of the gravitational constant might have roots in negative energy matter. The latter would affect the measured value of the effective gravitational constant, but would leave the equivalence principle untouched.

3.4. CONCLUSION

The problem of the cosmological constant is one of the toughest problems in theoretical physics. Its resolution would open the door to further understanding of the relation between quantum theory and general relativity. Since all more conventional approaches seem to have been more or less exploited without unambiguous success, the time is right for a more drastic novel approach. Such is the one which relies on the properties of the harmonic oscillator in a pseudo-Euclidean space. This can be applied to the field theory in which the fields behave as components of a harmonic oscillator. If the space of fields has a metric with signature $(+++\ldots ---)$ then the vacuum energy can be zero in the case in which the number of plus and minus signs is the same. As a consequence the expectation value of the stress–energy tensor, the source of the gravitational field, is finite, and there is no cosmological constant problem. However, the stress–energy tensor can be negative in certain circumstances and the matter then acquires exotic properties which are desirable for certain very important theoretical constructions, such as time machines [48] or a faster than light warp drive [49]. Negative energy matter, with repulsive gravitational field, is considered here as a prediction of the theory. On the contrary, in a more conventional approach just the opposite point of view is taken. It is argued that, since for all known forms of matter gravitation is attractive, certain energy conditions (weak, strong and dominant) must be satisfied [48]. But my point of view, advocated in this book, is that the existence of negative energy matter is necessary in order to keep the cosmological constant small (or zero). In addition to that, such exotic matter, if indeed present in the Universe, should manifest itself in various phenomena related to gravitation. Actually, we cannot claim to possess a complete knowledge and understanding of all those phenomena, especially when some of them are still waiting for a generally accepted explanation.

The theory of the pseudo-Euclidean signature harmonic oscillator is important for strings also. Since it eliminates the zero point energy it also eliminates the need for a critical dimension, as indicated in ref. [28]. We thus obtain a consistent string theory in an arbitrary even dimensional spacetime with suitable signature. Several exciting new possibilities of research are thus opened.

The question is only whether an n-dimensional spacetime V_n has indeed an equal number of time-like and space-like dimensions. Here we bring into the game the assumption that physical quantities are Clifford aggregates, or polyvectors, and that what we observe as spacetime is just a segment of a higher geometrical structure, called the *pandimensional continuum* [23]. Among the available elements of Clifford algebra one can choose n elements such that they serve as basis vectors of a pseudo-Euclidean space with arbitrary signature. Harmonic oscillator in pseudo-Euclidean space is only a tip of an iceberg. The iceberg is geometric calculus based on Clifford algebra which allows for any signature. Actually a signature is relative to a chosen basis e_μ, $e_{\mu\nu}$, $e_{\mu\nu\alpha}$, ... of Clifford algebra.

On the other hand, in conventional theory a signature with two or more time-like dimensions is problematic from the point of view of initial conditions: predictability is not possible. But such a problem does not arise in the theory with an invariant evolution parameter τ, where initial conditions are given in the entire spacetime, and the evolution then goes along τ. This is yet another argument in favor of a Stueckelberg-like theory. However, a question arises of how a theory in which initial conditions are given in the entire spacetime can be reconciled with the fact that we do not observe the entire spacetime "at once". This question will be answered in the remaining parts of the book.

II
EXTENDED OBJECTS

Chapter 4

GENERAL PRINCIPLES OF MEMBRANE KINEMATICS AND DYNAMICS

We are now going to extend our system. Instead of point particles we shall consider strings and higher-dimensional membranes. These objects are nowadays amongst the hottest topics in fundamental theoretical physics. Many people are convinced that strings and accompanying higher-dimensional membranes provide a clue to unify physics [51]. In spite of many spectacular successes in unifying gravity with other interactions, there still remain open problems. Amongst the most serious is perhaps the problem of a geometrical principle behind string theory [52]. The approach pursued in this book aims to shed some more light on just that problem. We shall pay much attention to the treatment of membranes as points in an infinite-dimensional space, called membrane space $\mathcal{M}$. When pursuing such an approach the researchers usually try to build in right from the beginning a complication which arises from reparametrization invariance (called also diffeomorphism invariance). Namely, the same n-dimensional membrane can be represented by different sets of parametric equations $x^\mu = X^\mu(\xi^a)$, where functions X^μ map a membrane's parameters (also called coordinates) ξ^a, $a = 1, 2, ..., n$, into spacetime coordinates x^μ, $\mu = 0, 1, 2, ..., N-1$. The problem is then what are coordinates of the membrane space $\mathcal{M}$? If there were no complication caused by reparametrization invariance, then the $X^\mu(\xi^a)$ would be coordinates of $\mathcal{M}$-space. But because of reparametrization invariance such a mapping from a point in $\mathcal{M}$ (a membrane) to its coordinates $X^\mu(\xi)$ is one-to-many. So far there is no problem: a point in any space can be represented by many different possible sets of coordinates, and all those different sets are related by coordinate transformations or diffeomorphisms. Since the latter transformations refer to the same point they are called *passive coordinate transformations* or *passive diffeomorphisms*. The problem occurs when one brings into the game *active diffeomorphisms*

108   *THE LANDSCAPE OF THEORETICAL PHYSICS: A GLOBAL VIEW*

which refer to different points of the space in question. In the case of membrane space $\mathcal{M}$ active diffeomorphisms would imply the existence of *tangentially deformed membranes*. But such objects are not present in a relativistic theory of membranes, described by the minimal surface action which is invariant under reparametrizations of ξ^a.

The approach pursued here is the following. We shall assume that *at the kinematic level such tangentially deformed membranes do exist* [53]. When considering dynamics it may happen that a certain action and its equations of motion exclude tangential motions within the membrane. This is precisely what happens with membranes obeying the relativistic minimal surface action. But the latter dynamical principle is not the most general one. We can extend it according to geometric calculus based on Clifford algebra. We have done so in Chapter 2 for point particles, and now we shall see how the procedure can be generalized to membranes of arbitrary dimension. And we shall find a remarkable result that *the polyvector generalization* of the membrane action allows for tangential motions of membranes. Because of the presence of an extra pseudoscalar variable entering the polyvector action, the membrane variables $X^\mu(\xi)$ and the corresponding momenta become *unconstrained*; tangentially deformed membranes are thus present not only at the kinematic, but also at the dynamical level. In other words, such a generalized dynamical principle allows for tangentially deformed membranes.

In the following sections I shall put the above description into a more precise form. But in the spirit of this book I will not attempt to achieve complete mathematical rigor, because for most readers this would be at the expense of seeing the main outline of my proposal of how to formulate membrane's theory

4.1.   MEMBRANE SPACE $\mathcal{M}$

The basic kinematically possible objects of the theory we are going to discuss are n-dimensional, arbitrarily deformable, and hence unconstrained, membranes $\mathcal{V}_n$ living in an N-dimensional space V_N. The dimensions n and N, as well as the corresponding signatures, are left unspecified at this stage. An unconstrained membrane $\mathcal{V}_n$ is represented by the embedding functions $X^\mu(\xi^a)$, $\mu = 0, 1, 2, ..., N - 1$, where ξ^a, $a = 0, 1, 2, ..., n - 1$, are local parameters (coordinates) on $\mathcal{V}_n$. The set of all possible membranes $\mathcal{V}_n$, with n fixed, forms an infinite-dimensional space $\mathcal{M}$. A membrane $\mathcal{V}_n$ can be considered as a point in $\mathcal{M}$ parametrized by coordinates $X^\mu(\xi^a) \equiv X^{\mu(\xi)}$ which bear a discrete index μ and n continuous indices ξ^a. To the discrete index μ we can ascribe arbitrary numbers: instead of $\mu = 0, 1, 2, ..., N - 1$

General principles of membrane kinematics and dynamics 109

we may set $\mu' = 1, 2, ..., N$ or $\mu' = 2, 5, 3, 1, ...$, etc.. In general,

$$\mu' = f(\mu), \tag{4.1}$$

where f is a transformation. Analogously, a continuous index ξ^a can be given arbitrary continuous values. Instead of ξ^a we may take ξ'^a which are functions of ξ^a :

$$\xi'^a = f^a(\xi). \tag{4.2}$$

As far as we consider, respectively, μ and ξ^a as a discrete and a continuous index of coordinates $X^{\mu(\xi)}$ in the infinite-dimensional space $\mathcal{M}$, reparametrization of ξ^a is analogous to a renumbering of μ. Both kinds of transformations, (4.1) and (4.2), refer to the same point of the space $\mathcal{M}$; they are *passive transformations*. For instance, under the action of (4.2) we have

$$X'^\mu(\xi') = X'^\mu(f(\xi)) = X^\mu(\xi) \tag{4.3}$$

which says that the same point $\mathcal{V}_n$ can be described either by functions $X^\mu(\xi)$ or $X'^\mu(\xi)$ (where we may write $X'^\mu(\xi)$ instead of $X'^\mu(\xi')$ since ξ' is a running parameter and can be renamed as ξ).

Then there also exist the *active transformations*, which transform one point of the space $\mathcal{M}$ into another. Given a parametrization of ξ^a and a numbering of μ, a point $\mathcal{V}_n$ of $\mathcal{M}$ with coordinates $X^\mu(\xi)$ can be transformed into another point $\mathcal{V}'_n$ with coordinates $X'^\mu(\xi)$. Parameters ξ^a are now considered as "body fixed", so that distinct functions $X^\mu(\xi)$, $X'^\mu(\xi)$ represent distinct points $\mathcal{V}_n$, $\mathcal{V}'_n$ of $\mathcal{M}$. Physically these are distinct membranes which may be deformed one with respect to the other. Such a membrane is *unconstrained*, since all coordinates $X^\mu(\xi)$ are necessary for its description [53]–[55]. In order to distinguish an unconstrained membrane $\mathcal{V}_n$ from the corresponding mathematical manifold V_n we use different symbols $\mathcal{V}_n$ and V_n.

It may happen, in particular, that two distinct membranes $\mathcal{V}_n$ and $\mathcal{V}'_n$ both lie on the same mathematical surface V_n, and yet they are physically distinct objects, represented by different points in $\mathcal{M}$.

The concept of an unconstrained membrane can be illustrated by imagining a rubber sheet spanning a surface V_2. The sheet can be deformed from one configuration (let me call it $\mathcal{V}_2$) into another configuration $\mathcal{V}'_2$ in such a way that both configurations $\mathcal{V}_2$, $\mathcal{V}'_2$ are spanning the same surface V_2. The configurations $\mathcal{V}_2$, $\mathcal{V}'_2$ are described by functions $X^i(\xi^1, \xi^2)$, $X'^i(\xi^1, \xi^2)$ ($i = 1, 2, 3$), respectively. The latter functions, from the mathematical point of view, both represent the same surface V_2, and can be transformed one into the other by a reparametrization of ξ^1, ξ^2. But from the physical point of view, $X^i(\xi^1, \xi^2)$ and $X'^i(\xi^1, \xi^2)$ represent two different configurations of the rubber sheet (Fig. 4.1).

110 *THE LANDSCAPE OF THEORETICAL PHYSICS: A GLOBAL VIEW*

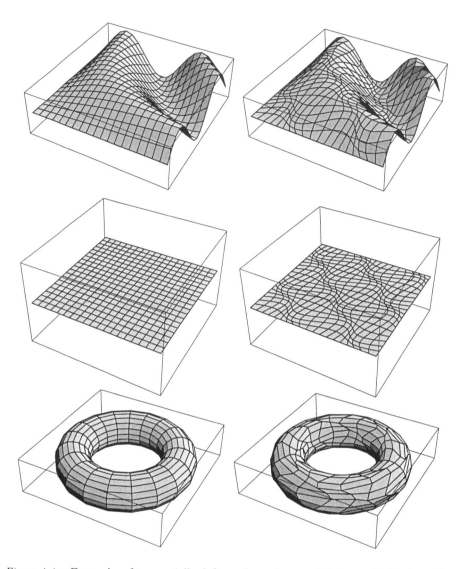

Figure 4.1. Examples of tangentially deformed membranes. Mathematically the surfaces on the right are the same as those on the left, but physically they are different.

The reasoning presented in the last few paragraphs implies that, since our membranes are assumed to be arbitrarily deformable, different functions $X^\mu(\xi)$ can always represent physically different membranes. This justifies use of the coordinates $X^\mu(\xi)$ for the description of points in $\mathcal{M}$. Later, when we consider a membrane's kinematics and dynamics we shall admit

General principles of membrane kinematics and dynamics 111

τ-dependence of coordinates $X^\mu(\xi)$. In this section all expressions refer to a fixed value of τ, therefore we omit it from the notation.

In analogy with the finite-dimensional case we can introduce the distance $d\ell$ in our infinite-dimensional space $\mathcal{M}$:

$$\begin{aligned} d\ell^2 &= \int d\xi\, d\zeta\, \rho_{\mu\nu}(\xi,\zeta)\, dX^\mu(\xi)\, dX^\nu(\zeta) \\ &= \rho_{\mu(\xi)\nu(\zeta)}\, dX^{\mu(\xi)}\, dX^{\nu(\zeta)} = dX^{\mu(\xi)} dX_{\mu(\xi)}, \end{aligned} \quad (4.4)$$

where $\rho_{\mu\nu}(\xi,\zeta) = \rho_{\mu(\xi)\nu(\zeta)}$ is the metric in $\mathcal{M}$. In eq. (4.4) we use a notation, similar to one that is usually used when dealing with more evolved functional expressions [56], [57]. In order to distinguish continuous indices from discrete indices, the former are written within parentheses. When we write $\mu(\xi)$ as a subscript or superscript this denotes a pair of indices μ and (ξ) (and not that μ is a function of ξ). We also use the convention that summation is performed over repeated indices (such as a, b) and integration over repeated continuous indices (such as $(\xi), (\zeta)$).

The tensor calculus in $\mathcal{M}$ [54, 55] is analogous to that in a finite-dimensional space. The differential of coordinates $dX^\mu(\xi) \equiv dX^{\mu(\xi)}$ is a vector in $\mathcal{M}$. The coordinates $X^{\mu(\xi)}$ can be transformed into new coordinates $X'^{\mu(\xi)}$ which are functionals of $X^{\mu(\xi)}$:

$$X'^{\mu(\xi)} = F^{\mu(\xi)}[X]. \quad (4.5)$$

The transformation (4.5) is very important. It says that if functions $X^\mu(\xi)$ represent a membrane $\mathcal{V}_n$ then any other functions $X'^\mu(\xi)$ obtained from $X^\mu(\xi)$ by a functional transformation also represent the same membrane $\mathcal{V}_n$. In particular, under a reparametrization of ξ^a the functions $X^\mu(\xi)$ change into new functions; a reparametrization thus manifests itself as a special functional transformation which belongs to a subclass of the general functional transformations (4.5).

Under a general coordinate transformation (4.5) a generic vector $A^{\mu(\xi)} \equiv A^\mu(\xi)$ transforms as[1]

$$A^{\mu(\xi)} = \frac{\partial X'^{\mu(\xi)}}{\partial X^{\nu(\zeta)}} A^{\nu(\zeta)} \equiv \int d\zeta\, \frac{\delta X'^\mu(\xi)}{\delta X^\nu(\zeta)} A^\nu(\zeta) \quad (4.6)$$

where $\delta/\delta X^\mu(\xi)$ denotes the functional derivative (see Box 4.1). Similar transformations hold for a covariant vector $A_{\mu(\xi)}$, a tensor $B_{\mu(\xi)\nu(\zeta)}$, etc..

---
[1] A similar formalism, but for a specific type of the functional transformations (4.5), namely the reparametrizations which functionally depend on string coordinates, was developed by Bardakci [56]

112 THE LANDSCAPE OF THEORETICAL PHYSICS: A GLOBAL VIEW

Indices are lowered and raised, respectively, by $\rho_{\mu(\xi)\nu(\zeta)}$ and $\rho^{\mu(\xi)\nu(\zeta)}$, the latter being the inverse metric tensor satisfying

$$\rho^{\mu(\xi)\alpha(\eta)}\rho_{\alpha(\eta)\nu(\zeta)} = \delta^{\mu(\xi)}{}_{\nu(\zeta)}. \tag{4.7}$$

Box 4.1: Functional derivative

Let $X^\mu(\xi)$ be a function of $\xi \equiv \xi^a$. The functional derivative of a functional $F[X^\mu(\xi)]$ is defined according to

$$\frac{\delta F}{\delta X^\nu(\xi')} = \lim_{\epsilon \to 0} \frac{F[X^\mu(\xi) + \epsilon\delta(\xi-\xi')\delta^\mu{}_\nu] - F[X^\mu(\xi)]}{\epsilon} \tag{4.8}$$

Examples

1) $F = X^\mu(\xi)$

$$\frac{\delta F}{\delta X^\nu(\xi)} = \delta(\xi-\xi')\delta^\mu{}_\nu$$

2) $F = \partial_a X^\mu(\xi)$

$$\frac{\delta F}{\delta X^\nu(\xi)} = \partial_a \delta(\xi-\xi')\delta^\mu{}_\nu$$

3) $F = \lambda(\xi)\partial_a X^\mu(\xi)$

$$\frac{\delta F}{\delta X^\nu(\xi)} = \lambda(\xi)\frac{\delta \partial_a X^\mu(\xi)}{\delta X^\nu(\xi')}$$

$$\left(\text{in general } \frac{\delta}{\delta X^\nu(\xi')}(\lambda(\xi)F[X]) = \lambda(\xi)\frac{\delta F}{\delta X^\nu(\xi)}\right)$$

4) $F = \partial_a X^\mu(\xi)\partial_b X_\mu(\xi)$

$$\frac{\delta F}{\delta X^\nu(\xi)} = \partial_a \delta(\xi-\xi')\partial_b X_\nu + \partial_b \delta(\xi-\xi')\partial_a X_\nu$$

General principles of membrane kinematics and dynamics 113

A suitable choice of the metric — assuring the invariance of the line element (4.4) under the transformations (4.2) and (4.5) — is, for instance,

$$\rho_{\mu(\xi)\nu(\zeta)} = \sqrt{|f|}\,\alpha\,g_{\mu\nu}\delta(\xi-\zeta), \tag{4.9}$$

where $f \equiv \det f_{ab}$ is the determinant of the induced metric

$$f_{ab} \equiv \partial_a X^\alpha \partial_b X^\beta\,g_{\alpha\beta} \tag{4.10}$$

on the sheet V_n, $g_{\mu\nu}$ is the metric tensor of the embedding space V_N, and α an arbitrary function of ξ^a.

With the metric (4.9) the line element (4.4) becomes

$$\mathrm{d}\ell^2 = \int \mathrm{d}\xi\sqrt{|f|}\,\alpha\,g_{\mu\nu}\,\mathrm{d}X^{\mu(\xi)}\mathrm{d}X^{\nu(\xi)}. \tag{4.11}$$

Rewriting the abstract formulas back into the usual notation, with explicit integration, we have

$$A^{\mu(\xi)} = A^\mu(\xi), \tag{4.12}$$

$$\begin{aligned}A_{\mu(\xi)} &= \rho_{\mu(\xi)\nu(\zeta)}A^{\nu(\zeta)} \\ &= \int \mathrm{d}\zeta\,\rho_{\mu\nu}(\xi,\zeta)A^\nu(\xi) = \sqrt{|f|}\,\alpha\,g_{\mu\nu}A^\nu(\xi).\end{aligned} \tag{4.13}$$

The inverse metric is

$$\rho^{\mu(\xi)\nu(\zeta)} = \frac{1}{\alpha\sqrt{|f|}}\,g^{\mu\nu}\delta(\xi-\zeta). \tag{4.14}$$

Indeed, from (4.7), (4.9) and (4.14) we obtain

$$\delta^{\mu(\xi)}{}_{\nu(\zeta)} = \int \mathrm{d}\eta\,g^{\mu\sigma}g_{\nu\sigma}\,\delta(\xi-\eta)\delta(\zeta-\eta) = \delta^\mu{}_\nu\delta(\xi-\zeta). \tag{4.15}$$

The invariant volume element (measure) of our membrane space $\mathcal{M}$ is [58]

$$\mathcal{D}X = (\mathrm{Det}\,\rho_{\mu\nu}(\xi,\zeta))^{1/2}\prod_{\xi,\mu}\mathrm{d}X^\mu(\xi). \tag{4.16}$$

Here Det denotes a continuum determinant taken over ξ,ζ as well as over μ,ν. In the case of the diagonal metric (4.9) we have

$$\mathcal{D}X = \prod_{\xi,\mu}\left(\sqrt{|f|}\,\alpha\,|g|\right)^{1/2}\mathrm{d}X^\mu(\xi) \tag{4.17}$$

As can be done in a finite-dimensional space, we can now also define the covariant derivative in $\mathcal{M}$. For a scalar functional $A[X(\xi)]$ the covariant functional derivative coincides with the ordinary functional derivative:

$$A_{;\mu(\xi)} = \frac{\delta A}{\delta X^\mu(\xi)} \equiv A_{,\mu(\xi)}. \qquad (4.18)$$

But in general a geometric object in $\mathcal{M}$ is a tensor of arbitrary rank, $A^{\mu_1(\xi_1)\mu_2(\xi_2)...}{}_{\nu_1(\zeta_1)\nu_2(\zeta_2)...}$, which is a functional of $X^\mu(\xi)$, and its covariant derivative contains the affinity $\Gamma^{\mu(\xi)}_{\nu(\zeta)\sigma(\eta)}$ composed of the metric (4.9) [54, 55]. For instance, for a vector we have

$$A^{\mu(\xi)}{}_{;\nu(\zeta)} = A^{\mu(\xi)}{}_{,\nu(\zeta)} + \Gamma^{\mu(\xi)}_{\nu(\zeta)\sigma(\eta)} A^{\sigma(\eta)}. \qquad (4.19)$$

Let the alternative notations for ordinary and covariant functional derivative be analogous to those used in a finite-dimensional space:

$$\frac{\delta}{\delta X^\mu(\xi)} \equiv \frac{\partial}{\partial X^\mu(\xi)} \equiv \partial_{\mu(\xi)} \quad , \quad \frac{\mathrm{D}}{\mathrm{D}X^\mu(\xi)} \equiv \frac{\mathrm{D}}{\mathrm{D}X^\mu(\xi)} \equiv \mathrm{D}_{\mu(\xi)}. \qquad (4.20)$$

4.2. MEMBRANE DYNAMICS

In the previous section I have considered arbitrary deformable membranes as *kinematically possible objects* of a membrane theory. A membrane, in general, is not static, but is assumed to move in an embedding space V_N. The parameter of evolution ("time") will be denoted τ. Kinematically every continuous trajectory $X^\mu(\tau, \xi^a) \equiv X^{\mu(\xi)}(\tau)$ is possible in principle. A particular dynamical theory then selects which amongst those kinematically possible membranes and trajectories are also dynamically possible. In this section I am going to describe the theory in which a dynamically possible trajectory $X^{\mu(\xi)}(\tau)$ is a *geodesic* in the membrane space $\mathcal{M}$.

MEMBRANE THEORY AS A FREE FALL IN $\mathcal{M}$-SPACE

Let $X^{\alpha(\xi)}$ be τ-dependent *coordinates* of a point in $\mathcal{M}$-space and $\rho_{\alpha(\xi')\beta(\xi'')}$ an arbitrary fixed metric in $\mathcal{M}$. From the point of view of a finite-dimensional space V_N the symbol $X^{\alpha(\xi)} \equiv X^\alpha(\xi)$ represents an n-dimensional membrane embedded in V_N. We assume that every dynamically possible trajectory

General principles of membrane kinematics and dynamics 115

$X^{\alpha(\xi)}(\tau)$ satisfies the variational principle given by the action

$$I[X^{\alpha(\xi)}] = \int d\tau' \left(\rho_{\alpha(\xi')\beta(\xi'')} \dot{X}^{\alpha(\xi')} \dot{X}^{\beta(\xi'')} \right)^{1/2}. \qquad (4.21)$$

This is just the action for a *geodesic* in $\mathcal{M}$-space.

The equation of motion is obtained if we functionally differentiate (4.21) with respect to $X^{\alpha(\xi)}(\tau)$:

$$\frac{\delta I}{\delta X^{\mu(\xi)}(\tau)} = \int d\tau' \frac{1}{\mu^{1/2}} \rho_{\alpha(\xi')\beta(\xi'')} \dot{X}^{\alpha(\xi'')} \frac{d}{d\tau'} \delta(\tau - \tau') \delta_{(\xi)}^{(\xi')}$$

$$+ \frac{1}{2} \int d\tau' \frac{1}{\mu^{1/2}} \left(\frac{\delta}{\delta X^{\mu(\xi)}(\tau)} \rho_{\alpha(\xi')\beta(\xi'')} \right) \dot{X}^{\alpha(\xi'')} \dot{X}^{\beta(\xi'')} = 0, \qquad (4.22)$$

where

$$\mu \equiv \rho_{\alpha(\xi')\beta(\xi'')} \dot{X}^{\alpha(\xi')} \dot{X}^{\beta(\xi'')} \qquad (4.23)$$

and

$$\delta_{(\xi)}^{(\xi')} \equiv \delta(\xi - \xi'). \qquad (4.24)$$

The integration over τ in the first term of eq. (4.22) can be easily performed and eq. (4.22) becomes

$$\frac{\delta I}{\delta X^{\mu(\xi)}(\tau)} = -\frac{d}{d\tau} \left(\frac{\rho_{\alpha(\xi')\mu(\xi)} \dot{X}^{\alpha(\xi')}}{\mu^{1/2}} \right)$$

$$+ \frac{1}{2} \int d\tau' \frac{1}{\mu^{1/2}} \left(\frac{\delta}{\delta X^{\mu(\xi)}(\tau)} \rho_{\alpha(\xi')\beta(\xi'')} \right) \dot{X}^{\alpha(\xi')} \dot{X}^{\beta(\xi'')} = 0. \qquad (4.25)$$

Some exercises with such a variation are performed in Box 4.2, where we use the notation $\partial_{\mu(\tau,\xi)} \equiv \delta/\delta X^{\mu}(\tau,\xi)$.

If the expression for the metric $\rho_{\alpha(\xi')\beta(\xi'')}$ does not contain the velocity $\dot{X}^{\mu}$, then eq. (4.25) further simplifies to

$$-\frac{d}{d\tau} \left(\dot{X}_{\mu(\xi)} \right) + \frac{1}{2} \partial_{\mu(\xi)} \rho_{\alpha(\xi')\beta(\xi'')} \dot{X}^{\alpha(\xi')} \dot{X}^{\beta(\xi'')} = 0. \qquad (4.26)$$

This can be written also in the form

$$\frac{d\dot{X}^{\mu(\xi)}}{d\tau} + \Gamma^{\mu(\xi)}{}_{\alpha(\xi')\beta(\xi'')} \dot{X}^{\alpha(\xi')} \dot{X}^{\beta(\xi'')} = 0, \qquad (4.27)$$

which is a straightforward generalization of the usual geodesic equation from a finite-dimensional space to an infinite-dimensional $\mathcal{M}$-space.

The metric $\rho_{\alpha(\xi')\beta(\xi'')}$ is arbitrary fixed background metric of $\mathcal{M}$-space. Choice of the latter metric determines, from the point of view of the embedding space V_N, a particular membrane theory. But from the viewpoint

Box 4.2: Excercises with variations and functional derivatives

1) $I[X^\mu(\tau)] = \dfrac{1}{2}\int d\tau'\, \dot X^\mu(\tau')\dot X^\nu(\tau')\eta_{\mu\nu}$

$$\dfrac{\delta I}{\delta X^\alpha(\tau)} = \int d\tau'\, \dot X^\mu(\tau')\dfrac{\delta \dot X^\nu(\tau')}{\delta X^\alpha(\tau)}\eta_{\mu\nu}$$

$$= \int d\tau'\, \dot X_\alpha(\tau')\dfrac{d}{d\tau'}\delta(\tau - \tau') = -\dfrac{d}{d\tau'}\dot X_\alpha$$

2) $I[X^\mu(\tau,\xi)] = \dfrac{1}{2}\int d\tau' d\xi'\, \sqrt{|f(\xi')|}\, \dot X^\mu(\tau',\xi')\dot X^\nu(\tau',\xi')\eta_{\mu\nu}$

$$\dfrac{\delta I}{\delta X^\alpha(\tau,\xi)} = \dfrac{1}{2}\int d\tau' d\xi'\, \dfrac{\delta\sqrt{|f(\tau',\xi')|}}{\delta X^\alpha(\tau',\xi')}\dot X^2(\tau',\xi')$$

$$+ \dfrac{1}{2}\int d\tau' d\xi'\, \sqrt{|f(\tau',\xi')|}\, 2\dot X^\mu \dfrac{\delta \dot X^\nu}{\delta X^\alpha(\tau,\xi)}\eta_{\mu\nu}$$

$$= \dfrac{1}{2}\int d\tau' d\xi'\, \sqrt{|f(\tau',\xi')|}\, \partial'^a X_\alpha \partial'_a \delta(\xi - \xi')\delta(\tau - \tau')\dot X^2(\tau',\xi')$$

$$+ \int d\tau' d\xi'\, \sqrt{|f(\tau',\xi')|}\, \dot X_\alpha(\tau',\xi')\dfrac{d}{d\tau'}\delta(\tau - \tau')\delta(\xi - \xi')$$

$$= -\dfrac{1}{2}\partial_a\left(\sqrt{|f|}\,\partial^a X_\alpha\, \dot X^2\right) - \dfrac{d}{d\tau}\left(\sqrt{|f|}\,\dot X_\alpha\right)$$

3) $I = \int d\tau'\, (\rho_{\alpha(\xi')\beta(\xi'')}\dot X^{\alpha(\xi')}\dot X^{\beta(\xi'')} + K)$

$$\dfrac{\delta I}{\delta X^{\mu(\xi)}(\tau)} = \dfrac{1}{2}\int d\tau'\,\Big[\partial_{\mu(\tau,\xi)}\rho_{\alpha(\xi')\beta(\xi'')}\dot X^{\alpha(\xi')}\dot X^{\beta(\xi'')}$$

$$+ 2\rho_{\alpha(\xi')\beta(\xi'')}\dfrac{d}{d\tau'}\delta(\tau - \tau')\delta(\xi - \xi')\delta_\mu{}^\alpha \dot X^{\beta(\xi'')} + \partial_{\mu(\tau,\xi)}K\Big]$$

$$= -\dfrac{d}{d\tau}(\rho_{\mu(\xi)\beta(\xi'')}\dot X^{\beta(\xi'')}) + \dfrac{1}{2}\int d\tau'\,\Big[\partial_{\mu(\tau,\xi)}\rho_{\alpha(\xi')\beta(\xi'')}\dot X^{\alpha(\xi')}\dot X^{\beta(\xi'')} + \partial_{\mu(\tau,\xi)}K\Big]$$

(*continued*)

Box 4.2 (*continued*)

a) $\quad \rho_{\alpha(\xi')\beta(\xi'')} = \dfrac{\kappa\sqrt{|f(\xi')|}}{\lambda(\xi')}\delta(\xi'-\xi'')\eta_{\alpha\beta}\,, \qquad K = \int d\xi \sqrt{|f|}\,\kappa\lambda$

$$\dfrac{\delta I}{\delta X^{\mu(\xi)}(\tau)} = -\dfrac{d}{d\tau}\left(\dfrac{\kappa\sqrt{|f|}}{\lambda}\dot{X}_\mu\right) - \dfrac{1}{2}\partial_a\left(\dfrac{\kappa\sqrt{|f|}\partial^a X_\mu \dot{X}^2}{\lambda}\right)$$

$$-\dfrac{1}{2}\partial_a(\kappa\sqrt{|f|}\partial^a X_\mu \lambda)$$

$$\dfrac{\delta I}{\lambda(\xi)} = 0 \quad\Rightarrow\quad \lambda^2 = \dot{X}^\alpha \dot{X}_\alpha$$

$$\Rightarrow\quad \dfrac{d}{d\tau}\left(\dfrac{\kappa\sqrt{|f|}}{\sqrt{\dot{X}^2}}\dot{X}_\mu\right) + \partial_a(\kappa\sqrt{|f|}\partial^a X_\mu \sqrt{\dot{X}^2}) = 0$$

b) $\quad \rho_{\alpha(\xi')\beta(\xi'')} = \dfrac{\kappa\sqrt{|f(\xi')|}}{\sqrt{\dot{X}^2(\xi')}}\delta(\xi'-\xi'')\eta_{\alpha\beta}\,, \qquad K = \int d\xi \sqrt{|f|}\,\kappa\sqrt{\dot{X}^2}$

$$\partial_{\mu(\tau,\xi)}\rho_{\alpha(\xi')\beta(\xi'')} = \kappa\dfrac{\delta\sqrt{|f(\tau',\xi')|}}{\delta X^\mu(\tau,\xi)}\dfrac{1}{\sqrt{\dot{X}^2(\tau',\xi')}}\eta_{\alpha\beta}\delta(\xi'-\xi'')$$

$$+\kappa\sqrt{|f(\tau',\xi')|}\dfrac{\delta}{\delta X^\mu(\tau,\xi)}\left(\dfrac{1}{\sqrt{\dot{X}^2(\tau',\xi')}}\right)\eta_{\alpha\beta}\delta(\xi'-\xi'')$$

$$= \kappa\sqrt{|f(\tau',\xi')|}\,\partial'^a X_\mu \dfrac{\partial'_a \delta(\xi-\xi')\delta(\tau-\tau')}{\sqrt{\dot{X}^2}}\eta_{\alpha\beta}\delta(\xi'-\xi'')$$

$$-\kappa\sqrt{|f(\tau',\xi')|}\dfrac{\dot{X}_\mu(\xi')}{\dot{X}^2(\xi'))^{3/2}}\delta(\xi-\xi')\delta(\xi'-\xi'')\dfrac{d}{d\tau'}\delta(\tau-\tau')\eta_{\alpha\beta}$$

$$\int d\tau' \dfrac{\delta\rho_{\alpha(\xi')\beta(\xi'')}}{\delta X^{\mu(\xi)}(\tau)}\dot{X}^{\alpha(\xi')}\dot{X}^{\beta(\xi'')}$$

$$= -\kappa\,\partial_a(\sqrt{|f|}\partial^a X_\mu \sqrt{\dot{X}^2}) + \kappa\dfrac{d}{d\tau}\left(\sqrt{|f|}\dfrac{\dot{X}_\mu}{\sqrt{\dot{X}^2}}\right)$$

$$\int d\tau' \dfrac{\delta K}{\delta X^{\mu(\xi)}(\tau)} = -\kappa\,\partial_a(\sqrt{|f|}\partial^a X_\mu \sqrt{\dot{X}^2}) - \kappa\dfrac{d}{d\tau}\left(\sqrt{|f|}\dfrac{\dot{X}_\mu}{\sqrt{\dot{X}^2}}\right)$$

of $\mathcal{M}$-space there is just one membrane theory in a background metric $\rho_{\alpha(\xi')\beta(\xi'')}$ which is an arbitrary functional of $X^{\mu(\xi)}(\tau)$.

Suppose now that the metric is given by the following expression:

$$\rho_{\alpha(\xi')\beta(\xi'')} = \kappa \frac{\sqrt{|f(\xi')|}}{\sqrt{\dot{X}^2(\xi')}} \delta(\xi' - \xi'')\eta_{\alpha\beta}, \qquad (4.28)$$

where $\dot{X}^2(\xi') \equiv \dot{X}^\mu(\xi')\dot{X}_\mu(\xi')$, and κ is a constant. If we insert the latter expression into the equation of geodesic (4.22) and take into account the prescriptions of Boxes 4.1 and 4.2, we immediately obtain the following equations of motion:

$$\frac{d}{d\tau}\left(\frac{1}{\mu^{1/2}}\frac{\sqrt{|f|}}{\sqrt{\dot{X}^2}}\dot{X}_\mu\right) + \frac{1}{\mu^{1/2}}\partial_a\left(\sqrt{|f|}\sqrt{\dot{X}^2}\partial^a X_\mu\right) = 0. \qquad (4.29)$$

The latter equation can be written as

$$\mu^{1/2}\frac{d}{d\tau}\left(\frac{1}{\mu^{1/2}}\right)\frac{\sqrt{|f|}}{\sqrt{\dot{X}^2}}\dot{X}_\mu + \frac{d}{d\tau}\left(\frac{\sqrt{|f|}}{\sqrt{\dot{X}^2}}\dot{X}_\mu\right) + \partial_a\left(\sqrt{|f|}\sqrt{\dot{X}^2}\partial^a X_\mu\right) = 0. \qquad (4.30)$$

If we multiply this by $\dot{X}^\mu$, sum over μ, integrate over ξ, and take into account that (4.23), we obtain

$$\begin{aligned}
\frac{1}{2}\frac{d\mu}{d\tau} &= \kappa\int d\xi \left[\frac{d}{d\tau}\left(\frac{\sqrt{|f|}}{\sqrt{\dot{X}^2}}\dot{X}_\mu\right)\dot{X}^\mu + \partial_a(\sqrt{|f|}\partial^a X_\mu\sqrt{\dot{X}^2})\dot{X}^\mu\right] \\
&= \kappa\int d\xi \left[\frac{d}{d\tau}\left(\frac{\sqrt{|f|}}{\sqrt{\dot{X}^2}}\dot{X}_\mu\right)\dot{X}^\mu - \sqrt{|f|}\sqrt{\dot{X}^2}\partial^a X_\mu\partial_a\dot{X}^\mu\right] \\
&= \kappa\int d\xi \left[\sqrt{\dot{X}^2}\frac{d\sqrt{|f|}}{d\tau} + \sqrt{|f|}\frac{d}{d\tau}\left(\frac{\dot{X}_\mu}{\sqrt{\dot{X}^2}}\right)\dot{X}^\mu\right. \\
&\qquad\qquad \left. - \sqrt{\dot{X}^2}\frac{d}{d\tau}\sqrt{|f|}\right] \\
&= 0. \qquad (4.31)
\end{aligned}$$

In the above calculation we have used the relations

$$\frac{d\sqrt{|f|}}{d\tau} = \frac{\partial\sqrt{|f|}}{\partial f_{ab}}\dot{f}_{ab} = \sqrt{|f|}f^{ab}\partial_a\dot{X}^\mu\partial_b X_\mu = \sqrt{|f|}\partial^a X_\mu\partial_a\dot{X}^\mu \qquad (4.32)$$

and

$$\frac{\dot{X}_\mu}{\sqrt{\dot{X}^2}}\frac{\dot{X}^\mu}{\sqrt{\dot{X}^2}} = 1 \quad\Rightarrow\quad \frac{d}{d\tau}\left(\frac{\dot{X}_\mu}{\sqrt{\dot{X}^2}}\right)\dot{X}^\mu = 0. \qquad (4.33)$$

We have thus seen that the equations of motion (4.29) automatically imply

$$\frac{d\mu}{d\tau} = 0 \quad \text{or} \quad \frac{d\sqrt{\mu}}{d\tau} = 0, \quad \mu \neq 0. \tag{4.34}$$

Therefore, instead of (4.29) we can write

$$\frac{d}{d\tau}\left(\frac{\sqrt{|f|}}{\sqrt{\dot{X}^2}}\dot{X}_\mu\right) + \partial_a\left(\sqrt{|f|}\sqrt{\dot{X}^2}\partial^a X_\mu\right) = 0. \tag{4.35}$$

This is precisely the equation of motion of the Dirac-Nambu-Goto membrane of arbitrary dimension. The latter objects are nowadays known as *p-branes*, and they include point particles (0-branes) and strings (1-branes). It is very interesting that the conventional theory of *p*-branes is just a particular case —with the metric (4.28)— of the membrane dynamics given by the action (4.21).

The action (4.21) is by definition invariant under reparametrizations of ξ^a. In general, it is not invariant under reparametrization of the evolution parameter τ. If the expression for the metric $\rho_{\alpha(\xi')\beta(\xi'')}$ does not contain the velocity $\dot{X}^\mu$ then the invariance of (4.21) under reparametrizations of τ is obvious. On the contrary, if $\rho_{\alpha(\xi')\beta(\xi'')}$ contains $\dot{X}^\mu$ then the action (4.21) is not invariant under reparametrizations of τ. For instance, if $\rho_{\alpha(\xi')\beta(\xi'')}$ is given by eq. (4.28), then, as we have seen, the equation of motion automatically contains the relation

$$\frac{d}{d\tau}\left(\dot{X}^{\mu(\xi)}\dot{X}_{\mu(\xi)}\right) \equiv \frac{d}{d\tau}\int d\xi\, \kappa\sqrt{|f|}\sqrt{\dot{X}^2} = 0. \tag{4.36}$$

The latter relation is nothing but *a gauge fixing relation*, where by "gauge" we mean here a choice of parameter τ. The action (4.21), which in the case of the metric (4.28) is not reparametrization invariant, contains the gauge fixing term. The latter term is not added separately to the action, but is implicit by the exponent $\frac{1}{2}$ of the expression $\dot{X}^{\mu(\xi)}\dot{X}_{\mu(\xi)}$.

In general the exponent in the Lagrangian is not necessarily $\frac{1}{2}$, but can be arbitrary:

$$I[X^{\alpha(\xi)}] = \int d\tau \left(\rho_{\alpha(\xi')\beta(\xi'')}\dot{X}^{\alpha(\xi')}\dot{X}^{\beta(\xi'')}\right)^a. \tag{4.37}$$

For the metric (4.28) the corresponding equation of motion is

$$\frac{d}{d\tau}\left(a\mu^{a-1}\frac{\kappa\sqrt{|f|}}{\sqrt{\dot{X}^2}}\dot{X}_\mu\right) + a\mu^{a-1}\partial_a\left(\kappa\sqrt{|f|}\sqrt{\dot{X}^2}\partial^a X_\mu\right) = 0. \tag{4.38}$$

For any a which is different from 1 we obtain a gauge fixing relation which is equivalent to (4.34), and the same equation of motion (4.35). When

$a=1$ we obtain directly the equation of motion (4.35), and no gauge fixing relation (4.34). For $a=1$ and the metric (4.28) the action (4.37) is invariant under reparametrizations of τ.

We shall now focus our attention to the action

$$I[X^{\alpha(\xi)}] = \int d\tau\, \rho_{\alpha(\xi')\beta(\xi'')} \dot X^{\alpha(\xi')} \dot X^{\beta(\xi')} = \int d\tau\, d\xi\, \kappa \sqrt{|f|}\sqrt{\dot X^2} \quad (4.39)$$

with the metric (4.28). It is invariant under the transformations

$$\tau \to \tau' = \tau'(\tau), \quad (4.40)$$

$$\xi^a \to \xi'^a = \xi'^a(\xi^a) \quad (4.41)$$

in which τ and ξ^a do not mix.

Invariance of the action (4.39) under reparametrizations (4.40) of the evolution parameter τ implies the existence of a constraint among the canonical momenta $p_{\mu(\xi)}$ and coordinates $X^{\mu(\xi)}$. Momenta are given by

$$\begin{aligned} p_{\mu(\xi)} &= \frac{\partial L}{\partial \dot X^{\mu(\xi)}} = 2\rho_{\mu(\xi)\nu(\xi')} \dot X^{\nu(\xi')} + \frac{\partial \rho_{\alpha(\xi')\beta(\xi'')}}{\partial \dot X^{\mu(\xi)}} \dot X^{\alpha(\xi')} \dot X^{\beta(\xi'')} \\ &= \frac{\kappa\sqrt{|f|}}{\sqrt{\dot X^2}} \dot X_\mu. \end{aligned} \quad (4.42)$$

By distinsguishing covariant and contravariant components one finds

$$p_{\mu(\xi)} = \dot X_{\mu(\xi)}, \quad p^{\mu(\xi)} = \dot X^{\mu(\xi)}. \quad (4.43)$$

We define

$$p_{\mu(\xi)} \equiv p_\mu(\xi) \equiv p_\mu, \quad \dot X^{\mu(\xi)} \equiv \dot X^\mu(\xi) \equiv \dot X^\mu. \quad (4.44)$$

Here p_μ and $\dot X^\mu$ have the meaning of the usual finite dimensional vectors whose components are lowered raised by the finite-dimensional metric tensor $g_{\mu\nu}$ and its inverse $g^{\mu\nu}$:

$$p^\mu = g^{\mu\nu} p_\nu, \quad \dot X_\mu = g_{\mu\nu} \dot X^\nu \quad (4.45)$$

Eq.(4.42) implies

$$p^\mu p_\mu - \kappa^2 |f| = 0 \quad (4.46)$$

which is satisfied at every ξ^a.

Multiplying (4.46) by $\sqrt{\dot X^2}/(\kappa\sqrt{|f|})$ and integrating over ξ we have

$$\frac{1}{2} \int d\xi \frac{\sqrt{\dot X^2}}{\kappa\sqrt{|f|}} (p^\mu p_\mu - \kappa^2|f|) = p_{\mu(\xi)} \dot X^{\mu(\xi)} - L = H = 0 \quad (4.47)$$

General principles of membrane kinematics and dynamics   121

where $L = \int \mathrm{d}\xi\, \kappa\sqrt{|f|}\sqrt{\dot{X}^2}$.

We see that the Hamiltonian belonging to our action (4.39) is identically zero. This is a well known consequence of the reparametrization invariance (4.40). The relation (4.46) is a constraint at ξ^a and the Hamiltonian (4.47) is a linear superposition of the constraints at all possible ξ^a.

An action which is equivalent to (4.39) is

$$I[X^{\mu(\xi)}, \lambda] = \frac{1}{2}\int \mathrm{d}\tau \mathrm{d}\xi\, \kappa\sqrt{|f|}\left(\frac{\dot{X}^\mu \dot{X}_\mu}{\lambda} + \lambda\right), \tag{4.48}$$

where λ is a Lagrange multiplier.

In the compact notation of $\mathcal{M}$-space eq. (4.48) reads

$$I[X^{\mu(\xi)}, \lambda] = \frac{1}{2}\int \mathrm{d}\tau \left(\rho_{\alpha(\xi')\beta(\xi'')}\dot{X}^{\alpha(\xi')}\dot{X}^{\beta(\xi'')} + K\right), \tag{4.49}$$

where

$$K = K[X^{\mu(\xi)}, \lambda] = \int \mathrm{d}\xi\, \kappa\sqrt{|f|}\lambda \tag{4.50}$$

and

$$\rho_{\alpha(\xi')\beta(\xi'')} = \rho_{\alpha(\xi')\beta(\xi'')}[X^{\mu(\xi)}, \lambda] = \frac{\kappa\sqrt{|f(\xi')|}}{\lambda(\xi')}\delta(\xi'-\xi'')\eta_{\alpha\beta}. \tag{4.51}$$

Variation of (4.49) with respect to $X^{\mu(\xi)}(\tau)$ and λ gives

$$\frac{\delta I}{\delta X^{\mu(\xi)}(\tau)} = -\frac{\mathrm{d}}{\mathrm{d}\tau}\left(\frac{\kappa\sqrt{|f|}}{\lambda}\dot{X}_\mu\right)$$

$$-\frac{1}{2}\partial_a\left(\kappa\sqrt{|f|}\partial^a X_\mu\left(\frac{\sqrt{\dot{X}^2}}{\lambda}+\lambda\right)\right) = 0, \tag{4.52}$$

$$\frac{\delta I}{\delta \lambda(\tau,\xi)} = -\frac{\dot{X}^\mu \dot{X}_\mu}{\lambda^2} + 1 = 0. \tag{4.53}$$

The system of equations (4.52), (4.53) is equivalent to (4.35). This is in agreement with the property that after inserting the λ "equation of motion" (4.53) into the action (4.48) one obtains the action (4.39) which directly leads to the equation of motion (4.35).

The invariance of the action (4.48) under reparametrizations (4.40) of the evolution parameter τ is assured if λ transforms according to

$$\lambda \to \lambda' = \frac{\mathrm{d}\tau'}{\mathrm{d}\tau}\lambda. \tag{4.54}$$

This is in agreement with the relations (4.53) which says that

$$\lambda = (\dot{X}^\mu \dot{X}_\mu)^{1/2}.$$

> **Box 4.3: Conservation of the constraint**
>
> Since the Hamiltonian $H = \int d\xi \lambda \mathcal{H}$ in eq. (4.67) is zero for any λ, it follows that the Hamiltonian density
>
> $$\mathcal{H}[X^\mu, p_\mu] = \frac{1}{2\kappa}\left(\frac{p_\mu p^\mu}{\sqrt{|f|}} - \kappa^2 \sqrt{|f|}\right) \qquad (4.55)$$
>
> vanishes for any ξ^a. The requirement that the constraint (6.1) is conserved in τ can be written as
>
> $$\dot{\mathcal{H}} = \{\mathcal{H}, H\} = 0, \qquad (4.56)$$
>
> which is satisfied if
>
> $$\{\mathcal{H}(\xi), \mathcal{H}(\xi')\} = 0. \qquad (4.57)$$
>
> That the Poisson bracket (4.57) indeed vanishes can be found as follows. Let us work in the language of the Hamilton–Jacobi functional $S[X^\mu(\xi)]$, in which one considers the momentum vector field $p_{\mu(\xi)}$ to be a function of position $X^{\mu(\xi)}$ in $\mathcal{M}$-space, i.e., a functional of $X^\mu(\xi)$ given by
>
> $$p_{\mu(\xi)} = p_{\mu(\xi)}(X^{\mu(\xi)}) \equiv p_\mu[X^\mu(\xi)] = \frac{\delta S}{\delta X^\mu(\xi)}. \qquad (4.58)$$
>
> Therefore $\mathcal{H}[X^\mu(\xi), p_\mu(\xi)]$ is a functional of $X^\mu(\xi)$. Since $\mathcal{H} = 0$, it follows that its functional derivative also vanishes:
>
> $$\frac{d\mathcal{H}}{dX^{\mu(\xi)}} = \frac{\partial \mathcal{H}}{\partial X^{\mu(\xi)}} + \frac{\partial \mathcal{H}}{\partial p_{\nu(\xi')}} \frac{\partial p_{\nu(\xi')}}{\partial X^{\mu(\xi)}}$$
>
> $$\equiv \frac{\delta \mathcal{H}}{\delta X^\mu(\xi)} + \int d\xi' \frac{\delta \mathcal{H}}{\delta p_\nu(\xi')} \frac{\delta p_\nu(\xi')}{\delta X^\mu(\xi)} = 0. \qquad (4.59)$$
>
> Using (4.59) and (4.58) we have
>
> $$\{\mathcal{H}(\xi), \mathcal{H}(\xi')\} = \int d\xi'' \left(\frac{\delta \mathcal{H}(\xi)}{\delta X^\mu(\xi'')}\frac{\delta \mathcal{H}(\xi')}{\delta p_\mu(\xi'')} - \frac{\delta \mathcal{H}(\xi')}{\delta X^\mu(\xi'')}\frac{\delta \mathcal{H}(\xi)}{\delta p_\mu(\xi'')}\right)$$
>
> $$= -\int d\xi'' d\xi''' \frac{\delta \mathcal{H}(\xi)}{\delta p_\nu(\xi''')}\frac{\delta \mathcal{H}(\xi')}{\delta p_\mu(\xi'')}\left(\frac{\delta p_\nu(\xi''')}{\delta X^\mu(\xi'')} - \frac{\delta p_\mu(\xi'')}{\delta X^\nu(\xi''')}\right) = 0. \qquad (4.60)$$
>
> *(continued)*

Box 4.3 (*continued*)

Conservation of the constraint (4.55) is thus shown to be automatically sastisfied.

On the other hand, we can calculate the Poisson bracket (4.57) by using the explicit expression (4.55). So we obtain

$$\{\mathcal{H}(\xi), \mathcal{H}(\xi')\} =$$

$$-\frac{\sqrt{|f(\xi)|}}{\sqrt{|f(\xi')|}} \partial_a \delta(\xi-\xi') p_\mu(\xi') \partial^a X^\mu(\xi) + \frac{\sqrt{|f(\xi')|}}{\sqrt{|f(\xi)|}} \partial'_a \delta(\xi-\xi') p_\mu(\xi) \partial'^a X^\mu(\xi')$$

$$= -\left(p_\mu(\xi) \partial^a X^\mu(\xi) + p_\mu(\xi') \partial'^a X^\mu(\xi')\right) \partial_a \delta(\xi - \xi') = 0, \quad (4.61)$$

where we have used the relation

$$F(\xi') \partial_a \delta(\xi - \xi') = \partial_a \left[F(\xi') \delta(\xi - \xi')\right] = \partial_a \left[F(\xi) \delta(\xi - \xi')\right]$$

$$= F(\xi) \partial_a \delta(\xi - \xi') + \partial_a F(\xi) \delta(\xi - \xi'). \quad (4.62)$$

Multiplying (4.61) by an arbitrary "test" function $\phi(\xi')$ and integrating over ξ' we obtain

$$2 p_\mu \partial^a X^\mu \partial_a \phi + \partial_a (p_\mu \partial^a X^\mu) \phi = 0. \quad (4.63)$$

Since ϕ and $\partial_a \phi$ can be taken as independent at any point ξ^a, it follows that

$$p_\mu \partial_a X^\mu = 0. \quad (4.64)$$

The "momentum" constraints (4.64) are thus shown to be automatically satisfied as a consequence of the conservation of the "Hamiltonian" constraint (4.55). This procedure was been discovered in ref. [59]. Here I have only adjusted it to the case of membrane theory.

If we calculate the Hamiltonian belonging to (4.49) we find

$$H = (p_{\mu(\xi)} \dot{X}^{\mu(\xi)} - L) = \tfrac{1}{2}(p_{\mu(\xi)} p^{\mu(\xi)} - K) \equiv 0, \quad (4.65)$$

where the canonical momentum is

$$p_{\mu(\xi)} = \frac{\partial L}{\partial \dot{X}^{\mu(\xi)}} = \frac{\kappa \sqrt{|f|}}{\lambda} \dot{X}_\mu . \quad (4.66)$$

Explicitly (4.65) reads

$$H = \frac{1}{2} \int d\xi \, \frac{\lambda}{\kappa \sqrt{|f|}} (p^\mu p_\mu - \kappa^2 |f|) \equiv 0. \quad (4.67)$$

The Lagrange multiplier λ is arbitrary. The choice of λ determines the choice of parameter τ. Therefore (4.67) holds for every λ, which can only be satisfied if we have

$$p^\mu p_\mu - \kappa^2 |f| = 0 \qquad (4.68)$$

at every point ξ^a on the membrane. Eq. (4.68) is a constraint at ξ^a, and altogether there are infinitely many constraints.

In Box 4.3 it is shown that the constraint (4.68) is conserved in τ and that as a consequence we have

$$p_\mu \partial_a X^\mu = 0. \qquad (4.69)$$

The latter equation is yet are another set of constraints[2] which are satisfied at any point ξ^a of the membrane manifold V_n

First order form of the action. Having the constraints (4.68), (4.69) one can easily write the first order, or phase space action,

$$I[X^\mu, p_\mu, \lambda, \lambda^a] = \int \mathrm{d}\tau\, \mathrm{d}\xi \left(p_\mu \dot{X}^\mu - \frac{\lambda}{2\kappa\sqrt{|f|}}(p^\mu p_\mu - \kappa^2|f|) - \lambda^a p_\mu \partial_a X^\mu \right), \qquad (4.70)$$

where λ and λ^a are Lagrange multipliers.

The equations of motion are

$$\delta X^\mu : \quad \dot{p}_\mu + \partial_a\left(\kappa\lambda\sqrt{|f|}\partial^a X_\mu - \lambda^a p_\mu\right) = 0, \qquad (4.71)$$

$$\delta p_\mu : \quad \dot{X}^\mu - \frac{\lambda}{\kappa\sqrt{|f|}} p_\mu - \lambda^a \partial_a X^\mu = 0, \qquad (4.72)$$

$$\delta\lambda : \quad p^\mu p_\mu - \kappa^2|f| = 0, \qquad (4.73)$$

$$\delta\lambda^a : \quad p_\mu \partial_a X^\mu = 0. \qquad (4.74)$$

Eqs. (4.72)–(4.74) can be cast into the following form:

$$p_\mu = \frac{\kappa\sqrt{|f|}}{\lambda}(\dot{X}_\mu - \lambda^a \partial_a X^\mu), \qquad (4.75)$$

$$\lambda^2 = (\dot{X}^\mu - \lambda^a \partial_a X^\mu)(\dot{X}_\mu - \lambda^b \partial_b X_\mu) \qquad (4.76)$$

---

[2]Something similar happens in canonical gravity. Moncrief and Teitelboim [59] have shown that if one imposes the Hamiltonian constraint on the Hamilton functional then the momentum constraints are automatically satisfied.

General principles of membrane kinematics and dynamics 125

$$\lambda_a = \dot{X}^\mu \partial_a X_\mu. \tag{4.77}$$

Inserting the last three equations into the phase space action (4.70) we have

$$I[X^\mu] = \kappa \int d\tau \, d\xi \sqrt{|f|} \left[\dot{X}^\mu \dot{X}^\nu (\eta_{\mu\nu} - \partial^a X_\mu \partial_a X_\nu) \right]^{1/2}. \tag{4.78}$$

The vector $\dot{X}(\eta_{\mu\nu} - \partial^a X_\mu \partial_a X_\nu)$ is normal to the membrane V_n; its scalar product with tangent vectors $\partial_a X^\mu$ is identically zero. The form $\dot{X}^\mu \dot{X}^\nu (\eta_{\mu\nu} - \partial^a X_\mu \partial_a X_\nu)$ can be considered as a 1-dimensional metric, equal to its determinant, on a line which is orthogonal to V_n. The product

$$f \dot{X}^\mu \dot{X}^\nu (\eta_{\mu\nu} - \partial^a X_\mu \partial_a X_\nu) = \det \partial_A X^\mu \partial_B X_\mu \tag{4.79}$$

is equal to the determinant of the induced metric $\partial_A X^\mu \partial_B X_\mu$ on the $(n+1)$-dimensional surface $X^\mu(\phi^A)$, $\phi^A = (\tau, \xi^a)$, swept by our membrane V_n. The action (4.78) is then *the minimal surface action* for the $(n+1)$-dimensional worldsheet V_{n+1}:

$$I[X^\mu] = \kappa \int d^{n+1}\phi \, (\det \partial_A X^\mu \partial_B X_\mu)^{1/2}. \tag{4.80}$$

This is the conventional Dirac–Nambu–Goto action, and (4.70) is one of its equivalent forms.

We have shown that from the point of view of $\mathcal{M}$-space a membrane of any dimension is just a point moving along a geodesic in $\mathcal{M}$. The metric of $\mathcal{M}$-space is taken to be an arbitrary fixed background metric. For a special choice of the metric we obtain the conventional p-brane theory. The latter theory is thus shown to be a particular case of the more general theory, based on the concept of $\mathcal{M}$-space.

Another form of the action is obtained if in (4.70) we use the replacement

$$p_\mu = \frac{\kappa \sqrt{|f|}}{\lambda} (\dot{X}_\mu - \lambda^a \partial_a X_\mu) \tag{4.81}$$

which follows from "the equation of motion" (4.72). Then instead of (4.70) we obtain the action

$$I[X^\mu, \lambda, \lambda^a] = \frac{\kappa}{2} \int d\tau \, d^n\xi \sqrt{|f|} \left(\frac{(\dot{X}^\mu - \lambda^a \partial_a X^\mu)(\dot{X}_\mu - \lambda^b \partial_b X_\mu)}{\lambda} + \lambda \right). \tag{4.82}$$

If we choose a gauge such that $\lambda^a = 0$, then (4.82) coincides with the action (4.48) considered before.

The analogy with the point particle. The action (4.82), and especially (4.48), looks like the well known Howe–Tucker action [31] for a point particle, apart from the integration over coordinates ξ^a of a space-like hypersurface Σ on the worldsheet V_{n+1}. Indeed, a worldsheet can be considered as a continuum collection or a bundle of worldlines $X^\mu(\tau, \xi^a)$, and (4.82) is an action for such a bundle. Individual worldlines are distinguished by the values of parameters ξ^a.

We have found a very interesting inter-relationship between various concepts:

1) membrane as a "point particle" moving along a geodesic in an infinite-dimensional membrane space $\mathcal{M}$;

2) worldsheet swept by a membrane as a minimal surface in a finite-dimensional embedding space V_N;

3) worldsheet as a bundle of worldlines swept by point particles moving in V_N.

MEMBRANE THEORY AS A MINIMAL SURFACE IN AN EMBEDDING SPACE

In the previous section we have considered a membrane as a point in an infinite-dimensional membrane space $\mathcal{M}$. Now let us change our point of view and consider a membrane as a surface in a finite-dimensional embedding space V_D When moving, a p-dimensional membrane sweeps a ($d = p + 1$)-dimensional surface which I shall call a *worldsheet*[3]. What is an action which determines the membrane dynamics, i.e., a possible worldsheet? Again the analogy with the point particle provides a clue. Since a point particle sweeps a worldline whose action is *the minimal length action*, it is natural to postulate that a membrane's worldsheet satisfies *the minimal surface action*:

$$I[X^\mu] = \kappa \int d^d\phi \, (\det \partial_A X^\mu \partial_B X_\mu)^{1/2}. \qquad (4.83)$$

This action, called also the Dirac–Nambu–Goto action, is invariant under reparametrizations of the worldsheet coordinates ϕ^A, $A = 0, 1, 2, ..., d - 1$. Consequently the dynamical variables X^μ, $\mu = 0, 1, 2, ..., D$, and the corresponding momenta are subjected to d primary constraints.

Another suitable form of the action (equivalent to (4.83)) is the Howe–Tucker action [31] generalized to a membrane of arbitrary dimension p

---

[3] In the literature on p-branes such a surface is often called "world volume" and sometimes world surface.

General principles of membrane kinematics and dynamics 127

(p-brane):

$$I[X^\mu, \gamma^{AB}] = \frac{\kappa_0}{2} \int \sqrt{|\gamma|}(\gamma^{AB}\partial_A X^\mu \partial_B X_\mu + 2 - d). \qquad (4.84)$$

Besides the variables $X^\mu(\phi)$, $\mu = 0, 1, 2, ..., D-1$, which denote the position of a d-dimensional ($d = p+1$) worldsheet V_d in the embedding spacetime V_D, the above action also contains the auxiliary variables γ^{AB} (with the role of Lagrange multipliers) which have to be varied independently from X^μ.

By varying (4.84) with respect to γ^{AB} we arrive at the equation for the induced metric on a worldsheet:

$$\gamma_{AB} = \partial_A X^\mu \partial_B X_\mu. \qquad (4.85)$$

Inserting (4.85) into (4.84) we obtain the Dirac–Nambu–Goto action (4.83).

In eq. (4.84) the γ^{AB} are the Lagrange multipliers, but they are not all independent. The number of worldsheet constraints is d, which is also the number of independent Lagrange multipliers. In order to separate out of γ^{AB} the independent multipliers we proceed as follows. Let Σ be a space-like hypersurface on the worldsheet, and n^A the normal vector field to Σ. Then the worldsheet metric tensor can be written as

$$\gamma^{AB} = \frac{n^A n^B}{n^2} + \bar\gamma^{AB}, \qquad \gamma_{AB} = \frac{n_A n_B}{n^2} + \bar\gamma_{AB}, \qquad (4.86)$$

where $\bar\gamma^{AB}$ is projection tensor, satisfying

$$\bar\gamma^{AB} n_B = 0, \quad \bar\gamma_{AB} n^B = 0. \qquad (4.87)$$

It projects any vector into the hypersurface to which n^a is the normal. For instance, using (4.86) we can introduce the tangent derivatives

$$\bar\partial_A X^\mu = \bar\gamma_A{}^B \partial_B X^\mu = \gamma_A{}^B \partial_B X^\mu - \frac{n_A n^B}{n^2} \partial_B X^\mu. \qquad (4.88)$$

An arbitrary derivative $\partial_A X^\mu$ is thus decomposed into a normal and tangential part (relative to Σ):

$$\partial_A X^\mu = n_A \partial X^\mu + \bar\partial_A X^\mu, \qquad (4.89)$$

where

$$\partial X^\mu \equiv \frac{n^A \partial_A X^\mu}{n^2}, \qquad n^A \bar\partial_A X^\mu = 0. \qquad (4.90)$$

Details about using and keeping the d-dimensional covariant notation as far as possible are given in ref. [61]. Here, following ref. [62], I shall present a shorter and more transparent procedure, but without the covariant notation in d-dimensions.

Let us take such a class of coordinate systems in which covariant components of normal vectors are

$$n_A = (1, 0, 0, ..., 0). \tag{4.91}$$

From eqs. (4.86) and (4.91) we have

$$n^2 = \gamma_{AB} n^A n^B = \gamma^{AB} n_A n_B = n^0 = \gamma^{00}, \tag{4.92}$$

$$\bar{\gamma}^{00} = 0, \quad \bar{\gamma}^{0a} = 0, \tag{4.93}$$

and

$$\gamma_{00} = \frac{1}{n^0} + \bar{\gamma}_{ab} \frac{n^a n^b}{(n^0)^2}, \tag{4.94}$$

$$\gamma_{0a} = -\frac{\bar{\gamma}_{ab} n^b}{n^0}, \tag{4.95}$$

$$\gamma_{ab} = \bar{\gamma}_{ab}, \tag{4.96}$$

$$\gamma^{00} = n^0, \tag{4.97}$$

$$\gamma^{0a} = n^a, \tag{4.98}$$

$$\gamma^{ab} = \bar{\gamma}^{ab} + \frac{n^a n^b}{n^0}, \quad a, b = 1, 2, ..., p. \tag{4.99}$$

The decomposition (4.89) then becomes

$$\partial_0 X^\mu = \partial X^\mu + \bar{\partial}_0 X^\mu, \tag{4.100}$$

$$\partial_a X^\mu = \bar{\partial}_a X^\mu, \tag{4.101}$$

where

$$\partial X^\mu = \dot{X}^\mu + \frac{n^a \partial_a X^\mu}{n^0}, \quad \dot{X}^\mu \equiv \partial_0 X^\mu \equiv \frac{\partial X^\mu}{\partial \xi^0}, \quad \partial_a X^\mu \equiv \frac{\partial x^\mu}{\partial \xi^a}, \tag{4.102}$$

$$\bar{\partial}_0 X^\mu = -\frac{n^a \partial_a X^\mu}{n^0}. \tag{4.103}$$

The $n^A = (n^0, n^a)$ can have the role of d independent Lagrange multipliers. We can now rewrite our action in terms of n^0, n^a, and $\bar{\gamma}^{ab}$ (instead of γ^{AB}). We insert (4.97)–(4.99) into (4.84) and take into account that

$$|\gamma| = \frac{\bar{\gamma}}{n^0}, \qquad (4.104)$$

where $\gamma = \det \gamma_{AB}$ is the determinant of the worldsheet metric and $\bar{\gamma} = \det \bar{\gamma}_{ab}$ the determinant of the metric $\bar{\gamma}_{ab} = \gamma_{ab}$, $a, b = 1, 2, ..., p$ on the hypersurface Σ.

So our action (4.84) after using (4.97)–(4.99) becomes

$$I[X^\mu, n^A, \bar{\gamma}^{ab}] = \frac{\kappa_0}{2} \int d^d\phi \, \frac{\sqrt{\bar{\gamma}}}{\sqrt{n^0}}$$

$$\times \left(n^0 \dot{X}^\mu \dot{X}_\mu + 2n^a \dot{X}^\mu \partial_a X_\mu + (\bar{\gamma}^{ab} + \frac{n^a n^b}{n^0}) \partial_a X^\mu \partial_b X_\mu + 2 - d \right). \qquad (4.105)$$

Variation of the latter action with respect to $\bar{\gamma}^{ab}$ gives the expression for the induced metric on the surface Σ:

$$\bar{\gamma}_{ab} = \partial_a X^\mu \partial_b X_\mu, \qquad \bar{\gamma}^{ab} \bar{\gamma}_{ab} = d - 1. \qquad (4.106)$$

We can eliminate $\bar{\gamma}^{ab}$ from the action (4.105) by using the relation (4.106):

$$I[X^\mu, n^a] = \frac{\kappa_0}{2} \int d^d\phi \, \frac{\sqrt{|f|}}{\sqrt{n^0}} \left(\frac{1}{n^0}(n^0 \dot{X}^\mu + n^a \partial_a X^\mu)(n^0 \dot{X}_\mu + n^b \partial_b X_\mu) + 1 \right), \qquad (4.107)$$

where $\sqrt{|f|} \equiv \det \partial_a X^\mu \partial_b X_\mu$. The latter action is a functional of the worldsheet variables X^μ and d independent Lagrange multipliers $n^A = (n^0, n^a)$. Varying (4.107) with respect to n^0 and n^a we obtain the worldsheet constraints:

$$\delta n^0 : \quad (\dot{X}^\mu + \frac{n^b \partial_b X^\mu}{n^0}) \dot{X}_\mu = \frac{1}{n^0}, \qquad (4.108)$$

$$\delta n^a : \quad (\dot{X}^\mu + \frac{n^b \partial_b X^\mu}{n^0}) \partial_a X_\mu = 0. \qquad (4.109)$$

Using (4.102) the constraints can be written as

$$\partial X^\mu \partial X_\mu = \frac{1}{n^0}, \qquad (4.110)$$

$$\partial X^\mu \partial_i X_\mu = 0. \qquad (4.111)$$

The action (4.107) contains the expression for the normal derivative ∂X^μ and can be written in the form

$$I = \frac{\kappa_0}{2} \int d\tau d^p\xi \sqrt{|f|}\left(\frac{\partial X^\mu \partial X_\mu}{\lambda} + \lambda\right), \quad \lambda \equiv \frac{1}{\sqrt{n^0}}, \quad (4.112)$$

where we have written $d^d\phi = d\tau\, d^p\xi$, since $\phi^A = (\tau, \xi^a)$.

So we arrived at an action which looks like the well known Howe–Tucker action for a point particle, except for the integration over a space-like hypersurface Σ, parametrized by coordinates $\xi^a, a = 1, 2, ..., p$. Introducing $\lambda^a = -n^a/n^0$ the normal derivative can be written as $\partial X^\mu \equiv \dot{X}^\mu - \lambda^a \partial_a X_\mu$. Instead of n^0, n^a we can take $\lambda \equiv 1/\sqrt{n^0}$, $\lambda^a \equiv -n^a/n^0$ as the Lagrange multipliers. In eq. (4.112) we thus recognize the action (4.82).

MEMBRANE THEORY BASED ON THE GEOMETRIC CALCULUS IN $\mathcal{M}$-SPACE

We have seen that a membrane's velocity $\dot{X}^{\mu(\xi)}$ and momentum $p_{\mu(\xi)}$ can be considered as components of vectors in an infinite-dimensional membrane space $\mathcal{M}$ in which every point can be parametrized by coordinates $X^{\mu(\xi)}$ which represent a membrane. In analogy with the finite-dimensional case considered in Chapter 2 we can introduce the concept of a vector in $\mathcal{M}$ and a set of basis vectors $e_{\mu(\xi)}$, such that any vector a can be expanded according to

$$a = a^{\mu(\xi)} e_{\mu(\xi)}. \quad (4.113)$$

From the requirement that

$$a^2 = a^{\mu(\xi)} e_{\mu(\xi)} a^{\nu(\xi')} e_{\nu(\xi')} = \rho_{\mu(\xi)\nu(\xi')} a^{\mu(\xi)} a^{\nu(\xi')} \quad (4.114)$$

we have

$$\frac{1}{2}(e_{\mu(\xi)} e_{\nu(\xi')} + e_{\nu(\xi')} e_{\mu(\xi)}) \equiv e_{\mu(\xi)} \cdot e_{\nu(\xi')} = \rho_{\mu(\xi)\nu(\xi')}. \quad (4.115)$$

This is the definition of *the inner product* and $e_{\mu(\xi)}$ are generators of *Clifford algebra* in $\mathcal{M}$-space.

A more complete elaboration of geometric calculus based on Clifford algebra in $\mathcal{M}$ will be provided in Chapter 6. Here we just use (4.113), (4.115) to extend the point particle polyvector action (2.56) to $\mathcal{M}$-space.

We shall start from the first order action (4.70). First we rewrite the latter action in terms of the compact $\mathcal{M}$-space notation:

$$I[X^\mu, p_\mu, \lambda, \lambda^a] = \int d\tau \left(p_{\mu(\xi)} \dot{X}^{\mu(\xi)} - \frac{1}{2}(p_{\mu(\xi)} p^{\mu(\xi)} - K) - \lambda^a p_{\mu(\xi)} \partial_a X^{\mu(\xi)}\right). \quad (4.116)$$

General principles of membrane kinematics and dynamics 131

In order to avoid introducing a new symbol, it is understood that the product $\lambda^a p_{\mu(\xi)}$ denotes covariant components of an $\mathcal{M}$-space vector. The Lagrange multiplier λ is included in the metric $\rho_{\mu(\xi)\nu(\xi')}$.

According to (4.113) we can write the momentum and velocity vectors as

$$p = p_{\mu(\xi)} e^{\mu(\xi)}, \qquad (4.117)$$

$$\dot{X} = \dot{X}^{\mu(\xi)} e_{\mu(\xi)}, \qquad (4.118)$$

where

$$e^{\mu(\xi)} = \rho^{\mu(\xi)\nu(\xi')} e_{\nu(\xi')} \qquad (4.119)$$

and

$$e^{\mu(\xi)} \cdot e_{\nu(\xi')} = \delta^{\mu(\xi)}{}_{\nu(\xi')}. \qquad (4.120)$$

The action (4.116) can be written as

$$I(X, p, \lambda, \lambda^a) = \int d\tau \left[p \cdot \dot{X} - \frac{1}{2}(p^2 - K) - \lambda^a \partial_a X \cdot p \right], \qquad (4.121)$$

where

$$\partial_a X = \partial_a X^{\mu(\xi)} e_{\mu(\xi)} \qquad (4.122)$$

are tangent vectors. We can omit the dot operation in (4.121) and write the action

$$I(X, p, \lambda, \lambda^a) = \int d\tau \left[p\dot{X} - \frac{1}{2}(p^2 - K) - \lambda^a p \, \partial_a X \right] \qquad (4.123)$$

which contains the scalar part and the bivector part. It is straightforward to show that the bivector part contains the same information about the equations of motion as the scalar part.

Besides the objects (4.113) which are 1-vectors in $\mathcal{M}$ we can also form 2-vectors, 3-vectors, etc., according to the analogous procedures as explained in Chapter 2. For instance, a 2-vector is

$$a \wedge b = a^{\mu(\xi)} b^{\nu(\xi')} e_{\mu(\xi)} \wedge e_{\nu(\xi')}, \qquad (4.124)$$

where $e_{\mu(\xi)} \wedge e_{\nu(\xi')}$ are basis 2-vectors. Since the index $\mu(\xi)$ has the discrete part μ and the continuous part (ξ), the wedge product

$$e_{\mu(\xi_1)} \wedge e_{\mu(\xi_2)} \wedge ... \wedge e_{\mu(\xi_k)} \qquad (4.125)$$

can have any number of terms with different values of ξ and the same value of μ. The number of terms in the wedge product

$$e_{\mu_1(\xi)} \wedge e_{\mu_2(\xi)} \wedge ... \wedge e_{\mu_k(\xi)}, \qquad (4.126)$$

with the same value of ξ, but with different values of μ, is limited by the number of discrete dimensions of $\mathcal{M}$-space. At fixed ξ the Clifford algebra of $\mathcal{M}$-space behaves as the Clifford algebra of a finite-dimensional space.

Let us write the pseudoscalar unit of the finite-dimensional subspace V_n of $\mathcal{M}$ as

$$I_{(\xi)} = e_{\mu_1(\xi)} \wedge e_{\mu_2(\xi)} \wedge ... \wedge e_{\mu_n(\xi)} \qquad (4.127)$$

A generic polyvector in $\mathcal{M}$ is a superposition

$$A = a_0 + a^{\mu(\xi)} e_{\mu(\xi)} + a^{\mu_1(\xi_1)\mu_2(\xi_2)} e_{\mu_1(\xi_1)} \wedge e_{\mu_2(\xi_2)} + ...$$
$$+ a^{\mu_1(\xi_1)\mu_2(\xi_2)...\mu_k(\xi_k)} e_{\mu_1(\xi_1)} \wedge e_{\mu_2(\xi_2)} \wedge ... \wedge e_{\mu_k(\xi_k)} + ... \qquad (4.128)$$

As in the case of the point particle I shall follow the principle that the most general physical quantities related to membranes, such as momentum P and velocity $\dot{X}$, are polyvectors in $\mathcal{M}$-space. I invite the interested reader to work out as an exercise (or perhaps as a research project) what physical interpretation[4] could be ascribed to all possible multivector terms of P and $\dot{X}$. For the finite-dimensional case I have already worked out in Chapter 2, Sec 3, to certain extent such a physical interpretation. We have also seen that at the classical level, momentum and velocity polyvectors which solve the equations of motion can have all the multivector parts vanishing except for the vector and pseudoscalar part. Let us assume a similar situation for the membrane momentum and velocity:

$$P = P^{\mu(\xi)} e_{\mu(\xi)} + m^{(\xi)} I_{(\xi)}, \qquad (4.129)$$

$$\dot{X} = \dot{X}^{\mu(\xi)} e_{\mu(\xi)} + \dot{s}^{(\xi)} . I_{(\xi)} \qquad (4.130)$$

In addition let us assume

$$\partial_a X = \partial_a X^{\mu(\xi)} e_{\mu(\xi)} + \partial_a s^{(\xi)} I_{(\xi)}. \qquad (4.131)$$

Let us assume the following general membrane action:

$$I(X, P, \lambda, \lambda^a) = \int \mathrm{d}\tau \left[P\dot{X} - \frac{1}{2}(P^2 - K) - \lambda^a \partial_a X P \right]. \qquad (4.132)$$

On the one hand the latter action is a generalization of the action (4.121) to arbitrary polyvectors $\dot{X}$, P, $\lambda^a \partial_a X$. On the other hand, (4.132) is a

---

[4]As a hint the reader is advised to look at Secs. 6.3 and 7.2.

generalization of the point particle polyvector action (2.56), where the polyvectors in a finite-dimensional space V_n are replaced by polyvectors in the infinite-dimensional space $\mathcal{M}$.

Although the polyvectors in the action (4.132) are arbitrary in principle (defined according to (4.128)), we shall from now on restrict our consideration to a particular case in which the polyvectors are given by eqs. (4.129)-(4.131). Rewriting the action action (4.132) in the component notation, that is, by inserting (4.129)-(4.131) into (4.132) and by taking into account (4.115), (4.120) and

$$e_{\mu(\xi)} \cdot e_{\nu(\xi')} = \rho_{\mu(\xi)\nu(\xi')} = \frac{\kappa\sqrt{|f|}}{\lambda}\delta(\xi-\xi')\eta_{\mu\nu}, \quad (4.133)$$

$$I_{(\xi)} \cdot I_{(\xi')} = \rho_{(\xi)(\xi')} = -\frac{\kappa\sqrt{|f|}}{\lambda}\delta(\xi-\xi'), \quad (4.134)$$

$$I_{(\xi)} \cdot e_{\nu(\xi')} = 0, \quad (4.135)$$

we obtain

$$\langle I \rangle_0 = \int d\tau \Bigl[p_{\mu(\xi)}\dot{X}^{\mu(\xi)} - \dot{s}^{(\xi)}m_{(\xi)} - \frac{1}{2}(p^{\mu(\xi)}p_{\mu(\xi)} + m^{(\xi)}m_{(\xi)} - K)$$

$$- \lambda^a (\partial_a X^{\mu(\xi)} p_{\mu(\xi)} - \partial_a s^{(\xi)} m_{(\xi)}) \Bigr]. \quad (4.136)$$

By the way, let us observe that using (4.133), (4.134) we have

$$-K = -\int d\xi\, \kappa\sqrt{|f|}\lambda = \lambda^{(\xi)}\lambda_{(\xi)}, \qquad \lambda^{(\xi)} \equiv \lambda(\xi), \quad (4.137)$$

which demonstrates that the term K can also be written in the elegant tensor notation.

More explicitly, (4.136) can be written in the form

$$I[X^\mu, s, p_\mu, m, \lambda, \lambda^a] = \int d\tau\, d^n\xi \Bigl[p_\mu \dot{X}^\mu - m\dot{s} - \frac{\lambda}{2\kappa\sqrt{|f|}}(p^\mu p_\mu - m^2 - \kappa^2|f|)$$

$$- \lambda^a(\partial_a X^\mu p_\mu - \partial_a s\, m) \Bigr] \quad (4.138)$$

This is a generalization of the membrane action (4.70) considered in Sec. 4.2. Besides the coordinate variables $X^\mu(\xi)$ we have now an additional variable $s(\xi)$. Besides the momentum variables $p_\mu(\xi)$ we also have the variable $m(\xi)$. We retain the same symbols p_μ and m as in the case of the point particle theory, with understanding that those variables are now ξ-dependent densities of weight 1.

The equations of motion derived from (4.138) are:

$$\delta s : \quad -\dot{m} + \partial_a(\lambda^a m) = 0, \tag{4.139}$$

$$\delta X^\mu : \quad \dot{p}_\mu + \partial_a\left(\kappa\lambda\sqrt{|f|}\partial^a X_\mu - \lambda^a p_\mu\right) = 0, \tag{4.140}$$

$$\delta m : \quad -\dot{s} + \lambda^a \partial_a s + \frac{\lambda}{\kappa\sqrt{|f|}}m = 0, \tag{4.141}$$

$$\delta p_\mu : \quad \dot{X}^\mu - \lambda^a \partial_a X^\mu - \frac{\lambda}{\kappa\sqrt{|f|}} p_\mu = 0, \tag{4.142}$$

$$\delta\lambda : \quad p^\mu p_\mu - m^2 - \kappa^2|f| = 0, \tag{4.143}$$

$$\delta\lambda^a : \quad \partial_a X^\mu p_\mu - \partial_a s\, m = 0. \tag{4.144}$$

Let us collect in the action those term which contain s and m and re-express them by using the equations of motion (4.141). We obtain

$$-m\dot{s} + \frac{\lambda}{2\sqrt{|f|}}m^2 + \lambda^a \partial_a s\, m = -\frac{\lambda}{2\kappa\sqrt{|f|}}m^2 \tag{4.145}$$

Using again (4.141) and also (4.139) we have

$$-\frac{\lambda}{2\kappa\sqrt{|f|}}m^2 = \frac{1}{2}(-m\dot{s} + \lambda^a \partial_a s\, m) = \frac{1}{2}\left(-\frac{\mathrm{d}(ms)}{\mathrm{d}\tau} + \partial_a(\lambda^a s\, m)\right). \tag{4.146}$$

Inserting (4.145) and (4.146) into the action (4.138) we obtain

$$I = \int \mathrm{d}\tau\, \mathrm{d}^n\xi \left[-\frac{1}{2}\frac{\mathrm{d}}{\mathrm{d}\tau}(ms) + \frac{1}{2}\partial_a(ms\lambda^a)\right.$$
$$\left. + p_\mu \dot{X}^\mu - \frac{\lambda}{2\kappa\sqrt{|f|}}(p^\mu p_\mu - \kappa^2|f|) - \lambda^a \partial_a X^\mu p_\mu \right]. \tag{4.147}$$

We see that the extra variables s, m occur only in the terms which are total derivatives. Those terms have no influence on the equations of motion, and can be omitted, so that

$$I[X^\mu, p_\mu] = \int \mathrm{d}\tau\, \mathrm{d}^n\xi \left[p_\mu \dot{X}^\mu - \frac{\lambda}{2\kappa\sqrt{|f|}}(p^\mu p_\mu - \kappa^2|f|) - \lambda^a \partial_a X^\mu p_\mu \right]. \tag{4.148}$$

This action action looks like the action (4.70) considered in Sec 2.2. However, now λ and λ^a are no longer Lagrange multipliers. They should be considered as fixed since they have already been "used" when forming the terms $-(\mathrm{d}/\mathrm{d}\tau)(ms)$ and $\partial_a(ms\lambda^a)$ in (4.147). Fixing of λ, λ^a means fixing

General principles of membrane kinematics and dynamics 135

the gauge, that is the choice of parameters τ and ξ^a. In (4.148) we have thus obtained a *reduced action* which is a functional of the reduced number of variables X^μ, p_μ. All X^μ or all p_μ are independent; there are no more constraints.

However, a choice of gauge (the fixing of λ, λ^a) must be such that the equations of motion derived from the reduced action are consistent with the equations of motion derived from the original constrained action. In our case we find that an admissible choice of gauge is given by

$$\frac{\lambda}{\kappa\sqrt{|f|}} = \Lambda, \quad \lambda^a = \Lambda^a, \qquad (4.149)$$

where Λ, Λ^a are arbitrary fixed functions of τ, ξ^a. So we obtained the following *unconstrained action*:

$$I[X^\mu, p_\mu] = \int d\tau\, d^n\xi \left[p_\mu \dot{X}^\mu - \frac{\Lambda}{2}(p^\mu p_\mu - \kappa^2|f|) - \Lambda^a \partial_a X^\mu p_\mu \right]. \quad (4.150)$$

The fixed function Λ does not transform as a scalar under reparametrizations of ξ^a, but a scalar density of weight -1, whereas Λ^a transforms as a vector. Under reparametrizations of τ they are assumed to transform according to $\Lambda' = (d\tau/d\tau')\Lambda$ and $\Lambda'^a = (d\tau/d\tau')\Lambda^a$. The action (4.150) is then *covariant* under reparametrizations of τ and ξ^a, i.e., it retains the same form. However, it is not *invariant* (Λ and Λ^a in a new parametrization are different functions of the new parameters), therefore there are no constraints.

Variation of the action (4.150) with respect to X^μ and p_μ gives

$$\delta X^\mu : \quad \dot{p}_\mu + \partial_a \left(\Lambda \kappa^2 |f| \partial^a X_\mu - \Lambda^a p_\mu \right) = 0, \qquad (4.151)$$

$$\delta p_\mu : \quad \dot{X}^\mu - \Lambda^a \partial_a X^\mu - \Lambda p^\mu = 0. \qquad (4.152)$$

The latter equations of motion are indeed equal to the equations of motion (4.140),(4.142) in which gauge is fixed according to (4.149).

Eliminating p_μ in (4.150) by using eq (4.152), we obtain

$$I[X^\mu] = \frac{1}{2} \int d\tau\, d^n\xi \left[\frac{(\dot{X}^\mu - \Lambda^a \partial_a X^\mu)(\dot{X}_\mu - \Lambda^b \partial_b X_\mu)}{\Lambda} + \Lambda \kappa^2 |f| \right]. \quad (4.153)$$

If $\Lambda^a = 0$ this simplifies to

$$I[X^\mu] = \frac{1}{2} \int d\tau\, d^n\xi \left(\frac{\dot{X}^\mu \dot{X}_\mu}{\Lambda} + \Lambda \kappa^2 |f| \right). \qquad (4.154)$$

In the static case, i.e., when $\dot X^\mu = 0$, we have

$$I[X^\mu] = \frac{1}{2}\int d\tau\, d^n\xi\, \Lambda\kappa^2|f|, \qquad (4.155)$$

which is the well known Schild action [63].

Alternative form of the $\mathcal{M}$-space metric. Let us now again consider the action (4.136). Instead of (4.133), (4.134) let us now take the following form of the metric:

$$\rho_{\mu(\xi)\nu(\xi')} = \frac{1}{\tilde\lambda}\delta(\xi-\xi')\eta_{\mu\nu}, \qquad (4.156)$$

$$\rho_{(\xi)(\xi')} = -\frac{1}{\tilde\lambda}\delta(\xi-\xi'), \qquad (4.157)$$

and insert it into (4.136). Then we obtain the action

$$I[X^\mu, s, p_\mu, m, \lambda', \lambda^a] = \qquad (4.158)$$

$$\int d\tau\, d^n\xi\, \left[-m\dot s + p_\mu \dot X^\mu - \frac{\tilde\lambda}{2}(p^\mu p_\mu - m^2 - \kappa^2|f|) - \lambda^a(\partial_a X^\mu p_\mu - \partial_a s\, m) \right],$$

which is equivalent to (4.138). Namely, we can easily verify that the corresponding equations of motion are equivalent to the equations of motion (4.139)–(4.144). From the action (4.158) we then obtain the unconstrained action (4.150) by fixing $\tilde\lambda = \Lambda$ and $\lambda^a = \Lambda^a$.

We have again found (as in the case of a point particle) that the polyvector generalization of the action naturally contains "time" and evolution of the membrane variables X^μ. Namely, in the theory there occurs an extra variable s whose derivative $\dot s$ with respect to the worldsheet parameter τ is the pseudoscalar part of the velocity polyvector. This provides a mechanism of obtaining the Stueckelberg action from a more basic principle.

Alternative form of the constrained action. Let us again consider the constrained action (4.138) which is a functional of the variables X^μ, s and the canonical momenta p_μ, m. We can use equation of motion (4.141) in order to eliminate the Lagrange multipliers λ from the action. By doing so we obtain

$$I = \int d\tau\, d^n\xi\, \left[-m(\dot s - \lambda^a\partial_a s) + p_\mu(\dot X^\mu - \lambda^a\partial_a X^\mu) \right.$$

$$\left. - \frac{\dot s - \lambda^a\partial_a s}{2m}(p^2 - m^2 - \kappa^2|f|) \right].$$

$$(4.159)$$

General principles of membrane kinematics and dynamics

We shall now prove that

$$\frac{\mathrm{d}s}{\mathrm{d}\tau} = \dot{s} - \lambda^a \partial_a s \quad \text{and} \quad \frac{\mathrm{d}X^\mu}{\mathrm{d}\tau} = \dot{X}^\mu - \lambda^a \partial_a X^\mu, \tag{4.160}$$

where $\dot{s} \equiv \partial s/\partial \tau$ and $\dot{X}^\mu = \partial X^\mu/\partial \tau$ are partial derivatives. The latter relations follow from the definitions of the total derivatives

$$\frac{\mathrm{d}s}{\mathrm{d}\tau} = \frac{\partial s}{\partial \tau} + \frac{\mathrm{d}\xi^a}{\mathrm{d}\tau} \quad \text{and} \quad \frac{\mathrm{d}X^\mu}{\mathrm{d}\tau} = \frac{\partial X^\mu}{\partial \tau} + \partial_a X^\mu \frac{\mathrm{d}\xi^a}{\mathrm{d}\tau}, \tag{4.161}$$

and the relation $\lambda^a = -\mathrm{d}\xi^a/\mathrm{d}\tau$, which comes from the momentum constraint $p_\mu \partial_a X^\mu = 0$.

Inserting (4.160) into eq. (4.159) we obtain yet another equivalent classical action

$$I[X^\mu, p_\mu, m] = \int \mathrm{d}s \, \mathrm{d}^n \xi \left[p_\mu \frac{\mathrm{d}X^\mu}{\mathrm{d}s} - \frac{m}{2} - \frac{1}{2m}(p^2 - \kappa^2 |f|) \right], \tag{4.162}$$

in which the variable s has disappeared from the Lagrangian, and it has instead become the evolution parameter. Alternatively, if in the action (4.159) we choose a gauge such that $\dot{s} = 1$, $\lambda^a = 0$, then we also obtain the same action (4.162).

The variable m in (4.162) acquired the status of a Lagrange multiplier leading to the constraint

$$\delta m: \quad p^2 - \kappa^2 |f| - m^2 = 0. \tag{4.163}$$

Using the constraint (4.163) we can eliminate m from the action (4.162) and we obtain the following reduced action

$$I[X^\mu, p_\mu] = \int \mathrm{d}s \, \mathrm{d}^n \xi \left(p_\mu \frac{\mathrm{d}X^\mu}{\mathrm{d}s} - \sqrt{p^2 - \kappa^2 |f|} \right) \tag{4.164}$$

which, of course, is unconstrained. It is straightforward to verify that the equations of motion derived from the unconstrained action (4.164) are the same as the ones derived from the original constrained action (4.138).

The extra variable s in the reparametrization invariant constrained action (4.138), after performing reduction of variables by using the constraints, has become the evolution parameter.

There is also a more direct derivation of the unconstrained action which will be provided in the next section.

Conclusion. Geometric calculus based on Clifford algebra in a finite-dimensional space can be generalized to the infinite-dimensional membrane space $\mathcal{M}$. Mathematical objects of such an algebra are Clifford numbers, also called Clifford aggregates, or polyvectors. It seems natural to assume that physical quantities are in general polyvectors in $\mathcal{M}$. Then, for instance, the membrane velocity $\dot{X}^\mu$ in general is not a vector, but a polyvector, and hence it contains all other possible r-vector parts, including a scalar and a pseudoscalar part. As a preliminary step I have considered here a model in which velocity is the sum of a vector and a pseudoscalar. The pseudoscalar component is $\dot{s}$, i.e., the derivative of an extra variable s. Altogether we thus have the variables X^μ and s, and the corresponding canonically conjugate momenta p_μ and m. The polyvector action is reparametrization invariant, and as a consequence there are constraints on those variables. Therefore we are free to choose appropriate number of extra relations which eliminate the redundancy of variables. We may choose relations such that we get rid of the extra variables s and m, but then the remaining variables X^μ, p_μ are *unconstrained*, and they evolve in the evolution parameter τ which, by choice of a gauge, can be made proportional to s.

Our model with the polyvector action thus allows for *dynamics* in spacetime. *It resolves the old problem of the conflict between our experience of the passage of time on the one hand, and the fact that the theory of relativity seems incapable of describing the flow of time at all:* past, present and future coexist in a four- (or higher-dimensional) "block" spacetime, with objects corresponding to worldlines (or worldsheets) within this block. And what, in my opinion, is very nice, the resolution is not a result of an *ad hoc procedure*, but is a necessary consequence of the existence of Clifford algebra as a general tool for the description of the geometry of spacetime!

Moreover, when we shall also consider dynamics of spacetime itself, we shall find out that the above model with the polyvector action, when suitably generalized, will provide a natural resolution of the notorious "problem of time" in quantum gravity.

4.3. MORE ABOUT THE INTERCONNECTIONS AMONG VARIOUS MEMBRANE ACTIONS

In the previous section we have considered various membrane actions. One action was just that of a *free fall in $\mathcal{M}$-space* (eq. (4.21)). For a special metric (4.28) which contains the membrane velocity we have obtained the equation of motion (4.35) which is identical to that of the *Dirac–Nambu–Goto membrane* described by *the minimal surface action* (4.80).

Instead of the free fall action in $\mathcal{M}$-space we have considered some *equivalent forms* such as *the quadratic actions* (4.39), (4.49) and the corresponding *first order* or *phase space action* (4.70).

General principles of membrane kinematics and dynamics 139

Then we have brought into the play *the geometric calculus based on Clifford algebra* and applied it to $\mathcal{M}$-space. The membrane velocity and momentum are promoted to *polyvectors*. The latter variables were then used to construct *the polyvector phase space action* (4.132), and its more restricted form in which the polyvectors contain the vector and the pseudoscalar parts only.

Whilst all the actions described in the first two paragraphs were equivalent to the usual minimal surface action which describes *the constrained membrane*, we have taken with the polyvectors a step beyond the conventional membrane theory. We have seen that the presence of a pseudoscalar variable results in *unconstraining* the rest of the membrane's variables which are $X^\mu(\tau, \xi)$. This has important consequences.

If momentum and velocity polyvectors are given by expressions (4.129)–(4.131), then the polyvector action (4.132) becomes (4.136) whose more explicit form is (4.138). Eliminating from the latter phase space action the variables P_μ and m by using their equations of motion (4.139), (4.142), we obtain

$$I[X^\mu, s, \lambda, \lambda^a]$$

$$= \frac{\kappa}{2} \int d\tau \, d^n\xi \sqrt{|f|} \tag{4.165}$$

$$\times \left(\frac{(\dot{X}^\mu - \lambda^a \partial_a X^\mu)(\dot{X}_\mu - \lambda^b \partial_b X_\mu) - (\dot{s} - \lambda^a \partial_a s)^2}{\lambda} + \lambda \right).$$

The choice of the Lagrange multipliers λ, λ^a fixes the parametrization τ and ξ^a. We may choose $\lambda^a = 0$ and action (4.165) simplifies to

$$I[X^\mu, s, \lambda] = \frac{\kappa}{2} \int d\tau \, d^n\xi \sqrt{|f|} \left(\frac{\dot{X}^\mu \dot{X}_\mu - \dot{s}^2}{\lambda} + \lambda \right), \tag{4.166}$$

which is an extension of the Howe–Tucker-like action (4.48) or (2.31) considered in the first two sections.

Varying (4.166) with respect to λ we have

$$\lambda^2 = \dot{X}^\mu \dot{X}_\mu - \dot{s}^2. \tag{4.167}$$

Using relation (4.167) in eq. (4.166) we obtain

$$I[X^\mu, s] = \kappa \int d\tau \, d^n\xi \sqrt{|f|} \sqrt{\dot{X}^\mu \dot{X}_\mu - \dot{s}^2}. \tag{4.168}$$

This reminds us of the relativistic point particle action (4.21). The difference is in the extra variable s and in that the variables depend not only on

140   THE LANDSCAPE OF THEORETICAL PHYSICS: A GLOBAL VIEW

the parameter τ but also on the parameters ξ^a, hence the integration over ξ^a with the measure $\mathrm{d}^n\xi\sqrt{|f|}$ (which is invariant under reparametrizations of ξ^a).

Bearing in mind $\dot X = \partial X^\mu/\partial \tau$, $\dot s = \partial s/\partial \tau$, and using the relations (4.160), (4.161), we can write (4.168) as

$$I[X^\mu] = \kappa \int \mathrm{d}s\, \mathrm{d}^n\xi\, \sqrt{|f|}\sqrt{\frac{\mathrm{d}X^\mu}{\mathrm{d}s}\frac{\mathrm{d}X_\mu}{\mathrm{d}s} - 1}. \qquad (4.169)$$

The step from (4.168) to (4.169) is equivalent to choosing the parametrization of τ such that $\dot s = 1$ for any ξ^a, which means that $\mathrm{d}s = \mathrm{d}\tau$.

We see that in (4.169) the extra variable s takes the role of the evolution parameter and that the variables $X^\mu(\tau,\xi)$ and the conjugate momenta $p_\mu(\tau,\xi) = \partial \mathcal{L}/\partial \dot X^\mu$ are *unconstrained* [5].

In particular, a membrane $\mathcal{V}_n$ which solves the variational principle (4.169) can have vanishing velocity

$$\frac{\mathrm{d}X^\mu}{\mathrm{d}s} = 0. \qquad (4.170)$$

Inserting this back into (4.169) we obtain the action[6]

$$I[X^\mu] = i\kappa \int \mathrm{d}s\, \mathrm{d}^n\xi\, \sqrt{|f|}, \qquad (4.171)$$

which governs the shape of such a *static* membrane $\mathcal{V}_n$.

In the action (4.168) or (4.169) the dimensions and signatures of the corresponding manifolds V_n and V_N are left unspecified. So action (4.169) contains many possible particular cases. Especially interesting are the following cases:

Case 1. The manifold V_n belonging to an unconstrained membrane $\mathcal{V}_n$ has the signature $(+ - - - ...)$ and corresponds to an *n-dimensional worldsheet* with one time-like and $n-1$ space-like dimensions. The index of the worldsheet coordinates assumes the values $a = 0, 1, 2, ..., n-1$.

Case 2. The manifold V_n belonging to our membrane $\mathcal{V}_n$ has the signature $(- - - - ...)$ and corresponds to a space-like *p-brane*; therefore we take $n = p$. The index of the membrane's coordinates ξ^a assumes the values $a = 1, 2, ..., p$.

Throughout the book we shall often use the single formalism and apply it, when convenient, either to the *Case 1* or to the *Case 2*.

---

[5]The invariance of action (4.169) under reparametrizations of ξ^a brings no constraints amongst the dynamical variables $X^\mu(\tau,\xi)$ and $p_\mu(\tau,\xi)$ which are related to motion in τ (see also [53]-[55]).

[6]The factor i comes from our inclusion of a *pseudoscalar* in the velocity polyvector. Had we instead included a scalar, the corresponding factor would then be 1.

When the dimension of the manifold V_n belonging to $\mathcal{V}_n$ is $n = p + 1$ and the signature is $(+ - - - ...)$, i.e. when we consider *Case 1*, then the action (4.171) is just that of the usual *Dirac–Nambu–Goto p-dimensional membrane* (well known under the name *p-brane*)

$$I = i\tilde{\kappa} \int d^n \xi \sqrt{|f|} \qquad (4.172)$$

with $\tilde{\kappa} = \kappa \int ds$.

The usual p-brane is considered here as a particular case of a more general membrane[7] which can *move* in the embedding spacetime (target space) according to the action (4.168) or (4.169). Bearing in mind two particular cases described above, our action (4.169) describes either

(i) *a moving worldsheet*, in the *Case I*; or
(ii) *a moving space like membrane*, in the *Case II*.

Let us return to the action (4.166). We can write it in the form

$$I[X^\mu, s, \lambda] = \frac{\kappa}{2} \int d\tau \, d^n \xi \left[\sqrt{|f|} \left(\frac{\dot{X}^\mu \dot{X}_\mu}{\lambda} + \lambda \right) - \frac{d}{d\tau} \left(\frac{\kappa \sqrt{|f|} \dot{s} s}{\lambda} \right) \right], \qquad (4.173)$$

where by the equation of motion

$$\frac{d}{d\tau} \left(\frac{\kappa \sqrt{|f|}}{\lambda} \dot{s} \right) = 0 \qquad (4.174)$$

we have

$$\frac{d}{d\tau} \left(\frac{\kappa \sqrt{|f|} \dot{s} s}{\lambda} \right) = \frac{\kappa \sqrt{|f|} \dot{s}^2}{\lambda}. \qquad (4.175)$$

The term with the total derivative does not contribute to the equations of motion and we may omit it, provided that we fix λ in such a way that the X^μ-equations of motion derived from the reduced action are consistent with those derived from the original constrained action (4.165). This is indeed the case if we choose

$$\lambda = \Lambda \kappa \sqrt{|f|}, \qquad (4.176)$$

where Λ is arbitrary fixed function of τ.

Using (4.167) we have

$$\sqrt{\dot{X}^\mu \dot{X}_\mu - \dot{s}^2} = \Lambda \kappa \sqrt{|f|}. \qquad (4.177)$$

---

[7] By using the name *membrane* we distinguish our moving extended object from the static object, which is called *p-brane*.

Inserting into (4.177) the relation

$$\frac{\kappa\sqrt{|f|}\dot{s}}{\sqrt{\dot{X}^\mu \dot{X}_\mu - \dot{s}^2}} = \frac{1}{C} = \text{constant}, \qquad (4.178)$$

which follows from the equation of motion (4.139), we obtain

$$\frac{\Lambda}{C} = \frac{\mathrm{d}s}{\mathrm{d}\tau} \quad \text{or} \quad \Lambda\,\mathrm{d}\tau = C\,\mathrm{d}s \qquad (4.179)$$

where the differential $\mathrm{d}s = (\partial s/\partial \tau)\mathrm{d}\tau + \partial_a s\,\mathrm{d}\xi^a$ is taken along a curve on the membrane along which $\mathrm{d}\xi^a = 0$ (see also eqs. (4.160), (4.161)). Our choice of parameter τ (given by a choice of λ in eq. (4.176)) is related to the variable s by the simple proportionality relation (4.179).

Omitting the total derivative term in action (4.173) and using the gauge fixing (4.176) we obtain

$$I[X^\mu] = \frac{1}{2}\int \mathrm{d}\tau\,\mathrm{d}^n\xi \left(\frac{\dot{X}^\mu \dot{X}_\mu}{\Lambda} + \Lambda\kappa^2|f|\right). \qquad (4.180)$$

This is the *unconstrained* membrane action that was already derived in previous section, eq. (4.154).

Using (4.179) we find that action (4.180) can be written in terms of s as the evolution parameter:

$$I[X^\mu] = \frac{1}{2}\int \mathrm{d}\tau\,\mathrm{d}^n\xi \left(\frac{\overset{\circ}{X}{}^\mu \overset{\circ}{X}_\mu}{C} + C\kappa^2|f|\right) \qquad (4.181)$$

where $\overset{\circ}{X}{}^\mu \equiv \mathrm{d}X^\mu/\mathrm{d}s$.

The equations of motion derived from the *constrained action* (4.166) are

$$\delta X^\mu: \quad \frac{\mathrm{d}}{\mathrm{d}\tau}\left(\frac{\kappa\sqrt{|f|}\dot{X}_\mu}{\sqrt{\dot{X}^2 - \dot{s}^2}}\right) + \partial_a\left(\kappa\sqrt{|f|}\sqrt{\dot{X}^2 - \dot{s}^2}\,\partial^a X_\mu\right) = 0, \qquad (4.182)$$

$$\delta s: \quad \frac{\mathrm{d}}{\mathrm{d}\tau}\left(\frac{\kappa\sqrt{|f|}\dot{s}}{\sqrt{\dot{X}^2 - \dot{s}^2}}\right) = 0, \qquad (4.183)$$

whilst those from the *reduced* or *unconstrained action* (4.180) are

$$\delta X^\mu: \quad \frac{\mathrm{d}}{\mathrm{d}\tau}\left(\frac{\dot{X}_\mu}{\Lambda}\right) + \partial_a\left(\kappa^2|f|\Lambda\partial^a X_\mu\right) = 0. \qquad (4.184)$$

By using the relation (4.177) we verify the equivalence of (4.184) and (4.182).

General principles of membrane kinematics and dynamics

The original, constrained action (4.168) implies the constraint

$$p^\mu p_\mu - m^2 - \kappa^2 |f| = 0, \qquad (4.185)$$

where

$$p_\mu - \kappa \sqrt{|f|} \dot{X}_\mu / \lambda \,, \quad m = \kappa \sqrt{|f|} \dot{s}/\lambda \,, \quad \lambda = \sqrt{\dot{X}^\mu \dot{X}_\mu - \dot{s}^2}.$$

According to the equation of motion (4.174) $\dot{m} = 0$, therefore

$$p^\mu p_\mu - \kappa^2 |f| = m^2 = \text{constant}. \qquad (4.186)$$

The same relation (4.186) also holds in the reduced, *unconstrained theory* based on the action (4.180). If, in particular, $m = 0$, then the corresponding solution $X^\mu(\tau, \xi)$ is identical with that for the ordinary Dirac–Nambu–Goto membrane described by the minimal surface action which, in a special parametrization, is

$$I[X^\mu] = \kappa \int d\tau d^n \xi \sqrt{|f|} \sqrt{\dot{X}^\mu \dot{X}_\mu} \qquad (4.187)$$

This is just a special case of (4.168) for $\dot{s} = 0$.

To sum up, *the constrained action* (4.168) has the two limits:

(i) *Limit* $\dot{X}^\mu = 0$. Then

$$I[X^\mu(\xi)] = i\tilde{\kappa} \int d^n \xi \sqrt{|f|}. \qquad (4.188)$$

This is the minimal surface action. Here the n-dimensional membrane (or the worldsheet in the *Case I*) is *static* with respect to the evolution[8] parameter τ.

(ii) *Limit* $\dot{s} = 0$. Then

$$I[X^\mu(\tau, \xi)] = \kappa \int d\tau \, d^n \xi \, \sqrt{|f|} \sqrt{\dot{X}^\mu \dot{X}_\mu}. \qquad (4.189)$$

This is an action for a *moving* n-dimensional membrane which sweeps an $(n+1)$-dimensional surface $X^\mu(\tau, \xi)$ subject to the constraint $p^\mu p_\mu - \kappa^2 |f| = 0$. Since the latter constraint is conserved in τ we have automatically also the constraint $p_\mu \dot{X}^\mu = 0$ (see Box 4.3). Assuming the *Case II* we have thus the motion of a conventional constrained p-brane, with $p = n$.

In general none of the limits (i) or (ii) is satisfied, and our membrane moves according to the action (4.168) which involves the constraint (4.185).

---

[8] The evolution parameter τ should not be confused with one of the worldsheet parameters ξ^a.

From the point of view of the variables X^μ and the conjugate momenta p_μ there is no constraint, and instead of (4.168) we can use *the unconstrained action* (4.180) or (4.181), where *the extra variable s has become the parameter of evolution* .

Chapter 5

MORE ABOUT PHYSICS IN $\mathcal{M}$-SPACE

In the previous chapter we have set up the general principle that a membrane of any dimension can be considered as a point in an infinite-dimensional membrane space $\mathcal{M}$. The metric of $\mathcal{M}$-space is arbitrary in principle and a membrane traces out a geodesic in $\mathcal{M}$-space. For a particular choice of $\mathcal{M}$-space metric, the geodesic in $\mathcal{M}$-space, when considered from the spacetime point of view, is a worldsheet obeying the dynamical equations derived from the Dirac–Nambu–Goto minimal surface action. We have thus arrived very close at a geometric principle behind the string (or, in general, the membrane) theory. In this chapter we shall explore how far such a geometric principle can be understood by means of the tensor calculus in $\mathcal{M}$.

From the $\mathcal{M}$-space point of view a membrane behaves as a point particle. It is minimally coupled to a fixed background metric field $\rho_{\mu(\xi)\nu(\xi')}$ of $\mathcal{M}$-space. The latter metric field has the role of a generalized gravitational field. Although at the moment I treat $\rho_{\mu(\xi)\nu(\xi')}$ as a fixed background field, I already have in mind a later inclusion of a kinetic term for $\rho_{\mu(\xi)\nu(\xi')}$. The membrane theory will thus become analogous to general relativity. We know that in the case of the usual point particles the gravitational interaction is not the only one; there are also other interactions which are elegantly described by gauge fields. One can naturally generalize such a gauge principle to the case of a point particle in $\mathcal{M}$ (i.e., to membranes).

5.1. GAUGE FIELDS IN $\mathcal{M}$-SPACE

GENERAL CONSIDERATIONS

One of the possible classically equivalent actions which describe the Dirac–Nambu–Goto membrane is the phase space action (4.70). In $\mathcal{M}$-space notation the latter action reads

$$I = \int d\tau \left[p_{\mu(\sigma)} \dot{X}^{\mu(\sigma)} - \tfrac{1}{2}(p_{\mu(\sigma)} p^{\mu(\sigma)} - K) - \lambda^a \partial_a X^{\mu(\sigma)} p_{\mu(\sigma)} \right], \qquad (5.1)$$

where the membrane's parameters are now denoted as $\sigma \equiv \sigma^a$, $a = 1, 2, ..., p$.

Since the action (5.1) is just like the usual point particle action, the inclusion of a gauge field is straightforward. First, let us add to (5.1) a term which contains a total derivative. The new action is then

$$I' = I + \int d\tau \, \frac{dW}{d\tau}, \qquad (5.2)$$

where W is a functional of X^μ. More explicitly,

$$\frac{dW}{d\tau} = \frac{\partial W}{\partial X^{\mu(\sigma)}} \dot{X}^{\mu(\sigma)}. \qquad (5.3)$$

If we calculate the canonical momentum belonging to I' we find

$$p'_{\mu(\sigma)} = p_{\mu(\sigma)} + \partial_{\mu(\sigma)} W. \qquad (5.4)$$

Expressed in terms of $p'_{\mu(\sigma)}$ the transformed action I' is a completely new functional of $p'_{\mu(\sigma)}$, depending on the choice of W. In order to remedy such an undesirable situation and obtain an action functional which would be invariant under the transformation (5.2) we introduce a compensating gauge field $\mathcal{A}_{\mu(\sigma)}$ in $\mathcal{M}$-space and, following the minimal coupling prescription, instead of (5.1) we write

$$I = \int d\tau \left[p_{\mu(\sigma)} \dot{X}^{\mu(\sigma)} - \tfrac{1}{2}(\pi_{\mu(\sigma)} \pi^{\mu(\sigma)} - K) - \lambda^a \partial_a X^{\mu(\sigma)} \pi_{\mu(\sigma)} \right], \qquad (5.5)$$

where

$$\pi_{\mu(\sigma)} \equiv p_{\mu(\sigma)} - q \mathcal{A}_{\mu(\sigma)}. \qquad (5.6)$$

q being the total electric charge of the membrane. If we assume that under the transformation (5.2)

$$\mathcal{A}_{\mu(\sigma)} \to \mathcal{A}'_{\mu(\sigma)} = \mathcal{A}_{\mu(\sigma)} + q^{-1} \partial_{\mu(\sigma)} W, \qquad (5.7)$$

the transformed action I' then has the same form as I:

$$I' = \int d\tau \left[p'_{\mu(\sigma)} \dot{X}^{\mu(\sigma)} - \tfrac{1}{2}(\pi'_{\mu(\sigma)} \pi'^{\mu(\sigma)} - K) - \lambda^a \partial_a X^{\mu(\sigma)} \pi'_{\mu(\sigma)} \right], \qquad (5.8)$$

where
$$\pi'_{\mu(\sigma)} \equiv p'_{\mu(\sigma)} - q\mathcal{A}'_{\mu(\sigma)} = p_{\mu(\sigma)} - q\mathcal{A}_{\mu(\sigma)} \equiv \pi_{\mu(\sigma)}.$$

Varying (5.5) with respect to $p_{\mu(\sigma)}$ we find
$$\pi^{\mu(\sigma)} = \dot{X}^{\mu(\sigma)} - \lambda^a \partial_a X^{\mu(\sigma)}. \tag{5.9}$$

Variation with respect to the Lagrange multipliers λ, λ^a gives the constraints[1], and the variation with respect to $X^{\mu(\sigma)}$ gives the equation of motion

$$\dot{\pi}_{\mu(\sigma)} - \partial_a(\lambda^a \pi_{\mu(\sigma)}) - \tfrac{1}{2}\partial_{\mu(\sigma)}K - \tfrac{1}{2}\partial_{\mu(\sigma)}\rho_{\alpha(\sigma')\beta(\sigma'')}\pi^{\alpha(\sigma')}\pi^{\beta(\sigma'')}$$
$$- \int d\tau'\, \mathcal{F}_{\mu(\sigma)\nu(\sigma')}(\tau,\tau')\dot{X}^{\nu(\sigma')}(\tau') = 0, \tag{5.10}$$

where
$$\mathcal{F}_{\mu(\sigma)\nu(\sigma')}(\tau,\tau') \equiv \frac{\delta \mathcal{A}_{\nu(\sigma')}(\tau')}{\delta X^{\mu(\sigma)}(\tau)} - \frac{\delta \mathcal{A}_{\mu(\sigma)}(\tau)}{\delta X^{\nu(\sigma')}(\tau')} \tag{5.11}$$

is the $\mathcal{M}$-space gauge field strength. A gauge field $\mathcal{A}_{\mu(\sigma)}$ in $\mathcal{M}$ is, of course, a function(al) of $\mathcal{M}$-space coordinate variables $X^{\mu(\sigma)}(\tau)$. In the derivation above I have taken into account the possibility that the expression for the functional $\mathcal{A}_{\mu(\sigma)}$ contains the velocity $\dot{X}^{\mu(\sigma)}(\tau)$ (see also Box 4.2).

For
$$K = \int d^p\sigma\, \kappa\sqrt{|f|}\lambda \tag{5.12}$$

and
$$\rho_{\alpha(\sigma')\beta(\sigma'')} = (\kappa\sqrt{|f|}/\lambda)\eta_{\alpha\beta}\delta(\sigma' - \sigma'') \tag{5.13}$$

eq. (5.10) becomes

$$\dot{\pi}_\mu + \partial_a(\kappa\sqrt{|f|}\partial^a X_\mu - \lambda^a \pi_\mu) - q\int d\tau'\, \mathcal{F}_{\mu(\sigma)\nu(\sigma')}(\tau,\tau')\dot{X}^{\nu(\sigma')}(\tau') = 0 \tag{5.14}$$

If we make use of (5.9) we find out that eq. (5.14) is indeed a generalization of the equation of motion (4.71) derived in the absence of a gauge field.

A SPECIFIC CASE

Now we shall investigate the transformation (5.2) in some more detail. Assuming
$$W = \int d^p\sigma\, (w(x) + \partial_a w^a(x)) \tag{5.15}$$

---

[1] Remember that $\pi_{\mu(\sigma)}\pi^{\mu(\sigma)} = \rho_{\mu(\sigma)\nu(\sigma')}\pi^{\mu(\sigma)}\pi^{\nu(\sigma')}$ and $\rho_{\mu(\sigma)\nu(\sigma')} = (\kappa\sqrt{|f|}/\lambda)\eta_{\mu\nu}\delta(\sigma - \sigma')$.

the total derivative (5.3) of the functional W can be written as

$$\frac{dW}{d\tau} = \partial_{\mu(\sigma)} W \dot{X}^{\mu(\sigma)} = \int d^p\sigma (\partial_\mu w - \partial_a \partial_\mu w^a) \dot{X}^\mu$$

$$= \int d^p\sigma (\partial_\mu w \dot{X}^\mu - \partial_a \dot{w}^a)$$

$$= \int d^p\sigma (\partial_\mu w \dot{X}^\mu - \partial_\mu \dot{w}^a \partial_a X^\mu). \quad (5.16)$$

In the above calculation we have used

$$\partial_a \partial_\mu w^a \dot{X}^\mu = \partial_\mu \partial_a w^a \dot{X}^\mu = (d/d\tau)\partial_a w^a = \partial_a \dot{w}^a.$$

In principle W is an arbitrary functional of $X^{\mu(\sigma)}$. Let us now choose the ansatz with $\chi = \chi(X^\mu)$:

$$w = e\chi, \qquad -\dot{w}^a = e^a \chi, \qquad (5.17)$$

subjected to the condition

$$\dot{e} + \partial_a e^a = 0 \qquad (5.18)$$

Then (5.16) becomes

$$\frac{dW}{d\tau} = \int d^p\sigma \, (e\dot{X}^\mu + e^a \partial_a X^\mu) \partial_\mu \chi \qquad (5.19)$$

and, under the transformation (5.2) which implies (5.4), (5.7), we have

$$p'_{\mu(\sigma)} \dot{X}^{\mu(\sigma)} = p_{\mu(\sigma)} \dot{X}^{\mu(\sigma)} + \int d^p\sigma (e\dot{X}^\mu + e^a \partial_a X^\mu) \partial_\mu \chi, \quad (5.20)$$

$$q\mathcal{A}'_{\mu(\sigma)} \dot{X}^{\mu(\sigma)} = q\mathcal{A}_{\mu(\sigma)} \dot{X}^{\mu(\sigma)} + \int d^p\sigma (e\dot{X}^\mu + e^a \partial_a X^\mu) \partial_\mu \chi. \quad (5.21)$$

The momentum $p_{\mu(\sigma)}$ thus consists of two terms

$$p_{\mu(\sigma)} = \hat{p}_{\mu(\sigma)} + \bar{p}_{\mu(\sigma)} \qquad (5.22)$$

which transform according to

$$\hat{p}'_{\mu(\sigma)} \dot{X}^{\mu(\sigma)} = \hat{p}_{\mu(\sigma)} \dot{X}^{\mu(\sigma)} + \int d^p\sigma \, e\dot{X}^\mu \partial_\mu \chi, \qquad (5.23)$$

$$\bar{p}'_{\mu(\sigma)} \dot{X}^{\mu(\sigma)} = \bar{p}_{\mu(\sigma)} \dot{X}^{\mu(\sigma)} + \int d^p\sigma \, e^a \partial_a X^\mu \partial_\mu \chi. \qquad (5.24)$$

Similarly for the gauge field $\mathcal{A}_{\mu(\sigma)}$:

$$\mathcal{A}_{\mu(\sigma)} = \hat{\mathcal{A}}_{\mu(\sigma)} + \bar{\mathcal{A}}_{\mu(\sigma)}, \qquad (5.25)$$

$$q\hat{\mathcal{A}}'_{\mu(\sigma)} \dot{X}^{\mu(\sigma)} = \hat{\mathcal{A}}_{\mu(\sigma)} \dot{X}^{\mu(\sigma)} + \int d^p\sigma \, e\dot{X}^\mu \partial_\mu \chi, \qquad (5.26)$$

$$q\bar{\mathcal{A}}'_{\mu(\sigma)} \dot{X}^{\mu(\sigma)} = \bar{\mathcal{A}}_{\mu(\sigma)} \dot{X}^{\mu(\sigma)} + \int d^p\sigma \, e^a \partial_a X^\mu \partial_\mu \chi. \qquad (5.27)$$

More about physics in $\mathcal{M}$-space

The last two relations are satisfied if we introduce a field A_μ which transforms according to
$$A'_\mu = A_\mu + \partial_\mu \chi \tag{5.28}$$
and satisfies the relations
$$q\widehat{\mathcal{A}}_{\mu(\sigma)} \dot{X}^{\mu(\sigma)} = \int \mathrm{d}^p\sigma \, e \dot{X}^\mu A_\mu, \tag{5.29}$$
$$q\bar{\mathcal{A}}_{\mu(\sigma)} \dot{X}^{\mu(\sigma)} = \int \mathrm{d}^p\sigma \, e^a \partial_a X^\mu A_\mu, \tag{5.30}$$

or

$$q\mathcal{A}_{\mu(\sigma)} \dot{X}^{\mu(\sigma)} = \int \mathrm{d}^p\sigma \, (e\dot{X}^\mu + e^a \partial_a X^\mu) A_\mu. \tag{5.31}$$

We now observe that from (5.9), which can be written as
$$\pi_\mu = \frac{\kappa \sqrt{|f|}}{\lambda} (\dot{X}_\mu - \lambda^a \partial_a X_\mu), \tag{5.32}$$
and from eq. (5.6) we have
$$p_\mu = \frac{\kappa \sqrt{|f|}}{\lambda} (\dot{X}_\mu - \lambda^a \partial_a X_\mu) + q\mathcal{A}_\mu. \tag{5.33}$$

Using the splitting (5.22) and (5.25) we find that the choice
$$\hat{p}_\mu = \frac{\kappa \sqrt{|f|}}{\lambda} (\dot{X}_\mu - \lambda^a \partial_a X_\mu) + q\widehat{\mathcal{A}}_\mu, \tag{5.34}$$
$$\bar{p}_\mu = q\bar{\mathcal{A}}_\mu \tag{5.35}$$

is consistent with the transformations (5.23), (5.24) and the relations (5.29), (5.30).

Thus the phase space action can be rewritten in terms of $\hat{p}_\mu$ by using (5.22), (5.35) and (5.30):

$$I = \int \mathrm{d}\tau \left[\hat{p}_{\mu(\sigma)} \dot{X}^{\mu(\sigma)} + e^a \partial_a X^{\mu(\sigma)} A_{\mu(\sigma)} \right.$$
$$\left. - \tfrac{1}{2} (\hat{\pi}_{\mu(\sigma)} \hat{\pi}^{\mu(\sigma)} - K) - \lambda^a \partial_a X^{\mu(\sigma)} \hat{\pi}_{\mu(\sigma)} \right]. \tag{5.36}$$

In a more explicit notation eq. (5.36) reads

$$I = \int d\tau\, d^p\sigma \left[\hat{p}_\mu \dot{X}^\mu + e^a \partial_a X^\mu A_\mu \right.$$

$$\left. - \frac{1}{2} \frac{\lambda}{\kappa\sqrt{|f|}} (\hat{\pi}_\mu \hat{\pi}^\mu - \kappa^2 |f|) - \lambda^a \partial_a X^\mu \hat{\pi}_\mu \right]. \tag{5.37}$$

This is an action [64] for a p-dimensional membrane in the presence of a fixed electromagnetic field A_μ. The membrane bears *the electric charge density* e and *the electric current density* e^a which satisfy the conservation law (5.18).

The Lorentz force term then reads

$$q \int d\tau'\, \mathcal{F}_{\mu(\sigma)\nu(\sigma')}(\tau,\tau') \dot{X}^{\nu(\sigma')}(\tau')$$

$$= q \int d\tau' \left(\frac{\delta \mathcal{A}_{\nu(\sigma')}(\tau')}{\delta X^{\mu(\sigma)}(\tau)} - \frac{\delta \mathcal{A}_{\mu(\sigma)}(\tau)}{\delta X^{\nu(\sigma')}(\tau')} \right) \dot{X}^{\nu(\sigma')}$$

$$= q \int d\tau' \left[\frac{\delta \left(\mathcal{A}_{\nu(\sigma')}(\tau') \dot{X}^{\nu(\sigma')}(\tau') \right)}{\delta X^{\mu(\sigma)}(\tau)} - \mathcal{A}_{\nu(\sigma')} \delta^{\nu(\sigma')}{}_{\mu(\sigma)} \frac{d}{d\tau'} \delta(\tau - \tau') \right.$$

$$\left. - \frac{\delta \mathcal{A}_{\mu(\sigma)}(\tau)}{\delta X^{\nu(\sigma')}(\tau')} \dot{X}^{\nu(\sigma')}(\tau') \right]$$

$$= \int d\tau'\, d^p\sigma' \left(\frac{\delta A_\nu(\tau',\sigma')}{\delta X^\mu(\tau,\sigma)} - \frac{\delta A_\mu(\tau,\sigma)}{\delta X^\nu(\tau',\sigma')} \right) \left(e\dot{X}^\nu(\tau',\sigma') + e^a \partial_a X^\nu(\tau',\sigma') \right)$$

$$= F_{\mu\nu}(e\dot{X}^\nu + e^a \partial_a X^\nu), \tag{5.38}$$

where we have used eq. (5.31). In the last step of the above derivation we have assumed that the expression for $A_\mu(\tau,\sigma)$ satisfies

$$\frac{\delta A_\mu(\tau,\sigma)}{\delta X^\nu(\tau',\sigma')} - \frac{\delta A_\nu(\tau',\sigma')}{\delta X^\mu(\tau,\sigma)} = \left(\frac{\partial A_\mu}{\partial X^\nu} - \frac{\partial A_\nu}{\partial X^\mu} \right) \delta(\tau - \tau') \delta(\sigma - \sigma'), \tag{5.39}$$

where $F_{\mu\nu} = \partial_\mu A_\nu - \partial_\nu A_\mu$.

Let us now return to the original phase space action (5.5). It is a functional of the canonical momenta $p_{\mu(\sigma)}$. The latter quantities can be eliminated from the action by using the equations of "motion" (5.9), i.e., for p_μ we use the substitution (5.33). After also taking into account the constraints $\pi^\mu \pi_\mu - \kappa^2 |f| = 0$ and $\partial_a X^\mu \pi_\mu = 0$, the action (5.5) becomes

$$I[X^\mu] = \int d\tau\, d^p\sigma \left[\frac{\kappa\sqrt{|f|}}{\lambda} (\dot{X}_\mu - \lambda^a \partial_a X_\mu) \dot{X}^\mu + q A_\mu \dot{X}^\mu \right]. \tag{5.40}$$

Using (5.31) we have

$$I[X^\mu] = \int d\tau d^p\sigma \left[\frac{\kappa\sqrt{|f|}}{\lambda}(\dot{X}_\mu - \lambda^a\partial_a X_\mu)\dot{X}^\mu + (e\dot{X}^\mu + e^a\partial_a X^\mu)A_\mu\right]. \tag{5.41}$$

From the constraints and from (5.32) we find the expressions (4.76), (4.77) for λ and λ^a which we insert into eq. (5.41). Observing also that

$$(\dot{X}_\mu - \lambda^a\partial_a X_\mu)\dot{X}^\mu = (\dot{X}_\mu - \lambda^a\partial_a X_\mu)(\dot{X}^\mu - \lambda^b\partial_b X^\mu)$$

we obtain

$$I[X^\mu] = \int d\tau\, d^p\sigma \left(\kappa\sqrt{|f|}\left[\dot{X}^\mu \dot{X}^\nu(\eta_{\mu\nu} - \partial^a X_\mu \partial_a X_\nu)\right]^{1/2}\right.$$

$$\left. + (e\dot{X}^\mu + e^a\partial_a X^\mu)A_\mu\right), \tag{5.42}$$

or briefly (see eq. (4.79))

$$I[X^\mu] = \int d^{p+1}\phi \left[\kappa(\det \partial_A X^\mu \partial_B X_\mu)^{1/2} + e^A\partial_A X^\mu A_\mu\right], \tag{5.43}$$

where

$$\partial_A \equiv \frac{\partial}{\partial \phi^A}, \quad \phi^A \equiv (\tau, \sigma^a) \quad \text{and} \quad e^A \equiv (e, e^a). \tag{5.44}$$

This form of the action was used in refs. [65], and is a generalization of the case for a 2-dimensional membrane considered by Dirac [66].

Dynamics for the gauge field A_μ can be obtained by adding the corresponding kinetic term to the action:

$$I[X^\mu(\phi), A_\mu]$$

$$= \int d^d\phi \left[\kappa(\det\partial_A X^\mu\partial_B X_\mu)^{1/2} + e^A\partial_A A_\mu\right]\delta^D(x - X(\phi))\, d^D x$$

$$+ \frac{1}{16\pi}\int F_{\mu\nu}F^{\mu\nu}\, d^D x, \tag{5.45}$$

where $F_{\mu\nu} = \partial_\mu A_\nu - \partial_\nu A_\mu$. Here $d = p+1$ and D are the worldsheet and the spacetime dimensions, respectively. The δ-function was inserted in order to write the corresponding Lagrangian as the *density* in the embedding spacetime V_D.

By varying (5.45) with respect to $X^\mu(\phi)$ we obtain the Lorentz force law for the membrane:

$$\kappa\,\partial_A(\sqrt{\det\partial_C X^\nu \partial_D X_\nu}\,\partial^A X^\mu) + e^A\partial_A X^\nu F_{\nu\mu} = 0. \tag{5.46}$$

The latter equation can also be derived directly from the previously considered equation of motion (5.14).

If, on the contrary, we vary (5.45) with respect to the electromagnetic potential A_μ, we obtain the Maxwell equations

$$F^{\mu\nu}{}_{,\nu} = -4\pi j^\mu, \qquad (5.47)$$

$$j^\mu = \int e^A \partial_A X^\mu \delta^D(x - X(\phi)) \mathrm{d}^d \phi. \qquad (5.48)$$

The field around our membrane can be expressed by the solution[2]

$$A^\mu(x) = \int e^A(\phi') \, \delta[(x - X(\phi'))^2] \, \partial_A X^\mu(\phi') \, \mathrm{d}^d \phi', \qquad (5.49)$$

which holds in the presence of the gauge condition $\partial_\mu A^\mu = 0$. This shows that $A^\mu(x)$ is indeed *a functional* of $X^\mu(\phi)$ as assumed in previous section. It is defined at every spacetime point x^μ. In particular, x^μ may be positioned on the membrane. Then instead of $A^\mu(x)$ we write $A^\mu(\phi)$, or $A^\mu(\tau, \sigma^a) \equiv A^{\mu(\sigma)}(\tau)$. If we use the splitting $\phi^A = (\tau, \sigma^a)$, $e^A = (e, e^a)$, $a = 1, 2, ..., p$, (5.49) then reads

$$A^\mu(\tau, \sigma) = \int \mathrm{d}\tau' \, \mathrm{d}^p\sigma' \, \delta^D\left[(X(\tau,\sigma) - X(\tau', \sigma'))^2\right] (e\dot{X}^\mu + e^a \partial_a X^\mu), \qquad (5.50)$$

which demonstrates that the expression for $A^{\mu(\sigma)}(\tau)$ indeed contains the velocity $\dot{X}^\mu$ as admitted in the derivation of the general equation of motion (5.10).

Moreover, by using the expression (5.50) for $A^{\mu(\sigma)}(\tau)$ we find that the assumed expression (5.39) is satisfied exactly.

$\mathcal{M}$-SPACE POINT OF VIEW AGAIN

In the previous subsection we have considered the action (5.45) which described the membrane dynamics from the point of view of spacetime. Let us now return to considering the membrane dynamics from the point of view of $\mathcal{M}$-space, as described by the action (5.5), or, equivalently, by the action

$$I = \int \mathrm{d}\tau \left(\dot{X}^{\mu(\sigma)} \dot{X}_{\mu(\sigma)} + q \mathcal{A}_{\mu(\sigma)} \dot{X}^{\mu(\sigma)} \right), \qquad (5.51)$$

which is a generalization of (4.39). In order to include the dynamics of the gauge field $\mathcal{A}_{\mu(\sigma)}$ itself we have to add its kinetic term. A possible choice

---

[2] This expression contains both the retarded and the advanced part, $A^\mu = \frac{1}{2}(A^\mu_{\text{ret}} + A^\mu_{\text{adv}})$. When initial and boundary conditions require so, we may use only the retarded part.

for the total action is then

$$I = \int d\tau \left(\dot{\tilde{X}}^{\mu(\sigma)} \dot{\tilde{X}}_{\mu(\sigma)} + q\mathcal{A}_{\mu(\sigma)}[X]\dot{\tilde{X}}^{\mu(\sigma)} \right) \delta^{(\mathcal{M})}(X - \tilde{X}(\tau))\mathcal{D}X$$

$$+ \frac{1}{16\pi} \int d\tau\, d\tau'\, \mathcal{F}_{\mu(\sigma)\nu(\sigma')}(\tau,\tau')\mathcal{F}^{\mu(\sigma)\nu(\sigma')}(\tau,\tau')\sqrt{|\rho|}\mathcal{D}X, \qquad (5.52)$$

where

$$\delta^{(\mathcal{M})}\left(X - \tilde{X}(\tau)\right) \equiv \prod_{\mu(\sigma)} \delta\left(X^{\mu(\sigma)} - \tilde{X}^{\mu(\sigma)}(\tau)\right) \qquad (5.53)$$

and

$$\sqrt{|\rho|}\mathcal{D}X \equiv \sqrt{|\rho|} \prod_{\mu(\sigma)} dX^{\mu(\sigma)} = \prod_{\mu(\sigma)} \frac{\kappa\sqrt{|f|}}{\sqrt{\dot{X}^2}} dX^{\mu(\sigma)} \qquad (5.54)$$

are respectively the δ-function and the volume element in $\mathcal{M}$-space.

Although (5.52) is a straightforward generalization of the point particle action in spacetime (such that the latter space is replaced by $\mathcal{M}$-space), I have encountered serious difficulties in attempting to ascribe a physical meaning to (5.52). In particular, it is not clear to me what is meant by the distinction between a point X at which the field $\mathcal{A}_{\mu(\sigma)}[X]$ is calculated and the point $\tilde{X}$ at which there is a "source" of the field. One might think that the problem would disappear if instead of a single membrane which is a source of its own field, a system of membranes is considered. Then, in principle one membrane could be considered as a "test membrane" (analogous to a "test particle") to probe the field "caused" by all the other membranes within our system. The field $\mathcal{A}_{\mu(\sigma)}[X]$ is then a functional of the test membrane coordinates $X^{\mu(\sigma)}$ (in analogy with $A_\mu(x)$ being a function of the test particle coordinates x^μ). Here there is a catch: how can a field $\mathcal{A}_{\mu(\sigma)}$ be a functional of a single membrane's coordinates, whilst it would be much more sensible (actually correct) to consider it as a functional of *all* membrane coordinates within our system. But then there would be no need for a δ-function in the action (5.52).

Let us therefore write the action without the δ-function. Also let us simplify our notation by writing

$$X^\mu(\tau,\sigma) \equiv X^{\mu(\tau,\sigma)} \equiv X^{\mu(\phi)}, \quad (\phi) \equiv (\tau,\sigma). \qquad (5.55)$$

Instead of the space of p-dimensional membranes with coordinates $X^{\mu(\sigma)}$ we thus obtain the space of $(p+1)$-dimensional membranes[3] with coordinates

---

[3] When a membrane of dimension p moves it sweeps a wordlsheet $X^{\mu(\tau,\sigma)}$, which is just another name for a *membrane* of dimension $p+1$.

$X^{\mu(\phi)}$. The latter space will also be called $\mathcal{M}$-space, and its metric is

$$\rho_{\mu(\phi)\nu(\phi')} = \frac{\kappa\sqrt{|f|}}{\sqrt{\dot X^2}}\eta_{\mu\nu}\delta(\phi-\phi'), \tag{5.56}$$

where $\delta(\phi-\phi') = \delta(\tau-\tau')\delta^p(\sigma-\sigma')$. A natural choice for the action is

$$I = \dot X_{\mu(\phi)}\dot X^{\mu(\phi)} + q\mathcal{A}_{\mu(\phi)}[X]\dot X^{\mu(\phi)} + \frac{\epsilon}{16\pi}\mathcal{F}_{\mu(\phi)\nu(\phi')}\mathcal{F}^{\mu(\phi)\nu(\phi')}, \tag{5.57}$$

where

$$\mathcal{F}_{\mu(\phi)\nu(\phi')} = \partial_{\mu(\phi)}\mathcal{A}_{\nu(\phi')} - \partial_{\nu(\phi')}\mathcal{A}_{\mu(\phi)}. \tag{5.58}$$

A constant ϵ is introduced for dimensional reasons (so that the units for $\mathcal{A}_{\mu(\phi)}$ in both terms of (5.57) will be consistent). Note that the integration over τ is implicit by the repetition of the index $(\phi) \equiv (\tau, \sigma^a)$.

Variation of the action (5.57) with respect to $X^{\mu(\phi)}$ gives

$$-2\frac{\mathrm{d}}{\mathrm{d}\tau}\dot X_{\mu(\phi)} + \partial_{\mu(\phi)}\rho_{\alpha(\phi')\beta(\phi'')}\dot X^{\alpha(\phi')}\dot X^{\beta(\phi'')} + q\mathcal{F}_{\mu(\phi)\nu(\phi')}\dot X^{\nu(\phi')}$$

$$+\frac{\epsilon}{8\pi}\left(\mathrm{D}_{\mu(\phi)}\mathcal{F}_{\alpha(\phi')\beta(\phi'')}\right)\mathcal{F}^{\alpha(\phi')\beta(\phi'')} = 0 \tag{5.59}$$

If we take the metric (5.56) we find that, apart from the last term, the latter equation of motion is equivalent to (5.14). Namely, as shown in Chapter 4, Sec. 2, the second term in eq. (5.59), after performing the functional derivative of the metric (5.56), brings an extra term $(\mathrm{d}\dot X_\mu/\mathrm{d}\tau)$ and also the term $\partial_A(\kappa\sqrt{|f|}\dot X^2\partial^A X_\mu)$.

If we raise the index $\mu(\phi)$ in eq. (5.59) we obtain

$$\frac{\mathrm{d}\dot X^{\mu(\phi)}}{\mathrm{d}\tau} + \Gamma^{\mu(\phi)}_{\alpha(\phi')\beta(\phi'')}\dot X^{\alpha(\phi')}\dot X^{\beta(\phi'')}$$

$$-\frac{q}{2}\mathcal{F}^{\mu(\phi)}{}_{\nu(\phi')}\dot X^{\nu(\phi')} - \frac{\epsilon}{4\pi}\left(\mathrm{D}^{\mu(\phi)}\mathcal{F}_{\alpha(\phi')\beta(\phi'')}\right)\mathcal{F}^{\alpha(\phi')\beta(\phi'')} = 0 \tag{5.60}$$

which contains the familiar term of the geodetic equation plus the terms resulting from the presence of the gauge field. The factor $\frac{1}{2}$ in front of the Lorentz force term compensates for such a factor also accourring in front of the acceleration after inserting the metric (5.56) into the affinity and summing over the first two terms of eq. (5.60).

If, on the contrary, we vary (5.57) with respect to the field $\mathcal{A}_{\mu(\phi)}[X]$, we obtain the equations which contain the $\mathcal{M}$-space δ-function. The latter is eliminated after integration over $\mathcal{D}X$. So we obtain the following field equations

$$-4\pi q\dot X^{\mu(\phi)} = \epsilon \mathrm{D}_{\nu(\phi')}\mathcal{F}^{\mu(\phi)\nu(\phi')}. \tag{5.61}$$

More about physics in $\mathcal{M}$-space 155

They imply the "conservation" law $D_{\mu(\phi)} \dot{X}^{\mu(\phi)} = 0$.

These equations describe the motion of a membrane interacting with its own electromagnetic field. First let us observe that the electromagnetic field $\mathcal{A}_{\mu(\phi)}[X]$ is a functional of the whole membrane. No explicit spacetime dependence takes place in $\mathcal{A}_{\mu(\phi)}[X]$. Actually, the concept of a field defined at spacetime points does not occur in this theory. However, besides the discrete index μ, the field bears the set of continuous indices $(\phi) = (\tau, \sigma^a)$, and $\mathcal{A}_{\mu(\phi)}$ means $\mathcal{A}_\mu(\phi) \equiv \mathcal{A}_\mu(\tau, \sigma^a)$, that is, the field is defined for all values of the worldsheet parameters τ, σ^a. At any values of τ, σ^a we can calculate the corresponding position x^μ in spacetime by making use of the mapping $x^\mu = X^\mu(\tau, \sigma)$ which, in principle, is determined after solving the system of equations (5.59), (5.61). The concept of a field which is a functional of the system configuration is very important and in our discussion we shall return to it whenever appropriate.

A SYSTEM OF MANY MEMBRANES

We have seen that that within the $\mathcal{M}$-space description a gauge field is defined for all values of membrane's parameters $\phi = (\tau, \sigma^a)$, or, in other words, the field is defined only at those points which are located on the membrane. If there are more than one membrane then the field is defined on all those membranes.

One membrane is described by the parametric equation $x^\mu = X^\mu(\tau, \sigma^a)$, where x^μ are target space coordinates and X^μ are functions of continuous parameters τ, σ^a. In order to describe a system of membranes we add one more parameter k, which has *discrete* values. The parametric equation for *system of membranes* is thus

$$x^\mu = X^\mu(\tau, \sigma^a, k). \tag{5.62}$$

A single symbol ϕ may now denote the set of those parameters:

$$\phi \equiv (\tau, \sigma^a, k), \tag{5.63}$$

and (5.62) can be written more compactly as

$$x^\mu = X^\mu(\phi) \quad \text{or} \quad x^\mu = X^{\mu(\phi)}. \tag{5.64}$$

With such a meaning of the index (ϕ) the equations (5.59), (5.61) describe the motion of a system of membranes and the corresponding gauge field.

The gauge field $\mathcal{A}_{\mu(\phi)}[X]$ is a *functional* of $X^{\mu(\phi)}$, i.e., it is a functional of the system's configuration, described by $X^{\mu(\phi)}$.

The gauge field $\mathcal{A}_{\mu(\phi)}[X]$ is also a function of the parameters $(\phi) \equiv (\tau, \sigma^a, k)$ which denote position on the k-th membrane. If the system of membranes "nearly" fills the target space, then the set of all the points x^μ which, by the mapping (5.62), corresponds to the set of all the parameters τ, σ^a, k approximately (whatever this words means) fills the target space.

Eq. (5.61) is a field equation in $\mathcal{M}$-space. We can transform it into a field equation in target space as follows. Multiplying (5.61) by $\delta(x - X(\phi))$ and integrating over ϕ we have

$$-4\pi q \int d\phi \, \dot{X}^{\mu(\phi)} \delta(x - X(\phi)) = \epsilon \int d\phi \, \delta(x - X(\phi)) \mathrm{D}_{\nu(\phi')} \mathcal{F}^{\mu(\phi)\nu(\phi')}. \quad (5.65)$$

The left hand side of the latter equation can be written explicitly as

$$-4\pi q \sum_k \int d\tau \, d^p\sigma \, \dot{X}^\mu(\tau, \sigma, k) \delta(x - X(\phi)(\tau, \sigma, k)) \equiv -4\pi j^\mu(x). \quad (5.66)$$

In order to calculate the right hand side of eq. (5.65) we write

$$\mathrm{D}_{\nu(\phi')} \mathcal{F}^{\mu(\phi)\nu(\phi)} = \rho^{\mu(\phi)\alpha(\phi'')} \rho^{\nu(\phi')\beta(\phi''')} \mathrm{D}_{\nu(\phi')} \mathcal{F}_{\alpha(\phi'')\beta(\phi''')} \quad (5.67)$$

and

$$\mathrm{D}_{\nu(\phi')} \mathcal{F}_{\alpha(\phi'')\beta(\phi''')} = \partial_{\nu(\phi')} \mathcal{F}_{\alpha(\phi'')\beta(\phi''')}$$

$$- \Gamma^{\epsilon(\tilde{\phi})}_{\nu(\phi')\alpha(\phi'')} \mathcal{F}_{\epsilon(\tilde{\phi})\beta(\phi''')} - \Gamma^{\epsilon(\tilde{\phi})}_{\nu(\phi')\beta(\phi''')} \mathcal{F}_{\alpha(\phi'')\epsilon(\tilde{\phi})}. \quad (5.68)$$

We also assume[4]

$$\mathcal{F}_{\mu(\phi)\nu(\phi')} = \mathcal{F}_{\mu\nu} \delta(\phi - \phi'), \quad \mathcal{F}_{\mu\nu} = \frac{\partial \mathcal{A}_\nu}{\partial X^\mu} - \frac{\partial \mathcal{A}_\mu}{\partial X^\nu}. \quad (5.69)$$

Then, for the metric (5.56), eq. (5.65) becomes

$$-4\pi j^\mu = \epsilon\delta(0) \sum_k \int d\tau d^p\sigma \, \delta(x - X(\tau, \sigma, k)) \frac{\dot{X}^2}{\kappa^2 |f|} \partial_\nu \mathcal{F}^{\mu\nu}(\tau, \sigma, k)$$

$$+ \text{(covariant derivative terms)}$$

$$= \partial_\nu F^{\mu\nu}(x) + \text{(covariant derivative terms)}, \quad (5.70)$$

where we have defined

$$\epsilon\delta(0) \sum_k \int d\tau d^p\sigma \, \delta(x - X(\tau, \sigma, k)) \frac{\dot{X}^2}{\kappa^2 |f|} \mathcal{A}^\mu(\tau, \sigma, k) \equiv A^\mu(x) \quad (5.71)$$

---

[4] If $\mathcal{A}^\mu(\phi)$ is given by an expression of the form (5.49) then eq. (5.69) is indeed satisfied.

and taken the normalization[5] such that $\epsilon\delta(0) = 1$. The factor $\dot{X}^2/\kappa|f|$ comes from the inverse of the metric 5.56. Apart from the 'covariant derivative terms', which I shall not discuss here, (5.70) are just the *Maxwell equations* in flat spacetime.

The gauge field $A^\mu(x)$ is formally defined at all spacetime points x, but because of the relation (5.71) it has non-vanishing values only at those points x^μ which are occupied by the membranes. A continuous gauge field over the target space is an approximate concept in this theory. Fundamentally, a gauge field is a functional of the system configuration.

One can also insert (5.69) directly into the action (5.57). We obtain

$$\frac{\epsilon}{16\pi}\mathcal{F}_{\mu(\phi)\nu(\phi')}\mathcal{F}^{\mu(\phi)\nu(\phi')}$$

$$= \frac{\epsilon\delta(0)}{16\pi}\int \frac{\dot{X}^2}{\kappa^2|f|}\mathcal{F}_{\mu\nu}\mathcal{F}^{\mu\nu}(\phi)\,\mathrm{d}\phi$$

$$= \frac{\epsilon\delta(0)}{16\pi}\int \mathcal{F}_{\mu\nu}\mathcal{F}^{\mu\nu}(\phi)\,\mathrm{d}\phi\,\delta(x - X(\phi))\,\mathrm{d}^N x$$

$$= \frac{1}{16\pi}\int F_{\mu\nu}F^{\mu\nu}(x)\,\mathrm{d}^N x \tag{5.72}$$

where we have used the definition (5.71) for A^μ and introduced

$$F_{\mu\nu} = \partial_\mu A_\nu - \partial_\nu A_\mu. \tag{5.73}$$

Similarly

$$\dot{X}_{\mu(\phi)}\dot{X}^{\mu(\phi)} = \int \mathrm{d}\phi\,\kappa\sqrt{\dot{X}^2}\sqrt{|f|} \tag{5.74}$$

and

$$q\mathcal{A}_{\mu(\phi)}\dot{X}^{\mu(\phi)} = \int \mathrm{d}\phi\,e^A\partial_A X_\mu A^\mu. \tag{5.75}$$

The action (5.45) is thus shown to be a special case of the $\mathcal{M}$-space action (5.57). This justifies use of (5.57) as the $\mathcal{M}$-space action. Had we used the form (5.52) with a δ-functional, then in the matter term the volume $\int \mathcal{D}X$ would not occur, while it would still occur in the field term.

---

[5] The infinity $\delta(0)$ in an expression such as (5.70) can be regularized by taking into account the plausible assumption that a generic physical object is actually a fractal (i.e., an object with detailed structure at all scales). The coordinates $X^\mu(\phi)$ which we are using in our calculations are well behaved functions with definite derivatives and should be considered as an approximate description of the actual physical object. This means that a description with a given $X^\mu(\phi)$ below a certain scale has no physical meaning. In order to make a physical sense of the expressions leading to (5.70), the δ function $\delta(\phi - \phi')$ should therefore be replaced by a function $F(a, \phi - \phi')$ which in the limit $a \to 0$ becomes $\delta(\phi - \phi')$. Instead of $\delta(0)$ we thus have $F(a, 0)$ which is finite for all finite values of a. We then take the normalization ϵ such that $\epsilon F(a, 0) = 1$ for any a, and assume that this is true in the limit $a \to 0$ as well.

5.2. DYNAMICAL METRIC FIELD IN $\mathcal{M}$-SPACE

So far I have considered the metric tensor of the membrane space $\mathcal{M}$ as a fixed background field. In previous section, for instance, I considered a fixed metric (5.56). As shown in Section 4.2, a consequence of such a metric is that a moving p-dimensional membrane sweeps a $(p+1)$-dimensional worldsheet, which is a solution to the usual minimal surface action principle. Such a worldsheet is a $(p+1)$-dimensional manifold parametrized by $\phi^A = (\tau, \sigma^a)$. All those parameters are on an equal footing, therefore we have used the compact notation when constructing the action (5.57).

Now I am going to ascribe the dynamical role to the $\mathcal{M}$-space metric as well. If the metric of $\mathcal{M}$-space is dynamical eq. (5.56) no longer holds. From the point of view of the infinite-dimensional $\mathcal{M}$-space we have the motion of a point "particle" in the presence of the metric field $\rho_{\mu(\phi)\nu(\phi')}$ which itself is dynamical. As a model for such a dynamical system let us consider the action

$$I[\rho] = \int \mathcal{D}X \sqrt{|\rho|} \left(\rho_{\mu(\phi)\nu(\phi')} \dot{X}^{\mu(\phi)} \dot{X}^{\nu(\phi')} + \frac{\epsilon}{16\pi} \mathcal{R} \right). \qquad (5.76)$$

where ρ is the determinant of the metric $\rho_{\mu(\phi)\nu(\phi')}$. Here $\mathcal{R}$ is the Ricci scalar in $\mathcal{M}$-space, defined according to

$$\mathcal{R} = \rho^{\mu(\phi)\nu(\phi')} \mathcal{R}_{\mu(\phi)\nu(\phi')}, \qquad (5.77)$$

where the Ricci tensor

$$\mathcal{R}_{\mu(\phi)\nu(\phi)} = \Gamma^{\rho(\phi'')}_{\mu(\phi)\rho(\phi''),\nu(\phi')} - \Gamma^{\rho(\phi'')}_{\mu(\phi)\nu(\phi'),\rho(\phi'')} \qquad (5.78)$$

$$+ \Gamma^{\rho(\phi'')}_{\sigma(\phi''')\nu(\phi')} \Gamma^{\sigma(\phi''')}_{\mu(\phi)\rho(\phi'')} - \Gamma^{\sigma(\phi''')}_{\mu(\phi)\nu(\phi')} \Gamma^{\rho(\phi'')}_{\sigma(\phi''')\rho(\phi'')}$$

is a contraction of the Riemann tensor in $\mathcal{M}$-space:

In order to perform the variation with respect to $X^{\mu(\phi)}$ we have to omit the integration over $\mathcal{D}X$ and consider the expression

$$I[\rho] = \left(\rho_{\mu(\phi)\nu(\phi')} \dot{X}^{\mu(\phi)} \dot{X}^{\nu(\phi')} + \frac{\epsilon}{16\pi} \mathcal{R} \right) \qquad (5.79)$$

Variation of (5.76) with respect to $X^{\mu(\phi)}$ gives

$$2 \left(\frac{\mathrm{d} \dot{X}^{\mu(\phi)}}{\mathrm{d}\tau} + \Gamma^{\mu(\phi)}_{\alpha(\phi')\beta(\phi'')} \dot{X}^{\alpha(\phi')} \dot{X}^{\beta(\phi'')} \right)$$

$$- \frac{\epsilon}{16\pi} \partial^{\mu(\phi)} \mathcal{R} - \frac{\partial^{\mu(\phi)} \sqrt{|\rho|}}{\sqrt{|\rho|}} \left(\dot{X}^2 + \frac{\epsilon}{16\pi} \mathcal{R} \right) = 0 \qquad (5.80)$$

More about physics in $\mathcal{M}$-space

Variation of the latter expression with respect to the metric gives the Einstein equations in $\mathcal{M}$-space:

$$\frac{\epsilon}{16\pi}\mathcal{G}_{\mu(\phi)\nu(\phi')} + \dot{X}_{\mu(\phi)}\dot{X}_{\nu(\phi')} - \tfrac{1}{2}\rho_{\mu(\phi)\nu(\phi')}\dot{X}^2 = 0 \tag{5.81}$$

where

$$\mathcal{G}_{\mu(\phi)\nu(\xi')} \equiv \mathcal{R}_{\mu(\xi)\nu(\xi')} - \tfrac{1}{2}\rho_{\mu(\xi)\nu(\xi')}\mathcal{R} \tag{5.82}$$

is the Einstein tensor in $\mathcal{M}$-space.

Contracting eq. (5.81) with $\rho^{\mu(\phi)\nu(\phi')}$ we have

$$\dot{X}^2 + \frac{\epsilon}{16\pi}\mathcal{R} = 0, \tag{5.83}$$

where $\dot{X}^2 \equiv \dot{X}^{\mu(\phi)}\dot{X}_{\mu(\phi)}$. Because of (5.83) the equations of motion (5.80), (5.81) can be written as

$$\frac{\mathrm{d}\dot{X}^{\mu(\phi)}}{\mathrm{d}\tau} + \Gamma^{\mu(\phi)}_{\alpha(\phi')\beta(\phi'')}\dot{X}^{\alpha(\phi')}\dot{X}^{\beta(\phi'')} - \frac{\epsilon}{32\pi}\partial^{\mu(\phi)}\mathcal{R} = 0 \tag{5.84}$$

$$\dot{X}^{\mu(\phi)}\dot{X}^{\nu(\phi)} + \frac{\epsilon}{16\pi}\mathcal{R}^{\mu(\phi)\nu(\phi')} = 0 \tag{5.85}$$

Using

$$\dot{X}^{\nu(\phi')}\mathrm{D}_{\nu(\phi')}\dot{X}^{\mu(\phi)} = \frac{\mathrm{D}\dot{X}^{\mu(\phi)}}{\mathrm{D}\tau}$$

$$\equiv \frac{\mathrm{d}\dot{X}^{\mu(\phi)}}{\mathrm{d}\tau} + \Gamma^{\mu(\phi)}_{\alpha(\phi')\beta(\phi'')}\dot{X}^{\alpha(\phi')}\dot{X}^{\beta(\phi'')}, \tag{5.86}$$

$$\dot{X}_{\mu(\phi)}\partial^{\mu(\phi)}\mathcal{R} = \frac{\mathrm{d}\mathcal{R}}{\mathrm{d}\tau}, \tag{5.87}$$

and multiplying (5.84) with $\det X_{\mu(\phi)}$ we have

$$\frac{1}{2}\frac{\mathrm{d}\dot{X}^2}{\mathrm{d}\tau} - \frac{\epsilon}{32\pi}\frac{\mathrm{d}\mathcal{R}}{\mathrm{d}\tau} = 0. \tag{5.88}$$

Equations (5.83) and (5.88) imply

$$\frac{\mathrm{d}\dot{X}^2}{\mathrm{d}\tau} = 0, \qquad \frac{\mathrm{d}\mathcal{R}}{\mathrm{d}\tau} = 0. \tag{5.89}$$

By the Bianchi identity

$$\mathrm{D}_{\nu(\phi')}\mathcal{G}^{\mu(\phi)\nu(\phi')} = 0 \tag{5.90}$$

eq. (5.81) implies

$$\left(D_{\nu(\phi')}\dot{X}^{\mu(\phi)}\right)\dot{X}^{\nu(\phi')} + \dot{X}^{\mu(\phi)}\left(D_{\nu(\phi')}\dot{X}^{\nu(\phi')}\right) - \tfrac{1}{2}\partial^{\mu(\phi)}\dot{X}^2 = 0 \qquad (5.91)$$

If we multiply the latter equation by $\dot{X}_{\mu(\phi)}$ (and sum over $\mu(\phi)$) we obtain

$$D_{\nu(\phi)}\dot{X}^{\nu(\phi)} = 0. \qquad (5.92)$$

After inserting this back into eq. (5.91) we have

$$\frac{D\dot{X}^{\mu(\phi)}}{D\tau} - \tfrac{1}{2}\partial^{\mu(\phi)}\dot{X}^2 = 0, \qquad (5.93)$$

or, after using (5.83),

$$\frac{D\dot{X}^{\mu(\phi)}}{D\tau} + \frac{\epsilon}{32\pi}\partial^{\mu(\phi)}\mathcal{R} = 0. \qquad (5.94)$$

The latter equation together with eq. (5.84) implies

$$\partial^{\mu(\phi)}\mathcal{R} = 0, \qquad (5.95)$$

or equivalently

$$\partial^{\mu(\phi)}\dot{X}^2 = 0. \qquad (5.96)$$

Equation of motion (5.84) or (5.93) thus becomes simply

$$\frac{D\dot{X}^{\mu(\phi)}}{D\tau} \equiv \frac{d\dot{X}^{\mu(\phi)}}{d\tau} + \Gamma^{\mu(\phi)}_{\alpha(\phi')\beta(\phi'')}\dot{X}^{\alpha(\phi')}\dot{X}^{\beta(\phi'')} = 0, \qquad (5.97)$$

which is *the geodesic equation* in $\mathcal{M}$-space.

Equation (5.96) is, in fact, the statement that the functional derivative of the quadratic form

$$\dot{X}^2 \equiv \dot{X}^{\mu(\phi)}\dot{X}_{\mu(\phi)} = \rho_{\mu(\phi)\nu(\phi')}\dot{X}^{\mu(\phi)}\dot{X}^{\nu(\phi')} \qquad (5.98)$$

with respect to $X^{\mu(\phi)}$ is zero. The latter statement is *the geodesic equation*.

Equation (5.95) states that the functional derivative of the curvature scalar does not matter. Formally, in our variation of the action (5.76) we have also considered the term $\partial^{\mu(\phi)}\mathcal{R}$, and from the consistency of the equations of motion we have found that $\partial^{\mu(\phi)}\mathcal{R} = 0$. Our equations of motion are therefore simply (5.97) and (5.85).

The metric $\rho_{\mu(\phi)\nu(\phi')}$ is a functional of the variables $X^{\mu(\phi)}$, and (5.85), (5.97) is a system of functional differential equations which determine the set of possible solutions for $X^{\mu(\phi)}$ and $\rho_{\mu(\phi)\nu(\phi')}$. Note that our membrane model is *independent of background* : there is no pre-existing space with

More about physics in M-space 161

a pre-existing metric, neither curved nor flat. The metric comes out as a solution of the system of functional differential equations, and at the same time we also obtain a solution for $X^\mu(\phi)$.

We can imagine a universe which consists of a single (n-dimensional) membrane whose configuration changes with τ according to the system of functional equations (5.85), (5.97). The metric $\rho_{\mu(\phi)\nu(\phi')}[X]$, at a given point $\phi^A = (\tau, \xi^a)$ on the membrane, depends on the membrane configuration described by the variables $X^\mu(\phi)$. In other words, *in the theory considered here the the metric $\rho_{\mu(\phi)\nu(\phi')}[X]$ is defined only on the membrane*, and it is meaningless to speak about a metric "outside" the membrane. Also there is no such thing as a metric of a (finite-dimensional) space into which the membrane is embedded. Actually, there is no embedding space. All what exists is the membrane configuration $X^\mu(\phi)$ and the corresponding metric $\rho_{\mu(\phi)\nu(\phi')}$ which determines the distance between two possible configurations $X^\mu(\phi)$ and $X'^\mu(\phi)$ in M-space. The latter space M itself together with the metric $\rho_{\mu(\phi)\nu(\phi')}$ is a solution to our dynamical system of equations (5.80),(5.81).

Instead of a one membrane universe we can imagine a many membrane universe as a toy model. If $X^\mu(\phi)$, with the set of continuous indices $\phi \equiv \phi^A$, represents a single membrane then $X^{\mu(\phi,k)}$ with an extra discrete index $k = 1, 2, ..., Z$ represents a system of Z membranes. If we replace (ϕ) with (ϕ, k), or, alternatively, if we interpret (ϕ) to include the index k, then *all equations (5.76)–(5.97) are also valid for a system of membranes*. The metric $\rho_{\mu(\phi,k)\nu(\phi',k')}$ as a solution to the system of functional equations of motion (5.80), (5.81) is defined only on those membranes. If there are many such membranes then the points belonging to all those membranes approximately sample a finite-dimensional manifold V_N into which the system of membranes is embedded. The metric $\rho_{\mu(\phi,k)\nu(\phi',k')}$, although defined only on the set of points (ϕ, k) belonging to the system of membranes, approximately sample the metric of a continuous finite-dimensional space V_N. See also Fig. 5.1 and the accompanying text.

The concepts introduced above are very important and I shall now rephrase them. Suppose we have a system of membranes. Each membrane is described by the variables[6] $X^\mu(\phi^A) = X^\mu(\xi^a, \tau)$, $\mu = 1, 2, ..., N$; $a = 1, 2, ..., n$; $\xi^a \in [0, \pi]$, $\tau \in [-\infty, \infty]$. In order to distinguish different

---
[6]Other authors also considered a class of membrane theories in which the embedding space has no prior existence, but is instead coded completely in the degrees of freedom that reside on the membranes. They take for granted that, as the background is not assumed to exist, there are no embedding coordinates (see e.g., [68]). This seems to be consistent with our usage of $X^\mu(\phi)$ which, at the fundamental level, are not considered as the embedding coordinates, but as the M-space coordinates. Points of M-space are described by coordinates $X^\mu(\phi)$, and the distance between the points is determined by the metric $\rho_{\mu(\phi)\nu(\phi')}$, which is dynamical.

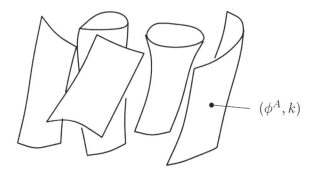

Figure 5.1. The system of membranes is represented as being embedded in a finite-dimensional space V_N. The concept of a continuous embedding space is only an approximation which, when there are many membranes, becomes good at large scales (i.e., at the "macroscopic" level). The metric is defined only at the points (ϕ, k) situated on the membranes. At large scales (or equivalently, when the membranes are "small" and densely packed together) the set of all the points (ϕ, k) is a good approximation to a continuous metric space V_N.

membranes we introduce an extra index k which assumes discrete values. The system of membranes is then described by variables $X^\mu(\phi^A, k)$, $k = 1, 2, ..., Z$. Each configuration of the system is considered as a point, with coordinates $X^{\mu(\phi,k)} \equiv X^\mu(\phi^A, k)$, in an infinite-dimensional space $\mathcal{M}$ with metric $\rho_{\mu(\phi,k)\nu(\phi',k')}$. The quantities $X^\mu(\phi^A, k)$ and $\rho_{\mu(\phi,k)\nu(\phi',k')}$ are not given *a priori*, they must be a solution to a dynamical system of equations. As a preliminary step towards a more realistic model I consider the action (5.76) which leads to the system of equations (5.85), (5.97). In this model everything that can be known about our toy universe is in the variables $X^{\mu(\phi,k)}$ and $\rho_{\mu(\phi,k)\nu(\phi',k')}$. The metric is defined for all possible values of the parameters (ϕ^A, k) assigned to the points on the membranes. At every chosen point (ϕ, k) the metric $\rho_{\mu(\phi,k)\nu(\phi',k')}$ is a functional of the system's configuration, represented by $X^{\mu(\phi,k)}$.

It is important to distinguish between the points, represented by the coordinates $X^{\mu(\phi,k)}$, of the infinite-dimensional $\mathcal{M}$-space, and the points, represented by (ϕ^A, k), of the finite-dimensional space spanned by the membranes. In the limit of large distances the latter space becomes the continuous target space into which a membrane is embedded. In this model the embedding space (or the "target space") is inseparable from the set of membranes. Actually, it is identified with the set of points (ϕ^A, k) situated on the membranes. Without membranes there is no target space. With membranes there is also the target space, and if the membranes are "small", and if their density (defined roughly as the number of membranes per volume) is high, then the target space becomes a good approximation to a continuous manifol V_N.

METRIC OF V_N FROM THE METRIC OF $\mathcal{M}$-SPACE

$\mathcal{M}$-space is defined here as the space of all possible configurations of a system of membranes. The metric $\rho_{\mu(\phi,k)\nu(\phi',k')}$ determines *the distance* between the points $X^{\mu(\phi,k)}$ and $X^{\mu(\phi,k)}+\mathrm{d}X^{\mu(\phi,k)}$ belonging to two different configurations (see Fig. 5.2) according to the relation

$$\mathrm{d}\ell^2 = \rho_{\mu(\phi,k)\nu(\phi',k')}\mathrm{d}X^{\mu(\phi,k)}\mathrm{d}X^{\nu(\phi',k')}, \tag{5.99}$$

where

$$\mathrm{d}X^{\mu(\phi,k)} = X'^{\mu(\phi,k)} - X^{\mu(\phi,k)}. \tag{5.100}$$

Besides (5.100), which is an $\mathcal{M}$-space vector joining two different membrane's configurations (i.e., two points of $\mathcal{M}$-space), we can define another quantity which connects two different points, in the usual sense of the word, within the same membrane configuration:

$$\widetilde{\Delta} X^{\mu}(\phi, k) \equiv X^{\mu(\phi',k')} - X^{\mu(\phi,k)}. \tag{5.101}$$

By using the $\mathcal{M}$-space metric let us define the quantity

$$\Delta s^2 = \rho_{\mu(\phi,k)\nu(\phi',k')}\widetilde{\Delta}X^{\mu}(\phi, k)\widetilde{\Delta}X^{\nu}(\phi', k'). \tag{5.102}$$

In the above formula summation over the repeated indices μ and ν is assumed, but no integration over ϕ, ϕ' and no summation over k, k'.

What is the geometric meaning of the quantity Δs^2 defined in (5.102)? To a point on a k-th membrane we can assign parameters, or *intrinsic coordinates*, (ϕ, k). Alternatively, we can assign to the same points the *extrinsic coordinates*[7] $X^{\mu(\phi,k)} \equiv X^{\mu}(\phi, k)$, $\mu = 1, 2, ..., N$, with ϕ, k fixed. It then seems very natural to interpret Δs^2 as *the distance* between the

---

[7] Obviously $X^{\mu(\phi,k)} \equiv X^{\mu}(\phi, k)$ has a double role. On the one hand it represents a *set* of all points with support on the system of membranes. On the other hand it represents a *single point* in an infinite-dimensional $\mathcal{M}$-space.

164   THE LANDSCAPE OF THEORETICAL PHYSICS: A GLOBAL VIEW

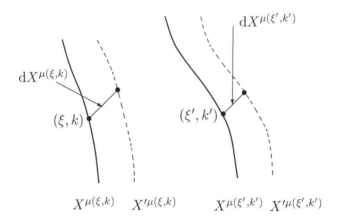

Figure 5.2. Two different membrane configurations, represented by the coordinates $X^{\mu(\xi,k)}$ and $X'^{\mu(\xi,k)}$, respectively. The infinitesimal $\mathcal{M}$-space vector is $\mathrm{d}X^{\mu(\xi,k)} = X'^{\mu(\xi,k)} - X^{\mu(\xi,k)}$.

points $X^\mu(\phi', k')$ and $X^\mu(\phi, k)$ within a given membrane configutations. The quantity $\rho_{\mu(\phi,k)\nu(\phi',k')}$ then has the role of a prototype of the *parallel propagator* [67] in V_N. Note that we do not yet have a manifold V_N, but we already have a *skeleton* S of it, formed by the set of all the points with support on our membranes. The quantity $\widetilde{\Delta}X^\mu(\phi', k')$ is a prototype of a *vector* in V_N, and if we act on a vector with the parallel propagator we perform a *parallel transport* of the vector along the geodesic joining the points (ϕ', k') and (ϕ, k). The latter terminology holds for a continuum space V_N, but we may retain it for our skeleton space S as well. The quantity $\rho_{\mu(\phi,k)\nu(\phi',k')}\widetilde{\Delta}X^\nu(\phi', k')$ is a vector, obtained from $\widetilde{\Delta}X^\nu(\phi', k')$ by parallel transport.

We define
$$\rho_{\mu(\phi,k)\nu(\phi',k')}\widetilde{\Delta}X^\nu(\phi', k') = \widetilde{\Delta}X_\mu(\phi, k). \qquad (5.103)$$

Then
$$\Delta s^2 = \widetilde{\Delta}X_\mu(\phi, k)\widetilde{\Delta}X^\mu(\phi, k). \qquad (5.104)$$

On the other hand, if in eq. (5.102) and (5.103) we take the coincidence limit $(\phi, k) \to (\phi', k')$, then

$$\Delta s^2 = \rho_{\mu\nu}(\phi, k)\widetilde{\Delta}x^\mu(\phi, k)\widetilde{\Delta}X^\nu(\phi, k), \qquad (5.105)$$

$$\rho_{\mu\nu}(\phi, k)\widetilde{\Delta}X^\nu(\phi, k) = \widetilde{\Delta}x^\mu(\phi, k), \qquad (5.106)$$

with
$$\rho_{\mu\nu}(\phi,k) \equiv \rho_{\mu(\phi,k)\nu(\phi,k)}. \qquad (5.107)$$

We see that the quantity $\rho_{\mu(\phi,k)\nu(\phi,k)}$ has the property of the metric in the skeleton target space S.

Now we can envisage a situation in which the index k is not discrete, but consists of a set of continiuous indices such that instead of the discrete membranes we have a *fluid* of membranes. We can arrange a situation such that the fluid fills an N-dimensional manifold V_N, and $\rho_{\mu\nu}(\phi, k)$ is the metric of V_N defined at the points (ϕ, k) of the fluid. Here the fluid is a reference system, or reference fluid (being itself a part of the dynamical system under consideration), with respect to which the points of the manifold V_N are defined.

The metric $\rho_{\mu\nu}(\phi, k)$ satisfies the functional differential equations (5.80) which presumably reduce to the usual Einstein equations, at least as an approximation. Later we shall show that this is indeed the case. The description with a metric tensor will be surpassed in Sec. 6.2, when we shall discuss a very promising description in terms of the Clifford algebra equivalent of the tetrad field which simplifies calculation significantly. The specific theory based on the action (5.76) should be considered as serving the purpose of introducing us to the concept of mathematical objects, based on $\mathcal{M}$-space, representing physical quantities at the fundamental level.

To sum up, we have taken *the membrane space $\mathcal{M}$* seriously as an arena for physics. The arena itself is also a part of the dynamical system, it is not prescribed in advance. The membrane space $\mathcal{M}$ is the space of all kinematically possible membrane configurations. Which particular configuration is actually realized is selected by the functional equations of motion together with the initial and boundary conditions. A configuration may consist of many membranes. The points located on all those membranes form a space which, at a sufficiently large scale, is a good approximation to a continuous N-dimensional space. The latter space is just *the embedding space* for a particular membrane.

We have thus formulated a theory in which an embedding space *per se* does not exist, but is intimately connected to the existence of membranes. Without membranes there is no embedding space. This approach is *background independent*. There is no pre-existing space and metric: they appear dynamically as solutions to the equations of motion.

The system, or condensate of membranes (which, in particular, may be so dense that the corrseponding points form a continuum), represents a *reference system* or *reference fluid* with respect to which the points of the target space are defined. Such a system was postulated by DeWitt [25], and recently reconsidered by Rovelli [26].

The famous *Mach principle* states that the motion of matter at a given location is determined by the contribution of all the matter in the universe and this provides an explanation for inertia (and inertial mass). Such a situation is implemented in the model of a universe consisting of a system of membranes: the motion of a k-th membrane, including its inertia, is determined by the presence of all the other membranes.

In my opinion the $\mathcal{M}$-space approach is also somehow related to *loop quantum gravity* proposed by Rovelli [69] and Smolin [70].

It seems to me that several concepts which have been around in theoretical physics for some time, and especially those sketched above, all naturally emerge from $\mathcal{M}$-space physics.

Chapter 6

EXTENDED OBJECTS AND CLIFFORD ALGEBRA

We have seen that geometric calculus based on Clifford algebra is a very useful tool for a description of geometry and point particle physics in *flat spacetime*. In Sec. 4.2 we have encountered a generalization of the concepts of vectors and polyvectors in an infinite-dimensional $\mathcal{M}$-space, which is in general a curved space. What we need now is a more complete description, firstly, of the usual finite-dimensional vectors and polyvectors in *curved space*, and secondly, of the corresponding objects in an infinite-dimensional space (which may be $\mathcal{M}$-space, in particular).

After such mathematical preliminaries I shall discuss a membrane description which employs $\mathcal{M}$-space basis vectors. A background independent approach will be achieved by proclaiming the basis vectors themselves as dynamical objects. The corresponding action has its parallel in the one considered in the last section of previous chapter. However, now the dynamical object is not the $\mathcal{M}$-space metric, but its "square root".

Until the techniques of directly solving the functional differential equations are more fully developed we are forced to make the transition from functional to the usual partial differential equations, i.e., from the infinite-dimensional to the finite-dimensional differential equations. Such a transition will be considered in Sec. 6.2, where a description in terms of position-dependent target space vectors will be provided. We shall observe that a membrane is a sort of a fluid localized in the target space, and exploit this concept in relation to the DeWitt–Rovelli reference fluid necessary for localization of spacetime points. Finally we shall realize that we have just touched the tip of an iceberg which is the full polyvector description of the membrane together with the target space in which the membrane is embedded.

168   THE LANDSCAPE OF THEORETICAL PHYSICS: A GLOBAL VIEW

6.1.   MATHEMATICAL PRELIMINARIES

I will now provide an intuitive description of vectors in curved spaces and their generalization to the infinite-dimensional (also curved) spaces. The important concept of the vector and multivector, or polyvector, derivative will also be explained. The usual functional derivative is just a component description of the vector derivative in an infinite-dimensional space. My aim is to introduce the readers into those very elegant mathematical concepts and give them a feeling about their practical usefulness. To those who seek a more complete mathematical rigour I advise consulting the literature [22]. In the case in which the concepts discussed here are not found in the existing literature the interested reader is invited to undertake the work and to develop the ideas initiated here further in order to put them into a more rigorous mathematical envelope. Such a development is beyond the scope of this book which aims to point out how various pieces of physics and mathematics are starting to merge before our eyes into a beautiful coherent picture.

VECTORS IN CURVED SPACES

Basis vectors need not be equal at all point of a space V_N. They may be position-dependent. Let $\gamma(x)$, $\mu = 1, 2, ..., N$, be N linearly independent vectors which depend on position x^μ. The inner products of the basis vectors form the metric tensor which is also position-dependent:

$$\gamma_\mu \cdot \gamma_\nu = g_{\mu\nu}, \tag{6.1}$$

$$\gamma^\mu \cdot \gamma^\nu = g^{\mu\nu}. \tag{6.2}$$

What is the relation between vectors at two successive infinitesimally separated points x^μ and $x^\mu + \mathrm{d}x^\mu$? Clearly vectors at $x^\mu + \mathrm{d}x^\mu$ cannot be linearly independent from the vectors at x^μ, since they both belong to the same vector space. Therefore the derivative of a basis vector must be a linear combination of basis vectors:

$$\partial_\alpha \gamma^\mu = -\Gamma^\mu_{\alpha\beta} \gamma^\beta. \tag{6.3}$$

Then

$$\Gamma^\mu(x + \mathrm{d}x) = \gamma^\mu(x) + \partial_\alpha \gamma^\mu(x)\mathrm{d}x^\alpha = (\delta^\mu{}_\alpha - \Gamma^\mu_{\alpha\beta}\mathrm{d}x^\beta)\gamma^\alpha,$$

which is indeed a linear conbination of $\gamma^\alpha(x)$.

The coefficients $\Gamma^\mu_{\alpha\beta}$ form *the affinity* of V_N. From (6.3) we have

$$\Gamma^\mu_{\alpha\beta} = -\gamma_\beta \cdot \partial_\alpha \gamma^\mu. \tag{6.4}$$

Extended objects and Clifford algebra 169

The relation $\gamma^\mu \cdot \gamma_\beta = \delta^\mu{}_\beta$ implies

$$\partial_\alpha \gamma^\mu \cdot \gamma_\beta + \gamma^\mu \cdot \partial_\alpha \gamma_\beta = 0, \tag{6.5}$$

from which it follows that

$$\Gamma^\mu_{\alpha\beta} = \gamma^\mu \cdot \partial_\alpha \gamma_\beta. \tag{6.6}$$

Multiplying the latter expression by γ_μ, summing over μ and using

$$\gamma_\mu (\gamma^\mu \cdot a) = a, \tag{6.7}$$

which holds for any vector a, we obtain

$$\partial_\alpha \gamma_\beta = \Gamma^\mu_{\alpha\beta} \gamma_\mu. \tag{6.8}$$

In general the affinity is *not symmetric*. In particular, when it is *symmetric*, we have

$$\gamma^\mu \cdot (\partial_\alpha \gamma_\beta - \partial_\beta \gamma_\alpha) = 0 \tag{6.9}$$

and also

$$\partial_\alpha \gamma_\beta - \partial_\beta \gamma_\alpha = 0 \tag{6.10}$$

in view of the fact that (6.9) holds for any γ^μ. For a symmetric affinity, after using (6.1), (6.2), (6.10) we find

$$\Gamma^\mu_{\alpha\beta} = \tfrac{1}{2} g^{\mu\nu} (g_{\nu\alpha,\beta} + g_{\nu\beta,\alpha} - g_{\alpha\beta,\nu}). \tag{6.11}$$

Performing the second derivative we have

$$\partial_\beta \partial_\alpha \gamma^\mu = -\partial_\beta \Gamma^\mu_{\alpha\sigma} \gamma^\sigma - \Gamma^\mu_{\alpha\sigma} \partial_\beta \gamma^\sigma = -\partial_\beta \Gamma^\mu_{\alpha\sigma} \gamma^\sigma + \Gamma^\mu_{\alpha\sigma} \Gamma^\sigma_{\beta\rho} \gamma^\rho \tag{6.12}$$

and

$$[\partial_\alpha, \partial_\beta] \gamma^\mu = R^\mu{}_{\nu\alpha\beta} \gamma^\nu, \tag{6.13}$$

where

$$R^\mu{}_{\nu\alpha\beta} = \partial_\beta \Gamma^\mu_{\nu\alpha} - \partial_\alpha \Gamma^\mu_{\nu\beta} + \Gamma^\mu_{\beta\rho} \Gamma^\rho_{\alpha\nu} - \Gamma^\mu_{\alpha\rho} \Gamma^\rho_{\beta\nu} \tag{6.14}$$

is *the curvature tensor*. In general the latter tensor does not vanish and we have a curved space.

SOME ILLUSTRATIONS

Derivative of a vector. Let a be an arbitrary position-dependent vector, expanded according to

$$a = a^\mu \gamma_\mu. \tag{6.15}$$

Taking the partial derivative with respect to coordinates x^μ we have

$$\partial_\nu a = \partial_\nu a^\mu \, \gamma_\mu + a^\mu \partial_\nu \gamma_\mu. \qquad (6.16)$$

Using (6.8) and renaming the indices we obtain

$$\partial_\nu a = \left(\partial_\nu a^\mu + \Gamma^\mu_{\nu\rho} a^\rho \right) \gamma_\mu, \qquad (6.17)$$

or

$$\gamma^\mu \cdot \partial_\nu a = \partial_\nu a^\mu + \Gamma^\mu_{\nu\rho} a^\rho \equiv \mathrm{D}_\nu a^\mu, \qquad (6.18)$$

which is the well known *covariant derivative*. The latter derivative is the projection of $\partial_\nu a$ onto one of the basis vectors.

Locally inertial frame. At each point of a space V_N we can define a set of N linearly independent vectors γ_a, $a = 1, 2, ..., N$, satisfying

$$\gamma_a \cdot \gamma_b = \eta_{ab}, \qquad (6.19)$$

where η_{ab} is the Minkowski tensor. The set of vector fields $\gamma_a(x)$ will be called the *locally inertial frame field*.

A coordinate basis vector can be expanded in terms of local basis vector

$$\gamma^\mu = e^\mu{}_a \gamma^a, \qquad (6.20)$$

where the expansion coefficients $e^\mu{}_a$ form the so called *fielbein field* (in 4-dimensions "fielbein" becomes "vierbein" or "tetrad").

$$e^\mu{}_a = \gamma^\mu \cdot \gamma_a. \qquad (6.21)$$

Also

$$\gamma_a = e^\mu{}_a \gamma_\mu, \qquad (6.22)$$

and analogous relations for the inverse vectors γ_μ and γ^a satisfying

$$\gamma^\mu \cdot \gamma_\nu = \delta^\mu{}_\nu, \qquad (6.23)$$

$$\gamma^a \cdot \gamma_b = \delta^a{}_b, \qquad (6.24)$$

From the latter relations we find

$$\gamma^\mu \cdot \gamma^\nu = (e^\mu{}_a \gamma^a) \cdot (e^\nu{}_b \gamma^b) = e^\mu{}_a e_\nu{}^a = g^{\mu\nu}, \qquad (6.25)$$

$$\gamma^a \cdot \gamma^b = (e_\mu{}^a \gamma^\mu) \cdot (e_\nu{}^b \gamma^\nu) = e_\mu{}^a e^{\mu b} = \eta^{ab}. \qquad (6.26)$$

A vector a can be expanded either in terms of γ_μ or γ_a:

$$a = a^\mu \gamma_\mu = a^\mu e_\mu{}^a \gamma_a = a^a \gamma_a , \qquad a^a = a^\mu e_\mu{}^a. \tag{6.27}$$

Differentiation gives

$$\partial_\mu \gamma^a = \omega^a{}_{b\mu} \gamma^b, \tag{6.28}$$

where $\omega^a{}_{b\mu}$ is the connection for the local frame field γ^a. Inserting (6.20) into the relation (6.3) we obtain

$$\partial_\nu \gamma^\mu = \partial_\nu (e^\mu{}_a \gamma^a) = \partial_\nu e^\mu{}_a \gamma^a + e^\mu{}_a \partial_\nu \gamma^a = -\Gamma^\mu_{\nu\sigma} \gamma^\sigma, \tag{6.29}$$

which, in view of (6.28), becomes

$$\partial_\nu e^{\mu a} + \Gamma^\mu_{\nu\sigma} e^{\sigma a} + \omega^{ab}{}_\nu e^\mu{}_b = 0. \tag{6.30}$$

Because of (6.25), (6.26) we have

$$\partial_\nu e_\mu{}^a - \Gamma^\sigma_{\nu\mu} e_\sigma{}^a + \omega^a{}_{b\mu} e_\mu{}^b = 0. \tag{6.31}$$

These are the well known relations for differentiation of the fielbein field.

Geodesic equation in V_N. Let p the momentum vector satisfying the equation of motion

$$\frac{\mathrm{d}p}{\mathrm{d}\tau} = 0. \tag{6.32}$$

Expanding $p = p^\mu \gamma_\mu$, where $p^\mu = m\dot{X}^\mu$, we have

$$\dot{p}^\mu + p^\mu \dot{\gamma}_\mu = \dot{p}^\mu \gamma_\mu + p^\mu \partial_\nu \gamma_\mu \dot{X}^\nu. \tag{6.33}$$

Using (6.8) we obtain, after suitably renaming the indices,

$$(\dot{p}^\mu \gamma_\mu + \Gamma^\mu_{\alpha\beta} p^\alpha \dot{X}^\beta) \gamma_\mu = 0, \tag{6.34}$$

which is *the geodesic equation* in component notation. The equation of motion (6.32) says that the vector p does not change during the motion. This means that vectors $p(\tau)$ for all values of the parameter τ remain parallel amongst themselves (and, of course, retain the same magnitude square p^2). After using the expansion $p = p^\mu \gamma_\mu$ we find that the change of the components p^μ is compensated by the change of basis vectors γ_μ.

Geometry in a submanifold V_n. In the previous example we considered a geodesic equation in spacetime V_N. Suppose now that a submanifold — a surface — V_n, parametrized by ξ^a, is embedded in V_N. Let[1] e^a, $a = 1, 2, ..., n$, be a set of tangent vectors to V_n. They can be expanded in

---

[1] Notice that the index a has now a different meaning from that in the case of a locally inertial frame considered before.

terms of basis vectors of V_N:

$$e_a = \partial_a X^\mu \gamma_\mu, \qquad (6.35)$$

where

$$\partial_a X^\mu = e_a \cdot \gamma^\mu \qquad (6.36)$$

are derivatives of the embedding functions of V_n. They satisfy

$$\partial_a X^\mu \, \partial_b X_\mu = (e_a \cdot \gamma^\mu)(e_b \cdot \gamma_\mu) = e_a \cdot e_b = \gamma_{ab}, \qquad (6.37)$$

which is the expression for the induced metric of V_n. Differentiation of e_a gives

$$\partial_b e_a = \partial_b \partial_a X^\mu \gamma_\mu + \partial_a X^\mu \partial_b . \gamma_\mu \qquad (6.38)$$

Using

$$\partial_b \gamma_\mu = \partial_\nu \gamma_\mu \partial_b X^\nu \qquad (6.39)$$

and the relation (6.8) we obtain from (6.38), after performing the inner product with e^c,

$$(e^c \cdot \partial_b e_a) = \partial_a \partial_b X^\mu \partial^c X_\mu + \Gamma^\sigma_{\mu\nu} \partial_a X^\mu \partial_b X^\nu \partial^c X_\sigma. \qquad (6.40)$$

On the other hand, the left hand side of eq. (6.38) involves the affinity of V_n:

$$\partial_b e_a = \Gamma^d_{ba} e_d , \quad \text{therefore} \quad e^c \cdot \partial_b e_a = \Gamma^c_{ba}, \qquad (6.41)$$

and so we see that eq. (6.40) is a relation between the affinity of V_n and V_N. Covariant derivative in the submanifold V_n is defined in terms of Γ^d_{ba}.

An arbitrary vector P in V_N can be expanded in terms of γ_μ:

$$P = P_\mu \gamma^\mu. \qquad (6.42)$$

It can be projected onto a tangent vector e_a:

$$P \cdot e_a = P_\mu \gamma^\mu \cdot e_a = P_\mu \partial_a X^\mu \equiv P_a \qquad (6.43)$$

In particular, a vector of V_N can be itself a tangent vector of a subspace V_n. Let p be such a tangent vector. It can be expanded either in terms of γ_μ or e_a:

$$p = p^\mu \gamma_\mu = p_a e^a, \qquad (6.44)$$

where

$$p_a = p_\mu \gamma^\mu \cdot e_a = p_\mu \partial_a X^\mu,$$

$$p_\mu = p_a e^a \cdot \gamma_\mu = p_a \partial^a X_\mu. \qquad (6.45)$$

Such symmetric relations between p_μ and p_a hold only for a vector p which is tangent to V_n.

Extended objects and Clifford algebra 173

Suppose now that p is tangent to a geodesic of V_n. Its derivative with respect to an invariant parameter τ along the geodesic is

$$\frac{\mathrm{d}}{\mathrm{d}\tau}(p^a e_a) = \frac{\mathrm{d}}{\mathrm{d}\tau}(p^a \partial_a X^\mu \gamma_\mu)$$
$$= (\dot{p}^a \partial_a X^\mu + \partial_a \partial_b X^\mu p^a \dot{\xi}^b + p^a \partial_a X^\alpha \Gamma^\mu_{\alpha\beta} \dot{X}^\beta)\gamma_\mu \quad (6.46)$$

where we have used eq. (6.8) and

$$\frac{\mathrm{d}}{\mathrm{d}\tau}\partial_a X^\mu = \partial_a \partial_b X^\mu \dot{\xi}^b.$$

The above derivative, in general, does not vanish: a vector p of V_N that is tangent to a geodesic in a subspace V_n changes with τ.

Making the inner product of the left and the right side of eq. (6.46) with e^c we obtain

$$\frac{\mathrm{d}p}{\mathrm{d}\tau}\cdot e^c = \dot{p}^c + \Gamma^c_{ab} p^a \dot{\xi}^b = 0. \quad (6.47)$$

Here Γ^c_{ab} is given by eq. (6.40). For a vector p tangent to a geodesic of V_n the right hand side of eq. (6.47) vanish.

On the other hand, starting from $p = p^\mu \gamma_\mu$ and using (6.8) the left hand side of eq. (6.47) gives

$$\frac{\mathrm{d}p}{\mathrm{d}\tau}\cdot e^c = \left(\frac{\mathrm{d}p^\mu}{\mathrm{d}\tau} + \Gamma^\mu_{\alpha\beta} p^\alpha \dot{X}^\beta\right) \partial^c X_\mu = 0. \quad (6.48)$$

Eqs.(6.47), (6.48) explicitly show that in general a geodesic of V_n is not a geodesic of V_N.

A warning is necessary. We have treated tangent vectors e_a to a subspace V_n as vectors in the embedding space V_N. As such they do *not* form a complete set of linearly independent vectors in V_N. An arbitrary vector of V_N, of course, cannot be expanded in terms of e_a; only a tangent vector to V_n can be expanded so. Therefore, the object $\left(\frac{\mathrm{d}p}{\mathrm{d}\tau}\cdot e^c\right) e_c$ should be distinguished from the object $\left(\frac{\mathrm{d}p}{\mathrm{d}\tau}\cdot \gamma^\mu\right) \gamma_\mu = \mathrm{d}p/\mathrm{d}\tau$. The vanishing of the former object does not imply the vanishing of the latter object.

DERIVATIVE WITH RESPECT TO A VECTOR

So far we have considered derivatives of position-dependent vectors with respect to (scalar) coordinates. We shall now consider the derivative with respect to a vector. Let $F(a)$ be a polyvector-valued function of a vector valued argument a which belongs to an n-dimensional vector space $\mathcal{A}_n$. For an arbitrary vector e in $\mathcal{A}_n$ the derivative of F in the direction e is given by

$$\left(e\cdot \frac{\partial}{\partial a}\right) F(a) = \lim_{\tau\to 0} \frac{F(a+e\tau) - F(a)}{\tau} = \frac{\partial F(a+e\tau)}{\partial \tau}. \quad (6.49)$$

For e we may choose one of the basis vectors. Expanding $a = a^\nu e_\nu$, we have

$$\left(e_\mu \cdot \frac{\partial}{\partial a}\right) F(a) \equiv \frac{\partial F}{\partial a^\mu} = \lim_{\tau \to 0} \frac{F(a^\nu e_\nu + e_\mu \tau) - F(a^\nu e_\nu)}{\tau}$$

$$= \lim_{\tau \to 0} \frac{F((a^\nu + \delta^\nu{}_\mu \tau) e_\nu) - F(a^\nu e_\nu)}{\tau}. \qquad (6.50)$$

The above derivation holds for an arbitrary function $F(a)$. For instance, for $F(a) = a = a^\nu e_\nu$ eq. (6.50) gives

$$\frac{\partial F}{\partial a^\mu} = \frac{\partial}{\partial a^\mu}(a^\nu e_\nu) = e_\mu = \delta_\mu{}^\nu e_\nu \qquad (6.51)$$

For the components $F \cdot e^\alpha = a^\alpha$ it is

$$\frac{\partial a^\alpha}{\partial a^\mu} = \delta_\mu{}^\alpha \qquad (6.52)$$

The derivative in the direction e_μ, as derived in (6.50), is the partial derivative with respect to the component a^μ of the vector argument a. (See Box 6.1 for some other examples.)

In eq. (6.50) we have derived the operator

$$e_\mu \cdot \frac{\partial}{\partial a} \equiv \frac{\partial}{\partial a^\mu}. \qquad (6.53)$$

For a running index μ these are components (or projections) of the operator

$$\frac{\partial}{\partial a} = e^\mu \left(e_\mu \cdot \frac{\partial}{\partial a}\right) = e^\mu \frac{\partial}{\partial a^\mu} \qquad (6.54)$$

which is *the derivative with respect to a vector* a.

The above definitions (6.49)–(6.53) hold for any vector a of $\mathcal{A}_n$. Suppose now that all those vectors are defined at a point a of an n-dimensional manifold V_n. They are said to be *tangent* to a point x in V_n [22]. If we allow the point x to vary we have thus a vector field $a(x)$. In components it is

$$a(x) = a^\mu(x) e_\mu(x), \qquad (6.55)$$

where $a^\mu(x)$ are arbitrary functions of x. A point x is parametrized by a set of n coordinates x^μ, hence $a^\mu(x)$ are functions of x^μ. In principle $a^\mu(x)$ are arbitrary functions of x^μ. In particular, we may choose

$$a^\mu(x) = x^\mu. \qquad (6.56)$$

Extended objects and Clifford algebra 175

Box 6.1: Examples of differentiation by a vector

1) *Vector valued function:*

$$F = x = x^\nu e_\nu \,; \qquad \frac{\partial F}{\partial x^\mu} = e_\mu = \delta_\mu{}^\nu e_\nu \,; \qquad \frac{\partial F}{\partial x} = e^\mu \frac{\partial F}{\partial x^\mu} = e^\mu e_\mu = n.$$

2) *Scalar valued function*

$$F = x^2 = x^\nu x_\nu \,; \qquad \frac{\partial F}{\partial x^\mu} = 2x_\mu \,; \qquad \frac{\partial F}{\partial x} = e^\mu 2 x_\mu = 2x.$$

3) *Bivector valued function*

$$F = b \wedge x = (b^\alpha e_\alpha) \wedge (x^\beta e_\beta) = b^\alpha x^\beta \, e_\alpha \wedge e_\beta,$$

$$\frac{\partial F}{\partial x^\mu} = \lim_{\tau \to 0} \frac{b^\alpha e_\alpha \wedge (x^\beta e_\beta + e_\mu \tau) - b^\alpha e_\alpha \wedge x^\beta e_\beta}{\tau} = b^\alpha e_\alpha \wedge e_\mu,$$

$$\frac{\partial F}{\partial x} = e^\mu \frac{\partial F}{\partial x^\mu} = b^\alpha e^\mu (e_\alpha \wedge e_\mu) = b^\alpha e^\mu \cdot (e_\alpha \wedge e_\mu) + b^\alpha e^\mu \wedge e_\alpha \wedge e_\mu,$$

$$= b^\alpha (\delta_\alpha{}^\mu e_\mu - \delta_\mu{}^\mu e_\alpha) = b^\alpha e_\alpha (1 - n) = b(1 - n).$$

Then
$$a(x) = x^\mu e_\mu(x). \tag{6.57}$$

Under a passive coordinate transformation the components $a^\mu(x)$ change according to
$$a'^\mu(x') = \frac{\partial x'^\mu}{\partial x^\nu} a^\nu(x). \tag{6.58}$$

This has to be accompanied by the corresponding (active) change of basis vectors,
$$e'^\mu(x') = \frac{\partial x^\nu}{\partial x'^\mu} e_\nu(x), \tag{6.59}$$

in order for a vector $a(x)$ to remain unchanged. In the case in which the components fields $a^\mu(x)$ are just coordinates themselves, the transformation (6.58) reads
$$a'^\mu(x') = \frac{\partial x'^\mu}{\partial x^\nu} x^\nu. \tag{6.60}$$

In new coordinates x'^μ the components a'^μ of a vector $a(x) = x^\mu e_\mu$ are, of course, not equal to the new coordinates x'^μ. The reader can check by performing some explicit transformations (e.g., from the Cartesian to spherical coordinates) that an object as defined in (6.56), (6.57) is quite a legitimate geometrical object and has, indeed, the required properties of a vector field, even in a curved space. The set of points of a curved space can then, at least locally[2], be considered as a vector field, such that its components in a certain basis are coordinates. In a given space, there are infinitely many fields with such a property, one field for every possible choice of coordinates. As an illustration I provide the examples of two such fields, denoted X and X':

$$a(x) = x^\mu e_\mu(x) = a'^\mu(x') e'_\mu(x') = X,$$
$$b(x) = b^\mu(x) e_\mu(x) = x'^\mu e'_\mu(x') = X'. \qquad (6.61)$$

Returning to the differential operator (6.49) we can consider a as a vector field $a(x)$ and the definition (6.49) is still valid at every point x of V_n. In particular we can choose

$$a(x) = x = x^\nu e_\nu. \qquad (6.62)$$

Then (6.50) reads

$$\left(e_\mu \cdot \frac{\partial}{\partial x}\right) F(x) \equiv \frac{\partial F}{\partial x^\mu} = \lim_{\tau \to 0} \frac{F\left((x^\nu + \delta_\mu{}^\nu \tau)e_\nu\right) - F(x^\nu e_\nu)}{\tau}. \qquad (6.63)$$

This is the partial derivative of a multivector valued function $F(x)$ of position x. The derivative with respect to the polyvector x is

$$\frac{\partial}{\partial x} = e^\mu \left(e'_\mu \cdot \frac{\partial}{\partial x}\right) = e^\mu \frac{\partial}{\partial x^\mu} \qquad (6.64)$$

Although we have denoted the derivative as $\partial/\partial x$ or $\partial/\partial a$, this notation should not be understood as implying that $\partial/\partial a$ can be defined as the limit of a difference quotient. The partial derivative (6.50) can be so defined, but not the derivative with respect to a vector.

DERIVATIVE WITH RESEPCT TO A POLYVECTOR

The derivative with respect to a vector can be generalized to polyvectors. Definition (6.49) is then replaced by

$$\left(E * \frac{\partial}{\partial A}\right) F(A) = \lim_{\tau \to 0} \frac{F(A + E\tau) - F(A)}{\tau} = \frac{\partial F(A + E\tau)}{\partial \tau}. \qquad (6.65)$$

---

[2] Globally this canot be true in general, since a single coordinate system cannot cover all the space.

Extended objects and Clifford algebra

Here $F(A)$ is a polyvector-valued function of a polyvector A, and E is an arbitrary polyvector. The star " $*$ " denotes the scalar product

$$A * B = \langle AB \rangle_0 \tag{6.66}$$

of two polyvectors A and B, where $\langle AB \rangle_0$ is the scalar part of the Clifford product AB. Let e_J be a complete set of basis vector of Clifford algebra satisfying[3]

$$e_J * e_K = \delta_{JK}, \tag{6.67}$$

so that any polyvector can be expanded as $A = A^J e_J$. For E in eq. (6.65) we may choose one of the basis vectors. Then

$$\left(e_K * \frac{\partial}{\partial A} \right) F(A) \equiv \frac{\partial F}{\partial A^K} = \lim_{\tau \to 0} \frac{F(A^J e_J + e_K \tau) - F(A^J e_J)}{\tau}. \tag{6.68}$$

This is the partial derivative of F with respect to the multivector components A_K. The derivative with respect to a polyvector A is the sum

$$\frac{\partial F}{\partial A} = e^J \left(e_J * \frac{\partial}{\partial A} \right) F = e^J \frac{\partial F}{\partial A^J}. \tag{6.69}$$

The polyvector A can be a polyvector field $A(X)$ defined over the position polyvector field X which is a generalizatin of the position vector field x defined in (6.61). In particular, the field $A(X)$ can be $A(X) = X$. Then (6.68), (6.69) read

$$\left(e_K * \frac{\partial}{\partial X} \right) F(X) \equiv \frac{\partial F}{\partial X^K} = \lim_{\tau \to 0} \frac{F(X^J e_J + e_K \tau) - F(X^J e_J)}{\tau}, \tag{6.70}$$

$$\frac{\partial F}{\partial X} = e^J \left(e_J * \frac{\partial}{\partial X} \right) F = e^J \frac{\partial F}{\partial X^J} \tag{6.71}$$

which generalizes eqs. (6.63),6.64).

VECTORS IN AN INFINITE-DIMENSIONAL SPACE

In functional analysis functions are considered as vectors in infinite dimensional spaces. As in the case of finite-dimensional spaces one can introduce a basis $h(x)$ in an infinite-dimensional space V_∞ and expand a vector of V_∞ in terms of the basis vectors. The expansion coefficients form a function $f(x)$:

$$f = \int \mathrm{d}x \, f(x) h(x). \tag{6.72}$$

---

[3] Remember that the set $\{e_J\} = \{1, e_\mu, e_\mu, e_\mu e_\nu, ...\}$.

The basis vectors $h(x)$ are elements of the Clifford algebra $\mathcal{C}_\infty$ of V_∞. The Clifford or geometric product of two vectors is

$$h(x)h(x') = h(x) \cdot h(x') + h(x) \wedge h(x'). \tag{6.73}$$

The symmetric part

$$h(x) \cdot h(x') = \tfrac{1}{2}\left(h(x)h(x') + h(x')h(x)\right) \tag{6.74}$$

is *the inner product* and *the antisymmetric part*

$$h(x) \wedge h(x') = \tfrac{1}{2}\left(h(x)h(x') - h(x')h(x)\right) \tag{6.75}$$

is *the outer, or wedge, product* of two vectors.

The inner product defines the metric $\rho(x, x')$ of V_∞:

$$h(x) \cdot h(x') = \rho(x, x') \tag{6.76}$$

The square or the norm of f is

$$f^2 = f \cdot f = \int \mathrm{d}x\, \mathrm{d}x'\, \rho(x,x') f(x) f(x'). \tag{6.77}$$

It is convenient to introduce notation with upper and lower indices and assume the convention of the integration over the repeated indices. Thus

$$f = f^{(x)} h_{(x)}, \tag{6.78}$$

$$f^2 = f^{(x)} f^{(x')} h_{(x)} \cdot h_{(x')} = f^{(x)} f^{(x')} \rho_{(x)(x')} = f^{(x)} f_{(x)}, \tag{6.79}$$

where (x) is the continuous index.

It is worth stressing here that $h_{(x)} \equiv h(x)$ are abstract elements satisfying the Clifford algebra relation (6.75) for a chosen metric $\rho(x, x') \equiv \rho_{(x)(x')}$. We do not need to worry here about providing an explicit representation of $h(x)$; the requirement that they satisfy the relation (6.75) is all that matters for our purpose[4].

The basis vectors $h_{(x)}$ are generators of Clifford algebra $\mathcal{C}_\infty$ of V_∞. An arbitrary element $F \in \mathcal{C}_\infty$, called a *polyvector*, can be expanded as

$$F = f_0 + f^{(x)} h_{(x)} + f^{(x)(x')} h_{(x)} \wedge h_{(x')} + f^{(x)(x')(x'')} h_{(x)} \wedge h_{(x')} \wedge h_{(x'')} + ..., \tag{6.80}$$

---

[4] Similarly, when introducing the imaginary number i, we do not provide an explicit representation for i. We remain satisfied by knowing that i satisfies the relation $i^2 = -1$.

Extended objects and Clifford algebra 179

i.e.,

$$F = f_0 + \int dx\, f(x)h(x) + \int dx\, dx'\, f(x,x')h(x)h(x')$$
$$+ \int dx\, dx'\, dx''\, f(x,x',x'')h(x)h(x')h(x'') + ...,\quad (6.81)$$

where the wedge product can be replaced by the Clifford product, if $f(x,x')$, $f(x,x',x'')$ are antisymmetric in arguments $x,x',\ldots$.

We see that once we have a space V_∞ of functions $f(x)$ and basis vectors $h(x)$, we also automatically have a larger space of antisymmetric functions $f(x,x')$, $f(x,x',x'')$. This has far reaching consequences which will be discussed in Sec.7.2.

DERIVATIVE WITH RESPECT TO AN INFINITE-DIMENSIONAL VECTOR

The definition (6.49), (6.50) of the derivative can be straightforwardly generalized to the case of polyvector-valued functions $F(f)$ of an infinite-dimensional vector argument f. The derivative in the direction of a vector g is defined according to

$$\left(g \cdot \frac{\partial}{\partial f}\right) F(f) = \lim_{\tau \to 0} \frac{F(f + g\tau) - F(f)}{\tau}. \quad (6.82)$$

If $g = h_{(x')}$ then

$$\left(h_{(x')} \cdot \frac{\partial}{\partial f}\right) F = \frac{\partial F}{\partial f^{(x')}} \equiv \frac{\delta F}{\delta f(x')}$$

$$= \lim_{\tau \to 0} \frac{F(f^{(x)} h_{(x)} + h_{(x')}\tau) - F(f^{(x)} h_{(x)})}{\tau}$$

$$= \lim_{\tau \to 0} \frac{F\left[(f^{(x)} + \tau \delta^{(x)}{}_{(x')})h_{(x')}\right] - F[f^{(x)} h_{(x)}]}{\tau},$$
$$(6.83)$$

where $\delta^{(x)}{}_{(x')} \equiv \delta(x - x')$. This is a definition of the *functional derivative*. A polyvector F can have a definite grade, e.g.,

$$r = 0 \;:\; F(f) = F_0[f^{(x)} h_{(x)}] = \Phi_0[f(x)],$$

$$r = 1 \;:\; F(f) = F^{(x)}[f^{(x)} h_{(x)}]h_{(x)} = \Phi^{(x)}[f(x)]h_{(x)},\quad (6.84)$$

$$r = 2 \;:\; F(f) = F^{(x)(x')}[f^{(x)} h_{(x)}]h_{(x)}h_{(x')} = \Phi^{(x)(x')}[f(x)]h_{(x)}h_{(x')}.$$

For a scalar-valued $F(f)$ the derivative (6.83) becomes

$$\frac{\delta \Phi_0}{\delta f(x')} = \lim_{\tau \to 0} \frac{\Phi_0[f(x) + \delta(x - x')\tau] - \Phi_0[f(x)]}{\tau}, \qquad (6.85)$$

which is the ordinary definition of the functional derivative. Analogously for an arbitrary r-vector valued field $F(f)$.

The derivative with respect to a vector f is then

$$h^{(x)} \left(h_{(x)} \cdot \frac{\partial}{\partial f} \right) F = \frac{\partial F}{\partial f} = h^{(x)} \frac{\partial F}{\partial f^{(x)}}. \qquad (6.86)$$

INCLUSION OF DISCRETE DIMENSIONS

Instead of a single function $f(x)$ of a single argument x we can consider a discrete set of functions $f^a(x^\mu)$, $a = 1, 2, ..., N$, of a multiple argument x^μ, $\mu = 1, 2, ..., n$. These functions can be considered as components of a vector f expanded in terms of the basis vectors $h_a(x)$ according to

$$f = \int \mathrm{d}x \, f^a(x) h_a(x) \equiv f^{a(x)} h_{a(x)}. \qquad (6.87)$$

Basis vectors $h_{a(x)} \equiv h_a(x)$ and components $f^{a(x)} \equiv f^a(x)$ are now labeled by a set of continuous numbers x^μ, $\mu = 1, 2, ..., n$, and by a set of discrete numbers a, such as, e.g., $a = 1, 2, ..., N$. All equations (6.72)–(6.85) considered before can be straightforwardly generalized by replacing the index (x) with $a(x)$.

6.2. DYNAMICAL VECTOR FIELD IN $\mathcal{M}$-SPACE

We shall now reconsider the action (5.76) which describes a membrane coupled to its own metric field in $\mathcal{M}$-space. The first term is the square of the velocity vector

$$\rho_{\mu(\phi)\nu(\phi)} \dot{X}^{\mu(\phi)} \dot{X}^{\nu(\phi')} \equiv \dot{X}^2. \qquad (6.88)$$

The velocity vector can be expanded in terms of $\mathcal{M}$-space basis vectors $h_{\mu(\phi)}$ (which are a particular example of generic basis vectors $h_a(x)$ considered at the end of Sec. 6.1):

$$\dot{X} = \dot{X}^{\mu(\phi)} h_{\mu(\phi)}. \qquad (6.89)$$

Extended objects and Clifford algebra

Basis vectors $h_{\mu(\phi)}$ are not fixed, but they depend on the membrane configuration. We must therefore include in the action not only a kinetic term for $\dot{X}^{\mu(\phi)}$, but also for $h_{\mu(\phi)}$. One possibility is just to rewrite the $\mathcal{M}$-space curvature scalar in terms of $h_{\mu(\phi)}$ by exploiting the relations between the metric and the basis vectors,

$$\rho_{\mu(\phi)\nu(\phi')} = h_{\mu(\phi)} \cdot h_{\nu(\xi')}, \tag{6.90}$$

and then perform variations of the action with respect to $h_{\mu(\phi)}$ instead of $\rho_{\mu(\phi)\nu(\phi')}$.

A more direct procedure is perhaps to exploit the formalism of Sec. 6.1. There we had the basis vectors $\gamma_\mu(x)$ which were functions of the coordinates x^μ. Now we have the basis vectors $h_{\mu(\phi)}[X]$ which are functionals of the membrane configuration X (i.e., of the $\mathcal{M}$-space coordinates $X^{\mu(\phi)}$). The relation (6.8) now generalizes to

$$\partial_{\alpha(\phi')} h_{\beta(\phi'')} = \Gamma^{\mu(\phi)}_{\alpha(\phi')\beta(\phi'')} h_{\mu(\phi)} \tag{6.91}$$

where

$$\partial_{\alpha(\phi')} \equiv \frac{\partial}{\partial X^{\alpha(\phi')}} \equiv \frac{\delta}{\delta X^\alpha(\phi')}$$

is the functional derivative. The commutator of two derivatives gives the curvature tensor in $\mathcal{M}$-space

$$[\partial_{\alpha(\phi')}, \partial_{\beta(\phi'')}] h^{\mu(\phi)} = \mathcal{R}^{\mu(\phi)}{}_{\nu(\bar{\phi})\alpha(\phi')\beta(\phi'')} h^{\nu(\bar{\phi})}. \tag{6.92}$$

The inner product of the left and the right hand side of the above equation with $h_{\nu(\bar{\phi}')}$ gives (after renaming the indices)

$$\mathcal{R}^{\mu(\phi)}{}_{\nu(\bar{\phi})\alpha(\phi')\beta(\phi'')} = \left([\partial_{\alpha(\phi')}, \partial_{\beta(\phi'')}] h^{\mu(\phi)}\right) \cdot h_{\nu(\bar{\phi})}. \tag{6.93}$$

The Ricci tensor is then

$$\mathcal{R}_{\nu(\bar{\phi})\beta(\phi'')} = \mathcal{R}^{\mu(\phi)}{}_{\nu(\bar{\phi})\mu(\phi)\beta(\phi'')} = \left([\partial_{\mu(\phi)}, \partial_{\beta(\phi'')}] h^{\mu(\phi)}\right) \cdot h_{\nu(\bar{\phi})}, \tag{6.94}$$

and the curvature scalar is

$$\begin{aligned}
\mathcal{R} &= \rho^{\nu(\bar{\phi})\beta(\phi'')} \mathcal{R}_{\nu(\bar{\phi})\beta(\phi'')} \\
&= \left([\partial_{\mu(\phi)}, \partial_{\nu(\phi')}]\right) \cdot h^{\nu(\phi')} \\
&= \left(\partial_{\mu(\phi)} \partial_{\nu(\phi')} h^{\mu(\phi)}\right) \cdot h^{\nu(\phi')} - \left(\partial_{\nu(\phi')} \partial_{\mu(\phi)} h^{\mu(\phi)}\right) \cdot h^{\nu(\phi')}.
\end{aligned} \tag{6.95}$$

A possible action is then

$$I[X^{\mu(\phi)}, h_{\mu(\phi)}] = \int \mathcal{D}X \sqrt{|\rho|} \left(\dot{X}^2 + \frac{\epsilon}{16\pi} \mathcal{R} \right), \qquad (6.96)$$

where ρ is the determinant of $\mathcal{M}$-space metric, and where $\mathcal{R}$ and ρ are now expressed in terms of $h_{\mu(\phi)}$. Variation of (6.96) with respect to $h_{\alpha(\phi)}$ gives their equations of motion. In order to perform such a variation we first notice that $\rho = \det \rho_{\mu(\phi)\nu(\phi')}$, in view of the relation (6.90), is now a function of $h_{\alpha(\phi)}$. Differentiation of $\sqrt{|\rho|}$ with respect to a vector $h_{\alpha(\phi)}$ follows the rules given in Sec. 6.1:

$$\frac{\partial \sqrt{|\rho|}}{\partial h^{\alpha(\phi)}} = \frac{\partial \sqrt{|\rho|}}{\partial \rho^{\mu(\phi')\nu(\phi'')}} \frac{\partial \rho^{\mu(\phi')\nu(\phi'')}}{\partial h^{\alpha(\phi)}}$$

$$= -\tfrac{1}{2}\sqrt{|\rho|}\rho_{\mu(\phi')\nu(\phi'')} (\delta^{\mu(\phi')}{}_{\alpha(\phi)} h^{\nu(\phi'')} + \delta^{\nu(\phi'')}{}_{\alpha(\phi)} h^{\mu(\phi')})$$

$$= -\sqrt{|\rho|} h_{\alpha(\phi)}. \qquad (6.97)$$

Since the vectors $h_{\alpha(\phi)}$ are functionals of the membrane's configuration $X^{\mu(\phi)}$, instead of the derivative we take the functional derivative

$$\frac{\delta \sqrt{|\rho[X]|}}{\delta h^{\alpha(\phi)}[X']} = -\sqrt{|\rho[X]|} h_{\alpha(\phi)}[X] \delta^{(\mathcal{M})}(X - X'), \qquad (6.98)$$

where

$$\delta^{(\mathcal{M})}(X - X') \equiv \prod_{\mu(\phi)} (X^{\mu(\phi)} - X'^{\mu(\phi)}) \qquad (6.99)$$

is the δ-functional in $\mathcal{M}$-space.

Functional derivative of $\mathcal{R}$ with respect to $h_{\alpha(\phi)}[X]$ gives

$$\frac{\delta \mathcal{R}[X]}{\delta h^{\alpha(\phi)}[X']} = \left([\partial_{\mu(\phi')}, \partial_{\nu(\phi'')}] \delta^{\mu(\phi')}{}_{\alpha(\phi)} \delta^{(\mathcal{M})}(X - X') \right) h^{\nu(\phi'')}[X]$$

$$+ [\partial_{\mu(\phi')}, \partial_{\nu(\phi'')}] h^{\mu(\phi')}[X] \delta^{\nu(\phi'')}{}_{\alpha(\phi)} \delta^{(\mathcal{M})}(X - X'). \qquad (6.100)$$

For the velocity term we have

$$\frac{\delta}{\delta h^{\alpha(\phi)}[X']} \left(\dot{X}_{\mu(\phi')} h^{\mu(\phi')} \dot{X}_{\nu(\phi'')} h^{\nu(\phi'')} \right)$$

$$= 2 \dot{X}_{\alpha(\phi)} \dot{X}_{\nu(\phi'')} h^{\nu(\phi'')} \delta^{(\mathcal{M})}(X - X'). \qquad (6.101)$$

We can now insert eqs. (6.97)–(6.101) into

$$\frac{\delta I}{\delta h^{\alpha(\phi)}[X']} = \int \mathcal{D}X \frac{\delta}{\delta h^{\alpha(\phi)}[X']} \left[\sqrt{|\rho|} \left(\dot{X}^2 + \frac{\epsilon}{16\pi} \mathcal{R} \right) \right] = 0. \qquad (6.102)$$

We obtain

$$\dot{X}_{\alpha(\phi)}\dot{X}_{\nu(\phi')}h^{\nu(\phi')} - \tfrac{1}{2}h_{\alpha(\phi)}\dot{X}^2$$
$$+ \frac{\epsilon}{16\pi}\left(-\tfrac{1}{2}h_{\alpha(\phi)}\mathcal{R} + [\partial_{\nu(\phi')},\partial_{\alpha(\phi'')}]h^{\nu(\phi')}\right) = 0. \quad (6.103)$$

These are the equations of "motion" for the variables $h_{\mu(\phi)}$. The equations for $X^{\mu(\phi)}$ are the same equations (5.80).

After performing the inner product of eq. (6.103) with a basis vector $h_{\beta(\phi')}$ we obtain the $\mathcal{M}$-space Einstein equations (5.81). This justifies use of $h_{\mu(\phi)}$ as dynamical variables, since their equations of motion contain the equations for the metric $\rho_{\mu(\phi)\nu(\phi')}$.

DESCRIPTION WITH THE VECTOR FIELD IN SPACETIME

The set of $\mathcal{M}$-space basis vectors $h_{\mu(\phi)}$ is an arbitrary solution to the dynamical equations (6.103). In order to find a connection with the usual theory which is formulated, not in $\mathcal{M}$-space, but in a finite-dimensional (spacetime) manifold V_N, we now assume a particular Ansatz for $h^{\mu(\phi)}$:

$$h^{\mu(\phi)} = h^{(\phi)}\gamma^{\mu}(\phi), \quad (6.104)$$

where

$$h^{(\phi)} \cdot h^{(\phi')} = \frac{\sqrt{\dot{X}^2}}{\kappa\sqrt{|f|}}\delta(\phi - \phi') \quad (6.105)$$

and

$$\gamma^{\mu}(\phi) \cdot \gamma^{\nu}(\phi) = g^{\mu\nu}. \quad (6.106)$$

Altogether the above Ansatz means that

$$h^{\mu(\phi)} \cdot h^{\nu(\phi')} = \frac{\sqrt{\dot{X}^2}}{\sqrt{|f|}} g^{\mu\nu}(\phi)\delta(\phi - \phi'). \quad (6.107)$$

Here $g^{\mu\nu}(\phi)$ is *the proto-metric*, and $\gamma^{\mu}(\phi)$ are *the proto-vectors* of spacetime. The symbol f now meens

$$f \equiv \det f_{ab}, \quad f_{ab} \equiv \partial_a X^{\mu}\partial_b X^{\nu}\gamma_{\mu}\gamma_{\nu} = \partial_a X_{\mu}\partial_b X_{\nu}\gamma^{\mu}\gamma^{\nu}. \quad (6.108)$$

Here the $\mathcal{M}$-space basis vectors are factorized into the vectors $h^{(\phi)}$, $\phi = (\phi^A, k)$, $\phi^A \in [0, 2\pi]$, $k = 1, 2, ..., Z$, which are independent for all values of μ, and are functions of parameter ϕ. Loosely speaking, $h(\phi)$ bear the task of being basis vectors of the infinite-dimensional part (index (ϕ) of $\mathcal{M}$-space, while $\gamma^{\mu}(\phi)$ are basis vectors of the finite-dimensional part (index μ)

of $\mathcal{M}$-space. As in Sec. 6.1, γ^μ here also are functions of the membrane's parameters $\phi = (\phi^A, k)$.

The functional derivative of $h^{\mu(\phi)}$ with respect to the membrane coordinates $X^{\mu(\phi)}$ is

$$\partial_{\nu(\phi')} h^{\mu(\phi)} = \frac{\delta h(\phi)}{\delta X^\nu(\phi')} \gamma^\mu(\phi) + h(\phi) \frac{\delta \gamma^\mu(\phi)}{\delta X^\nu(\phi')}. \tag{6.109}$$

Let us assume a particular case where

$$[\partial_{\mu(\phi)}, \partial_{\nu(\phi')}] \gamma^\mu(\phi'') = \left[\frac{\partial}{\partial X^\mu}, \frac{\partial}{\partial X^\nu} \right] \gamma^\mu(\phi'') \delta(\phi - \phi'') \delta(\phi' - \phi'')$$

$$\neq 0 \tag{6.110}$$

and

$$[\partial_{\mu(\phi)}, \partial_{\nu(\phi')}] h(\phi'') = 0. \tag{6.111}$$

Inserting the relations (6.104)–(6.111) into the action (6.96), and omitting the integration over $\mathcal{D}X$ we obtain

$$\epsilon \mathcal{R} = \epsilon \int \frac{\sqrt{\dot{X}^2}}{\kappa \sqrt{|f(\phi)|}} \delta^2(0) \mathrm{d}\phi \left(\left[\frac{\partial}{\partial X^\mu}, \frac{\partial}{\partial X^\nu} \right] \gamma^\mu(\phi) \right) \cdot \gamma^\nu(\phi)$$

$$= \epsilon \int \frac{\sqrt{\dot{X}^2}}{\kappa \sqrt{|f(\phi)|}} \delta^2(0) \mathrm{d}\phi$$

$$\times \left(\left[\frac{\partial}{\partial X^\mu}, \frac{\partial}{\partial X^\nu} \right] \gamma^\mu(\phi) \right) \cdot \gamma^\nu(\phi) \frac{\delta(x - X(\phi))}{\sqrt{|g|}} \sqrt{|g|} \, \mathrm{d}^N x$$

$$= \frac{1}{G} \int \mathrm{d}^N x \sqrt{|g|} \, \tilde{\mathcal{R}}, \tag{6.112}$$

where we have set

$$\epsilon \int \mathrm{d}\phi \, \delta^2(0) \frac{\sqrt{\dot{X}^2}}{\kappa \sqrt{|f(\phi)|}} \left(\left[\frac{\partial}{\partial X^\mu}, \frac{\partial}{\partial X^\nu} \right] \gamma^\mu(\phi) \right) \cdot \gamma^\nu(\phi) \frac{\delta(x - X(\phi))}{\sqrt{|g|}}$$

$$= \frac{1}{G} \tilde{\mathcal{R}}(x), \tag{6.113}$$

G being the gravitational constant. The expression $\tilde{\mathcal{R}}(x)$ is defined formally at all points x, but because of $\delta(x - X(\phi))$ it is actually different from zero only on the set of membranes. If we have a set of membranes filling spacetime, then $\tilde{\mathcal{R}}(x)$ becomes a continuous function of x:

$$\tilde{\mathcal{R}}(x) = ([\partial_\mu, \partial_\nu] \gamma^\mu(x)) \cdot \gamma^\nu(x) = R(x) \tag{6.114}$$

Extended objects and Clifford algebra 185

which is actually a *Ricci scalar*.
For the first term in the action (6.96) we obtain

$$h_{\mu(\phi)}h_{\nu(\phi')}\dot X^{\mu(\phi)}\dot X^{\nu(\phi')}$$

$$= \int \frac{\kappa\sqrt{|f|}}{\sqrt{\dot X^2}}\,\mathrm{d}\phi\,\gamma_\mu(\phi)\gamma_\nu(\phi)\dot X^\mu(\phi)\dot X^\nu(\phi)$$

$$= \int \mathrm{d}^n\phi\,\frac{\kappa\sqrt{|f|}}{\sqrt{\dot X^2}}\,\gamma_\mu(x)\gamma_\nu(x)\dot X^\mu(\phi)\dot X^\nu(\phi)\,\delta\left(x-X(\phi)\right)\,\mathrm{d}^N x$$

$$= \int \mathrm{d}^n\phi\,\frac{\kappa\sqrt{|f|}}{\sqrt{\dot X^2}}\,g_{\mu\nu}(x)\dot X^\mu(\phi)\dot X^\nu(\phi)\,\delta\left(x-X(\phi)\right)\,\mathrm{d}^N x. \quad (6.115)$$

Altogether we have

$$I[X^\mu(\phi),\gamma^\mu(x)] = \kappa\int \mathrm{d}\phi\,\sqrt{|f|}\sqrt{\gamma_\mu(x)\gamma_\nu(x)\dot X^\mu(\phi)\dot X^\nu(\phi)}\,\delta^N\left(x-X(\phi)\right)\,\mathrm{d}^N x$$

$$+\frac{1}{16\pi G}\int \mathrm{d}^N x\sqrt{|g|}\,([\partial_\mu,\partial_\nu]\gamma^\mu(x))\cdot\gamma^\nu(x). \quad (6.116)$$

This is an action for the spacetime vector field $\gamma^\mu(x)$ in the presence of a membrane configuration filling spacetime. It was derived from the action (6.96) in which we have omitted the integration over $\mathcal{D}X\sqrt{\rho}$.

Since $\gamma_\mu\cdot\gamma_\nu = g_{\mu\nu}$, and since according to (6.13)

$$([\partial_\mu,\partial_\nu]\gamma^\mu)\cdot\gamma^\nu = R, \quad (6.117)$$

the action (6.116) is equivalent to

$$I[X^\mu(\phi),g_{\mu\nu}] = \kappa\int \mathrm{d}\phi\,\sqrt{|f|}\sqrt{\dot X^2}\,\delta\left(x-X(\phi)\right)\,\mathrm{d}^N x$$

$$+\frac{1}{16\pi G}\int \mathrm{d}^N x\sqrt{|g|}\,R \quad (6.118)$$

which is an action for the gravitational field $g_{\mu\nu}$ in the presence of membranes.

Although (6.118) formally looks the same as the usual gravitational action in the presence of matter, there is a significant difference. In the conventional general relativity the matter part of the action may vanish and we thus obtain the Einstein equations in vacuum. On the contrary, in the theory based on $\mathcal{M}$-space, the metric of $\mathcal{M}$-space is intimately connected to the existence of a membrane configuration. Without membranes there is no $\mathcal{M}$-space and no $\mathcal{M}$-space metric. When considering the $\mathcal{M}$-space action (5.76) or (6.96) from the point of view of an effective spacetime (defined in our case by the Ansatz (6.104)–(6.107), we obtain the spacetime action (6.118) in which the matter part cannot vanish. There is always present a set of membranes filling spacetime. Actually, the points of spacetime are identified with the points on the membranes.

The need to fill spacetime with a reference fluid (composed of a set of reference particles) has been realized recently by Rovelli [26], following an earlier work by DeWitt [25]. According to Rovelli and DeWitt, because of the Einstein "hole argument" [42], spacetime points cannot be identified at all. This is a consequence of the invariance of the Einstein equations under active diffeomorphisms. One can identify spacetime points if there exists a material reference fluid with respect to which spacetime points are identified.

We shall now vary the action (6.116) with respect to the vector field $\gamma^\alpha(x)$:

$$\frac{\delta I}{\delta \gamma^\alpha(x)} = \int \mathrm{d}\phi \, \frac{\sqrt{|f|}}{\sqrt{\dot{X}^2}} \dot{X}_\alpha \dot{X}_\nu \, \gamma^\nu \delta\left(x - X(\phi)\right) \tag{6.119}$$

$$+ \int \mathrm{d}x' \left[\frac{\delta \sqrt{|g(x')|}}{\delta \gamma^\alpha(x)} R(x') + \sqrt{|g(x')|} \frac{\delta R(x')}{\delta \gamma^\alpha(x)} \right].$$

For this purpose we use

$$\frac{\delta \gamma^\mu(x)}{\delta \gamma^\nu(x')} = \delta^\mu{}_\nu (x - x') \tag{6.120}$$

and

$$\frac{\delta \sqrt{|g(x)|}}{\delta \gamma^\nu(x')} = - \sqrt{|g(x)|} \, \gamma_\nu(x) \delta(x - x'). \tag{6.121}$$

In (6.121) we have taken into account that $g \equiv \det g_{\mu\nu}$ and $g_{\mu\nu} = \gamma_\mu \cdot \gamma_\nu$. Using (6.120), we have for the gravitational part

$$\frac{\delta I_g}{\delta \gamma^\alpha(x)} = \frac{1}{16\pi G} \int \mathrm{d}^N x' \sqrt{|g(x')|} \bigg[-\gamma_\alpha(x') R(x') \delta(x - x')$$

$$+ \left([\partial'_\mu, \partial'_\nu] \delta^\mu{}_\alpha \delta(x - x')\right) \gamma^\nu(x') + [\partial'_\mu, \partial'_\nu] \gamma^\mu(x') \delta^\nu{}_\alpha \delta(x - x') \bigg]$$

$$= \frac{1}{16\pi} \sqrt{|g|} \left(-\gamma_\alpha R + 2[\partial_\mu, \partial_\alpha]\gamma^\mu\right). \tag{6.122}$$

The equations of motion for $\gamma_\alpha(x)$ are thus

$[\partial_\mu, \partial_\alpha]\gamma^\mu - \frac{1}{2}R\gamma_\alpha$

$$= -8\pi G \int \mathrm{d}\phi \sqrt{|f|} \sqrt{\dot{X}^2} \left(\frac{\dot{X}_\alpha \dot{X}_\nu}{\dot{X}^2} + \partial^a X_\alpha \partial_a X_\nu \right)$$

$$\times \gamma^\nu \frac{\delta(x - X(\phi))}{\sqrt{|g|}}, \qquad (6.123)$$

where now $\dot{X}^2 \equiv \dot{X}^\mu \dot{X}_\mu$.

After performing the inner product with γ_β the latter equations become the Einstein equations

$$R_{\alpha\beta} - \tfrac{1}{2}R g_{\alpha\beta} = -8\pi G T_{\alpha\beta}, \qquad (6.124)$$

where

$$R_{\alpha\beta} = ([\partial_\mu, \partial_\alpha])\gamma^\mu \cdot \gamma_\beta \qquad (6.125)$$

and

$$T_{\alpha\beta} = \kappa \int \mathrm{d}\phi \sqrt{|f|} \sqrt{\dot{X}^2} \left(\frac{\dot{X}_\alpha \dot{X}_\beta}{\dot{X}^2} + \partial^a X_\alpha \partial_a X_\beta \right) \frac{\delta(x - X(\phi))}{\sqrt{|g|}} \qquad (6.126)$$

is the stress–energy tensor of the membrane configuration. It is the ADW split version of the full stress–energy tensor

$$T_{\mu\nu} = \kappa \int \mathrm{d}\phi \, (\det \partial_A X^\alpha \partial_B X_\alpha)^{1/2} \, \partial^A X_\mu \partial_A X_\nu \, \delta(x - X(\phi)). \qquad (6.127)$$

The variables $\gamma^\mu(x)$ appear much easier to handle than the variables $g_{\mu\nu}$. The expressions for the curvature scalar (6.117) and the Ricci tensor (6.125) are very simple, and it is easy to vary the action (6.116) with respect to $\gamma(x)$.[5]

We should not forget that the matter stress–energy tensor on the right hand side of the Einstein equations (6.124) is present everywhere in spacetime and it thus represents a sort of background matter[6] whose origin is in the original $\mathcal{M}$-space formulation of the theory. The ordinary matter is then expected to be present in addition to the background matter. This will be discussed in Chapter 8.

---

[5] At this point we suggest the interested reader study the Ashtekar variables [74], and compare them with $\gamma(x)$.

[6] It is tempting to speculate that this background is actually *the hidden mass* or *dark matter* postulated in astrophysics and cosmology [73].

6.3. FULL COVARIANCE IN THE SPACE OF PARAMETERS ϕ^A

So far we have exploited the fact that, according to (4.21), a membrane moves as a point particle in an infinite-dimensional $\mathcal{M}$-space. For a special choice of $\mathcal{M}$-space metric (5.4) which involves the membrane velocity we obtain equations of motion which are identical to the equations of motion of the conventional constrained membrane, known in the literature as the *p-brane*, or simply the *brane*[7]. A moving brane sweeps a surface V_n which incorporates not only the brane parameters ξ^a, $a = 1, 2, ..., p$, but also an extra, time-like parameter τ. Altogether there are $n = p + 1$ parameters $\phi^A = (\tau, \xi^a)$ which denote a point on the surface V_n. The latter surface is known in the literature under names such as *world surface*, *world volume* (now the most common choice) and *world sheet* (my favorite choice).

Separating the parameter τ from the rest of the parameters turns out to be very useful in obtaining *the unconstrained membrane* out of the Clifford algebra based polyvector formulation of the theory.

On the other hand, when studying interactions, the separate treatment of τ was a nuisance, therefore in Sec. 5.1 we switched to a description in terms of the variables $X^\mu(\tau, \xi^a) \equiv X^\mu(\phi^A)$ which were considered as $\mathcal{M}$-space coordinates $X^\mu(\phi^A) \equiv X^\mu(\phi)$. So $\mathcal{M}$-space was enlarged from that described by coordinates $X^\mu(\xi)$ to that described by $X^\mu(\phi)$. In the action and in the equations of motion there occurred the $\mathcal{M}$-space velocity vector $\dot{X}^\mu(\phi) \equiv \partial X^\mu(\phi)/\partial \tau$. Hence manifest covariance with respect to reparametrizations of ϕ^A was absent in our formulation. In my opinion such an approach was good for introducing the theory and fixing the development of the necessary concepts. This is now to be superceded. We have learnt enough to be able to see a way how a fully reparametrization covariant theory should possibly be formulated.

DESCRIPTION IN SPACETIME

As a first step I now provide a version of the action (6.116) which is invariant under arbitrary reparametrizations $\phi^A \to \phi'^A = f^A(\tau)$. In the form as it stands (6.116) (more precisely, its "matter" term) is invariant under reparametrizations of ξ^a and τ separately. A fully invariant action

---

[7] For this reason I reserve the name *p-brane* or *brane* for the extended objetcs described by the conventional theory, while the name *membrane* stands for the extended objects of the more general, $\mathcal{M}$-space based theory studied in this book (even if the same name "membrane" in the conventional theory denotes 2-branes, but this should not cause confusion).

Extended objects and Clifford algebra

(without a kinetic term for the gravitational field) is

$$I[X^\mu] = \kappa \int d^n\phi \, (\det \partial_A X^\mu \partial_B X^\nu g_{\mu\nu})^{1/2}. \tag{6.128}$$

An equivalent form (considered in Sec. 4.2) is

$$I[X^\mu, \gamma^{AB}] = \frac{\kappa}{2} \int d^n\phi \, \sqrt{|\gamma|} (\gamma^{AB} \partial_A X^\mu \partial_B X^\nu g_{\mu\nu} + 2 - n), \tag{6.129}$$

where X^μ and γ^{AB} are to be varied independently. Remember that the variation of (6.129) with respect to γ^{AB} gives the relation

$$\gamma_{AB} = \partial_A X^\mu \partial_B X_\mu, \tag{6.130}$$

which says that γ_{AB} is the induced metric on V_n.

DESCRIPTION IN TERMS OF VECTORS e^A

Let us now take into account the basic relation

$$\gamma^{AB} = e^A \cdot e^B, \tag{6.131}$$

where e^A, $A = 1, 2, ..., n$, form a set of basis vectors in an n-dimensional space V_n whose points are described by coordinates ϕ^A. After inserting (6.131) into (6.129) we obtain the action

$$I[X^\mu, e^A] = \frac{\kappa}{2} \int d^n\phi \, \sqrt{|\gamma|} (e^A \partial_A X^\mu e^B \partial_B X^\nu g_{\mu\nu} + 2 - n) \tag{6.132}$$

which is a functional of $X^\mu(\phi^A)$ and e^A. The vectors e^A now serve as the auxiliary variables. The symbol $\gamma \equiv \det \gamma_{AB}$ now depends on e_A. We have

$$\frac{\partial \sqrt{|\gamma|}}{\partial e^A} = \frac{\partial \sqrt{|\gamma|}}{\partial \gamma^{CD}} \frac{\partial \gamma^{CD}}{\partial e^A} = -\tfrac{1}{2}\sqrt{|\gamma|}\, \gamma_{CD}\, \tfrac{1}{2}(\delta^C{}_A e^D + \delta^D{}_A e^C)$$

$$= -\tfrac{1}{2}\sqrt{|\gamma|}\, e_A. \tag{6.133}$$

Variation of (6.132) with respect to e^C gives

$$-\frac{1}{2} e_C (e^A \partial_A X^\mu e^B \partial_B X_\mu + 2 - n) + \partial_C X^\mu \partial_D X_\mu e^D = 0. \tag{6.134}$$

Performing the inner product with e^C and using $e^C \cdot e_C = n$ we find

$$e^A \partial_A X^\mu e^B \partial_B X_\mu = n. \tag{6.135}$$

Eq. (6.134) then becomes

$$e_C = \partial_C X^\mu \partial_D X_\mu e^D, \tag{6.136}$$

or
$$e_D(\delta^D{}_C - \partial_C X^\mu \partial_D X_\mu) = 0. \tag{6.137}$$

Forming the inner product of the left hand side and the right hand side of eq. (6.136) with e_A we obtain

$$e_C \cdot e_A = \partial_C X^\mu \partial_D X_\mu\, e^D \cdot e_A. \tag{6.138}$$

Since $e_C \cdot e_A = \gamma_{CA}$ and $e^D \cdot e_A = \delta^D{}_A$ the equation (6.138) becomes (after renaming the indices)

$$\gamma_{AB} = \partial_A X^\mu \partial_B X_\mu. \tag{6.139}$$

This is the same relation as resulting when varying $I[X^\mu, \gamma^{AB}]$ (eq. (6.129)) with respect to γ^{AB}. So we have verified that $I[X^\mu, e^A]$ is equivalent to $I[X^\mu, \gamma^{AB}]$.

In the action (6.132) $e^A(\phi)$ are auxiliary fields serving as Lagrange multipliers. If we add a kinetic term for e^A then e^A become dynamical variables:

$$I[X^\mu, e^A] = \frac{\kappa}{2} \int d^n\phi\, \sqrt{|\gamma|}(e^A \partial_A X^\mu\, e^B \partial_B X_\mu + 2 - n)$$

$$+ \frac{1}{16\pi G^{(n)}} \int d^n\phi\, \sqrt{|\gamma|}\, R^{(n)} \tag{6.140}$$

Here, according to the theory of Sec. 6.2,

$$R^{(n)} = \bigl([\partial_A, \partial_B]e^A\bigr) \cdot e^B. \tag{6.141}$$

Variation of (6.140) with respect to e^C leads to the Einstein equations in V_n which can be cast in the form

$$\gamma_{AB} = \partial_A X^\mu \partial_B X_\mu + \frac{1}{16\pi G^{(n)}} R^{(n)}_{AB}. \tag{6.142}$$

In general, therefore, γ_{AB}, satisfying (6.140), is not the induced metric on V_n.

DESCRIPTION IN TERMS OF γ^μ

In eqs. (6.132)–(6.140) we have written the metric of V_n as $\gamma^{AB} = e^A \cdot e^B$ and considered the action as a functional of the vectors e^A, but we have kept the metric $g_{\mu\nu}$ of the embedding space V_N fixed. Now let us write

$$g_{\mu\nu} = \gamma_\mu \cdot \gamma_\nu, \tag{6.143}$$

where γ_μ, $\mu = 1, 2, ..., N$, form a complete set of basis vectors in V_N, and write an action

$$I[X^\mu, \gamma^{AB}, \gamma^\mu]$$
$$= \frac{\kappa}{2} \int d^n\phi \sqrt{|\gamma|} (\gamma^{AB}\partial_A X^\mu \partial_B X^\nu \gamma_\mu \gamma_\nu + 2 - n)\delta^N(x - X(\phi))d^N x$$
$$+ \frac{1}{16\pi G^{(N)}} \int d^N x \sqrt{|g|}\, R^{(N)}. \tag{6.144}$$

Variation of the latter action with respect to γ^α goes along similar lines to eqs. (6.119)–(6.122). We obtain

$$[\partial_\mu, \partial_\alpha]\gamma^\mu - \frac{1}{2}R^{(N)}\gamma_\alpha \tag{6.145}$$

$$= -8\pi G^{(N)} \int d^n\phi \sqrt{|\gamma|}\, \partial^A X_\alpha \partial_A X_\nu \gamma^\nu \frac{\delta(x - X(\phi))}{\sqrt{|g|}}.$$

These equations of motion for γ^μ are a fully reparametrization covariant form of eq. (6.123). The inner product with γ_β gives the Einstein equations (6.124) with

$$T_{\alpha\beta} = \kappa \int d^n\phi \sqrt{|\gamma|}\, \partial^A X_\alpha \partial_A X_\beta \frac{\delta(x - X(\phi))}{\sqrt{|g|}}, \tag{6.146}$$

which is a fully reparametrization covariant form of the stress–energy tensor (6.126).

DESCRIPTION IN TERMS OF e^A AND γ^μ

Combining the descriptions as given in eqs. (6.140) and (6.144) we obtain the following action:

$$I[X^\mu, e^A, \gamma^\mu] = \int d\phi \sqrt{|\gamma|} \left(\frac{\kappa}{2} e^A e^B \partial_A X^\mu \partial_B X^\nu \gamma_\mu \gamma_\nu + \frac{\kappa}{2}(2-n) \right.$$
$$\left. + \frac{1}{16\pi G^{(n)}} R^{(n)} \right) d^N x\, \delta(x - X(\phi))$$
$$+ \frac{1}{16\pi G^{(N)}} \int d^N x \sqrt{|g|} R^{(N)}. \tag{6.147}$$

The first term in (6.147) can be written as

$$(e^A \cdot e^B + e^A \wedge e^B)(\gamma_\mu \cdot \gamma_\nu + \gamma_\mu \wedge \gamma_\nu)\partial_A X^\mu \partial_B X^\nu, \tag{6.148}$$

which gives the usual scalar part and a multivector part:

$$(e^A \cdot e^B)\partial_A X^\mu \partial_B X^\nu (\gamma_\mu \cdot \gamma_\nu) + (e^A \wedge e^B)\partial_A X^\mu \partial_B X^\nu (\gamma_\mu \wedge \gamma_\nu), \tag{6.149}$$

Now, considering *the string* (worldsheet dimension $n = 2$), we have $e^A \wedge e^B = I_{(2)}\epsilon^{AB}$, where $I_{(2)}$ is the pseudoscalar unit, $I_{(2)} = e^1 e^2$, in 2 dimensions, and ϵ^{AB} is the antisymmetric tensor density in 2 dimensions. So we have

$$(e^A \wedge e^B)(\gamma_\mu \wedge \gamma_\nu) = \epsilon^{AB} I_{(2)}(\gamma_\mu \wedge \gamma_\nu) \tag{6.150}$$

$$\langle e^1 e^2 (\gamma_\mu \wedge \gamma_\nu) \rangle_0 = e^1 \left[(e^2 \cdot \gamma_\mu)\gamma_\nu - (e^2 \cdot \gamma_\nu)\gamma_\mu \right] \tag{6.151}$$

$$= (e^2 \cdot \gamma_\mu)(e^1 \cdot \gamma_\nu) - (e^2 \cdot \gamma_\nu)(e^1 \cdot \gamma_\mu)$$

$$\equiv B_{\mu\nu} \tag{6.152}$$

where $\langle \ \rangle_0$ means the scalar part. For the string case the antisymmetric field $B_{\mu\nu}$ thus naturally occurs in the action (6.147) whose Lagrangian contains the terms

$$\gamma^{AB} \partial_A X^\mu \partial_B X^\nu g_{\mu\nu} + \epsilon^{AB} \partial_A X^\mu \partial_B X^\nu B_{\mu\nu} , \tag{6.153}$$

well known in the usual theory of strings. However, in the theory of p-branes (when $n = p + 1$ is arbitrary), instead of a 2-form field there occurs an n-form field and the action contains the term

$$\epsilon^{A_1 A_2 \ldots A_n} \partial_{A_1} X^{\mu_1} \ldots \partial_{A_n} X^{\mu_n} B_{\mu_1 \ldots \mu_n} . \tag{6.154}$$

If we generalize the action (6.147) to a corresponding *polyvector action*

$$I[X^\mu, e^A, \gamma^\mu]$$

$$= \frac{\kappa}{2} \int d^n \phi \sqrt{|\gamma|} \Big(e^{A_1} e^{A_2} \partial_{A_1} X^{\mu_1} \partial_{A_2} X^{\mu_2} \gamma_{\mu_1} \gamma_{\mu_2}$$

$$+ e^{A_1} e^{A_2} e^{A_3} \partial_{A_1} X^{\mu_1} \partial_{A_2} X^{\mu_2} \partial_{A_3} x^{\mu_3} \gamma_{\mu_1} \gamma_{\mu_2} \gamma_{\mu_3}$$

$$+ \ldots + e^{A_1} e^{A_2} \ldots e^{A_n} \partial_{A_1} X^{\mu_1} \partial_{A_2} x^{\mu_2} \ldots \partial_{A_n} X^{\mu_n} \gamma_{\mu_1} \gamma_{\mu_2} \ldots \gamma_{\mu_n}$$

$$+ (2 - n) + \frac{1}{16\pi G^{(n)}} \frac{2}{\kappa} R^{(n)} \Big) d^N x \, \delta(x - X(\phi))$$

$$+ \frac{1}{16 G^{(N)}} \int d^N x \sqrt{|g|} \, R^{(N)} , \tag{6.155}$$

then in addition to the other multivector terms we also obtain a term (6.154), since the highest rank multivector term gives

Extended objects and Clifford algebra 193

$$\langle(e^{A_1} \wedge ... \wedge e^{A_n})(\gamma_{\mu_1} \wedge ... \wedge \gamma_{\mu_n})\rangle_0 = \langle \epsilon^{A_1...A_n} I_{(n)}(\gamma_{\mu_1} \wedge ... \wedge \gamma_{\mu_n})\rangle_0$$
$$\equiv \epsilon^{A_1...A_n} B_{\mu_1...\mu_n}. \quad (6.156)$$

Alternatively, for the *p*-brane Lagrangian we can take

$$\mathcal{L}(X^\mu) = \kappa \sqrt{|\gamma|} \left[\left\langle \sum_{r=1}^{2n} \prod_{i=1}^{r} e^{A_i} \partial_{A_i} X^{\mu_i} \gamma_{\mu_i} \right\rangle_0 \right]^{1/2} \quad (6.157)$$

The latter expression is obtained if we square the brane part of the Lagrangian (6.155), take the scalar part, and apply the square root.

Although the action (6.155) looks quite impressive, it is not the most general polyvector action for a membrane. Some preliminary suggestions of how to construct such an action will now be provided.

POLYVECTOR ACTION

Polyvector action in terms of e^A. The terms $e^A \partial_A X^\mu$, $\mu = 1, 2, ..., N$ which occur in the action (6.140) can be generalized to *world sheet polyvectors*:

$$V^\mu = s^\mu + e^A \partial_A X^\mu + e^{A_1} \wedge e^{A_2} v^\mu_{A_1 A_2} + ... + e^{A_1} \wedge ... \wedge e^{A_n} . v^\mu_{A_1....A_n} \quad (6.158)$$

The action can then be

$$I[X^\mu, e^A, g_{\mu\nu}] = \frac{\kappa}{2} \int d^n \phi \sqrt{|\gamma|} V^\mu V^\nu g_{\mu\nu} + I[e^A] + I[g_{\mu\nu}]. \quad (6.159)$$

The kinetic term for e^A includes, besides $R^{(n)}$, the term with $(2-n)$ and possible other terms required for consistency.

Polyvector action in terms of γ^μ. Similarly we can generalize the terms $\partial_A X^\mu \gamma_\mu$, $A = 1, 2, ..., n$, in eq. (6.144), so as to become spacetime polyvectors:

$$C_A = c_A + \partial_A X^\mu \gamma_\mu + c_A^{\mu_1 \mu_2} \gamma_{\mu_1} \wedge \gamma_{\mu_2} + ... + c_A^{\mu_1...\mu_N} \gamma_{\mu_1} \wedge \gamma_{\mu_1} \wedge ... \wedge \gamma_{\mu_N} \quad (6.160)$$

A corresponding action is then

$$I[X^\mu, \gamma^{AB}, \gamma^\mu] = \tfrac{1}{2} \int d^n \phi \sqrt{|\gamma|} \gamma^{AB} C_A C_B + I[\gamma^{AB}] + I[\gamma^\mu]. \quad (6.161)$$

where

$$I[\gamma^\mu] = \frac{1}{16\pi G^{(N)}} \int d^N x \sqrt{|g|} \left([\partial_\mu, \partial_\nu] \gamma^\mu(x)\right) \cdot \gamma^\nu(x). \quad (6.162)$$

Polyvector action in terms of e^A and γ^μ. From the expression V^μ in eq. (6.158) we can form another expression

$$V^\mu \gamma_\mu = s^\mu \gamma_\mu + e^A \partial_A X^\mu \gamma_\mu + e^{A_1} \wedge e^{A_2} v^\mu_{A_1 A_2} \gamma_\mu + ...$$
$$+ e^{A_1} \wedge ... \wedge e^{A_n} v^\mu_{A_1....A_n} \gamma_\mu. \tag{6.163}$$

The latter expression can be generalized accoridng to

$$V = s + V^\mu \gamma_\mu + V^{\mu_1 \mu_2} \gamma_{\mu_1} \wedge \gamma_{\mu_2} + ... + V^{\mu_1 ... \mu_N} \gamma_{\mu_1} \wedge ... \wedge \gamma_{\mu_N}, \tag{6.164}$$

where we assume

$$V^{\mu_1 \mu_2 ... \mu_k} \equiv V^{\mu_1} V^{\mu_2} ... V^{\mu_k} \tag{6.165}$$

and where V^{μ_1}, V^{μ_2}, etc., are given by (6.158). For instance,

$$V^{\mu_1 \mu_2} = V^{\mu_1} V^{\mu_2}$$
$$= (s^{\mu_1} + e^A \partial_A X^{\mu_1} + e^{A_1} \wedge e^{A_2} v^{\mu_1}_{A_1 A_2} + ... + e^{A_1} \wedge ... \wedge e^{A_n} v^{\mu_1}_{A_1....A_n})$$
$$\times (s^{\mu_2} + e^{A'} \partial_{A'} X^{\mu_2} + e^{A'_1} \wedge e^{A'_2} v^{\mu_2}_{A'_1 A'_2} + ...$$
$$+ e^{A'_1} \wedge ... \wedge e^{A'_n} v^{\mu_2}_{A'_1....A'_n}). \tag{6.166}$$

The action is then

$$I[X^\mu, v_{A_1...A_r}, e^A, \gamma^\mu] = \tfrac{1}{2} \int d^n \phi \sqrt{|\gamma|}\, V^2 + I[e^A] + I[\gamma^\mu]. \tag{6.167}$$

Another possibility is to use (6.161) and form the expression

$$e^A C_A = e^A c_A + e^A \partial_A X^\mu \gamma_\mu + e^A c^{\mu_1 \mu_2}_A \gamma_{\mu_1} \wedge \gamma_{\mu_2} + ...$$
$$+ e^A c^{\mu_1 ... \mu_N}_A \gamma_{\mu_1} \wedge ... \wedge \gamma_{\mu_N}, \tag{6.168}$$

which is a spacetime polyvector, but a worldsheet 1-vector. We can generalize (6.168) so as to become a world sheet polyvector:

$$C = c + e^A C_A + e^{A_1} \wedge e^{A_2} C_{A_1 A_2} + ... + e^{A_1} \wedge ... \wedge e^{A_n} C_{A_1...A_n}, \tag{6.169}$$

where

$$C_{A_1 A-2...A_k} = C_{A_1} C_{A_2}...C_{A_k} \tag{6.170}$$

and where C_{A_1}, C_{A_2},... are given by the expression (6.168).

A corresponding action is

$$I[X^\mu, c^{\mu_1...\mu_k}_A, e^A, \gamma^\mu] = \tfrac{1}{2} \int d^n \phi \sqrt{|\gamma|}\, C^2 + I[c^{\mu_1...\mu_k}_A] + I[e^A] + I[\gamma^\mu] \tag{6.171}$$

Extended objects and Clifford algebra

Whichever action, (6.167) or (6.171), we choose, we have an action which is a *polyvector both in the target space and on the world sheet*. According to the discussion of Chapter 2, Section 5, particular polyvectors are *spinors*. Our action therefore contains the target space and and the worldsheet spinors. In order to obtain dynamics of such spinors the coefficients $c_A^{\mu_1...\mu_k}$ or $v_{A_1...A_k}^{\mu}$ and the corresponding kinetic terms should be written explicitly. This will come out naturally within an $\mathcal{M}$-space description of our membrane.

DESCRIPTION IN $\mathcal{M}$-SPACE

So far in this section we have considered various possible formulations of the membrane action, but without the powerful $\mathcal{M}$-space formalism. In Sec. 6.2, throughout eqs. (6.88)–(6.103), we have used a description with $\mathcal{M}$-space basis vectors $h_{\mu(\phi)}$, but the equations were not fully invariant or covariant with respect to arbitrary reparametrizations of the worldsheet parameters $\phi^A \equiv (\tau, \xi^a)$. One parameter, namely τ, was singled out, and the expressions contained the velocity $\dot{X}^{\mu(\phi)}$. On the contrary, throughout eqs. (6.129)–(6.171) of this section we have used a fully reparametrization covariant formalism.

We shall now unify both formalisms. First we will consider a straightforward reformulation of the action (6.140) and then proceed towards a "final" proposal for a polyvector action in $\mathcal{M}$-space.

By introducing the metric

$$\rho_{\mu(\phi)\nu(\phi')} = \kappa\sqrt{|\gamma|}\,\delta(\phi-\phi')g_{\mu\nu} \tag{6.172}$$

the first term in the action (6.140) can be written as the quadratic form in $\mathcal{M}$-space

$$\frac{\kappa}{2}\int d^n\phi\sqrt{|\gamma|}\,g_{\mu\nu}e^A\partial_A X^\mu e^B\partial_B X^\nu = \frac{1}{2}\rho_{\mu(\phi)\nu(\phi')}v^{\mu(\phi)}v^{\nu(\phi')}, \tag{6.173}$$

where

$$v^{\mu(\phi)} = e^A\partial_A X^{\mu(\phi)}. \tag{6.174}$$

Here $v^{\mu(\phi)}$ is the uncountable (infinite) set of vectors in the n-dimensional worldsheet manifold V_n.

The projection

$$e_0 \cdot v^{\mu(\phi)} = \partial_0 X^{\mu(\phi)} \equiv \dot{X}^{\mu(\phi)} \tag{6.175}$$

is the membrane velocity considered in Chapter 4.

More generally,

$$e_B \cdot v^{\mu(\phi)} = \partial_B X^{\mu(\phi)}. \tag{6.176}$$

POLYVECTOR ACTION IN TERMS OF e^A

The vectors (6.174) can be promoted to *polyvectors* in V_n:

$$V^{\mu(\phi)} = s^{\mu(\phi)} + e^A \partial_A X^{\mu(\phi)} + e^{A_1} \wedge e^{A_2} v^{\mu(\phi)}_{A_1 A_2} + \ldots + e^{A_1} \wedge \ldots \wedge v^{\mu(\phi)}_{A_1 \ldots A_n}. \quad (6.177)$$

This is the same expression (6.158), $V^{\mu(\phi)} \equiv V^\mu(\phi) = V^\mu$.

The first term in the action (6.159) can be written as

$$V^{\mu(\phi)} V_{\mu(\phi)} = \rho_{\mu(\phi)\nu(\phi')} V^{\mu(\phi)} V^{\mu(\phi')}. \quad (6.178)$$

Instead of the action (6.159) we can consider its $\mathcal{M}$-space generalization

$$I[X^{\mu(\phi)}, e^A, \rho_{\mu(\phi)\nu(\phi')}] \quad (6.179)$$

$$= \int \mathcal{D} X \sqrt{|\rho|} \left(\frac{1}{2} \rho_{\mu(\phi)\nu(\phi')} V^{\mu(\phi)} V^{\mu(\phi')} + \mathcal{L}[e^A] + \mathcal{L}[\rho_{\mu(\phi)\nu(\phi')}] \right),$$

where

$$\mathcal{L}[e^A] = \int d^n \phi \sqrt{|\gamma|} \left(\frac{\kappa}{2}(2-n) + \frac{1}{16\pi G^{(n)}} R^{(n)} \right) \quad (6.180)$$

and

$$\mathcal{L}[\rho_{\mu(\phi)\nu(\phi')}] = \frac{\epsilon}{16\pi} \mathcal{R} \quad (6.181)$$

are now $\mathcal{M}$-space Lagrangians.

POLYVECTOR ACTION IN TERMS OF e^A AND $h_{\mu(\phi)}$

The $\mathcal{M}$-space metric can be written as the inner product of two $\mathcal{M}$-space basis vectors:

$$\rho_{\mu(\phi)\nu(\phi')} = h_{\mu(\phi)} \cdot h_{\nu(\phi')}. \quad (6.182)$$

Hence (6.173) becomes

$$v^{\mu(\phi)} v_{\mu(\phi)} = v^{\mu(\phi)} v^{\nu(\phi')} h_{\mu(\phi)} h_{\nu(\phi')}$$

$$= e^A \partial_A X^{\mu(\phi)} e^B \partial_B X^{\nu(\phi')} h_{\mu(\phi)} h_{\nu(\phi')}. \quad (6.183)$$

It is understood that $e^A \partial_A X^{\mu(\phi)}$ is considered as a single object with index $\mu(\phi)$.

Using (6.182) the quadratic form (6.178) becomes

$$V^{\mu(\phi)} V_{\mu(\phi)} = V^{\mu(\phi)} h_{\mu(\phi)} V^{\nu(\phi')} h_{\mu(\phi')}. \quad (6.184)$$

Extended objects and Clifford algebra 197

The object $V^{\mu(\phi)} h_{\mu(\phi)}$ is a *vector* from the $\mathcal{M}$-space point of view and a *polyvector* from the point of view of the worldsheet V_n. We can promote $V^{\mu(\phi)} h_{\mu(\phi)}$ to an $\mathcal{M}$-space polyvector:

$$V = s + V^{\mu(\phi)} h_{\mu(\phi)} + V^{\mu_1(\phi_1)\mu_2(\phi_2)} h_{\mu_1(\phi_1)} \wedge h_{\mu_2(\phi_2)} + ... $$
$$+ V^{\mu_1(\phi_1)...\mu_N(\phi_N)} h_{\mu_1(\phi_1)} \wedge ... \wedge h_{\mu_N(\phi_N)}, \qquad (6.185)$$

where $V^{\mu_1(\phi_1)...\mu_k(\phi_k)}$ are generalizations of $V^{\mu(\phi)}$:

$$V^{\mu_1(\phi_1)...\mu_k(\phi_k)} = s^{\mu_1(\phi_1)...\mu_k(\phi_k)} + e^A \partial_A X^{\mu_1(\phi_1)...\mu_k(\phi_k)} + ...$$
$$+ e^{A_1} \wedge ... \wedge e^{A_n} v^{\mu_1(\phi_1)...\mu_k(\phi_k)}_{A_1...A_n}. \qquad (6.186)$$

A corresponding action is

$$I[X^{\mu(\phi)}, v^{\mu_1(\phi_1)...\mu_k(\phi_k)}_{A_1...A_p}, e^A, h^{\mu(\phi)}]$$
$$= \int \mathcal{D}X \sqrt{|\rho|} \left(\frac{1}{2} V^2 + \mathcal{L}[v^{\mu_1(\phi_1)...\mu_k(\phi_k)}_{A_1...A_p}] + \mathcal{L}[e^A] + \mathcal{L}[h^{\mu(\phi)}] \right). \qquad (6.187)$$

Although in any single term in an expression which involves products of e^A and $h_{\mu(\phi)}$ (or γ_μ) there appears an ordering ambiguity because e^A and $h_{\mu(\phi)}$ (or γ_μ) do not commute, there is no such ordering ambiguity in any expression which invloves worldsheet and target space polyvectors. A change of order results in a reshuffling of various multivector terms, so that only the coefficients become different from those in the original expression, but the form of the expression remains the same.

A question now arises of what are the coefficients $v^{\mu_1(\phi_1)...\mu_k(\phi_k)}_{A_1...A_p}$ in the expansion (6.186). There is no simple and straightforward answer to this question if $\phi \equiv \phi^A$ is a set of scalar coordinate functions (parameters) on the worldsheet V_n. In the following we shall find out that the theory becomes dramatically clear if we generalize ϕ to polyvector coordinate functions.

DESCRIPTION IN THE ENLARGED $\mathcal{M}$-SPACE

So far we have considered a set of N scalar-valued functions $X^\mu(\phi^A)$ of parameters ϕ^A. This can be written as a set of scalar-valued functions of a vector valued parameter $\phi = \phi^A e_A$, where ϕ^A are coordinate functions in the coordinate basis e_A (see also Sec. 6.1).

Let us now promote ϕ to a polyvector Φ in V_n:

$$\Phi = \phi_0 + \phi^A e_A + \phi^{A_1 A_2} e_{A_1} e_{A_2} + ... \phi^{A_1...A_n} e_{A_1}...e_{A_n} \equiv \phi^J e_J, \qquad (6.188)$$

where e_J are basis vectors of the Clifford algebra generated by the vectors e_A. The worldsheet polyvector $V^{\mu(\phi)} \equiv V^\mu(\phi)$, considered in (6.177), is an extension of the velocity vector $e^A \partial_A X^\mu$. The latter expression is nothing but the derivative of $X^\mu(\phi)$ with respect to the coordinate vector:

$$v^\mu(\phi) \equiv \frac{\partial X^\mu}{\partial \phi} = e^A \frac{\partial X^\mu}{\partial \phi^A}. \tag{6.189}$$

If instead of $X^\mu(\phi)$ we have $X^\mu(\Phi)$, i.e., a function of the polyvector (6.188), then (6.189) generalizes to

$$V^\mu(\Phi) \equiv \frac{\partial X^\mu}{\partial \Phi} = \frac{\partial X^\mu}{\partial \phi_0} + e^A \frac{\partial X^\mu}{\partial \phi^A} + e^{A_1} e^{A_2} \frac{\partial X^\mu}{\partial \phi^{A_1 A_2}} + \ldots$$

$$+ e^{A_1} \ldots e^{A_n} \frac{\partial X^\mu}{\partial \phi^{A_1 \ldots A_n}}$$

$$\equiv e^J \partial_J X^\mu. \tag{6.190}$$

This is just like (6.177), except that the argument is now Φ instead of ϕ.

In eq. (6.177) $V^\mu(\Phi)$, $\mu = 1, 2, ..., N$, are *polyvectors* in the worldsheet V_n, but *scalars* in target space V_N. By employing the basis vectors γ_μ, $\mu = 1, 2, ..., N$, we can form a *vector* $V^\mu \gamma_\mu$ in V_N (and also generalize it to a polyvector in V_N).

What about $\mathcal{M}$-space? Obviously $\mathcal{M}$-*space is now enlarged*: instead of $X^\mu(\phi) \equiv X^{\mu(\phi)}$ we have $X\mu(\Phi) \equiv X^{\mu(\Phi)}$. Our basic object is no longer a simple worldsheet (n-dimensional membrane or surface) described by $X^{\mu(\phi)}$, but a more involved geometrical object. It is described by $X^{\mu(\Phi)}$ which are coordinates of a point in an infinite-dimensional space which I will denote $\mathcal{M}_{(\Phi)}$, whereas the $\mathcal{M}$-space considered so far will from now on be denoted $\mathcal{M}_{(\phi)}$. Whilst the basis vectors in $\mathcal{M}_{(\phi)}$-space are $h_{\mu(\phi)}$, the basis vectors in $\mathcal{M}_{(\Phi)}$-space are $h_{\mu(\Phi)}$. The objects $V^\mu \equiv V^{\mu(\Phi)}$ are *components* of a vector $V^{\mu(\Phi)} h_{\mu(\Phi)}$ in $\mathcal{M}_{(\Phi)}$-space.

The generalization of (6.190) is now at hand:

$$V = v_0 + e^J \partial_J X^{\mu(\Phi)} h_{\mu(\Phi)} + e^J \partial_J X^{\mu_1(\Phi_1)\mu_2(\Phi_2)} h_{\mu_1(\Phi_1)} h_{\mu_2(\Phi_2)} + \ldots. \tag{6.191}$$

It is instructive to write the latter object more explicitly

$$V = v_0 + \left(\frac{\partial X^{\mu(\Phi)}}{\partial \phi_0} + e^A \frac{\partial X^{\mu(\Phi)}}{\partial \phi^A} + e^{A_1} e^{A_2} \frac{\partial X^{\mu(\Phi)}}{\partial \phi^{A_1 A_2}} + \ldots\right.$$

$$\left. + e^{A_1}\ldots e^{A_n} \frac{\partial X^{\mu(\Phi)}}{\partial \phi^{A_1\ldots A_n}}\right) h_{\mu(\Phi)}$$

$$+ \left(\frac{\partial X^{\mu_1(\Phi_1)\mu_2(\Phi_2)}}{\partial \phi_0} + e^A \frac{\partial X^{\mu_1(\Phi_1)\mu_2(\Phi_2)}}{\partial \phi^A} + e^{A_1} e^{A_2} \frac{\partial X^{\mu_1(\Phi_1)\mu_2(\Phi_2)}}{\partial \phi^{A_1 A_2}} + \ldots\right.$$

$$\left. + e^{A_1}\ldots e^{A_n} \frac{\partial X^{\mu_1(\Phi_1)\mu_2(\Phi_2)}}{\partial \phi^{A_1\ldots A_n}}\right) h_{\mu_1(\Phi_1)} h_{\mu_2(\Phi_2)} + \ldots, \qquad (6.192)$$

and more compactly,

$$V = e^J \partial_J X^{[Z]} h_{[Z]}, \qquad (6.193)$$

where $h_{[Z]} = h_{\mu(\phi)}, h_{\mu_1(\Phi_1)} h_{\mu_2(\Phi_2)}, \ldots$ form a basis of Clifford algebra in $\mathcal{M}_{(\Phi)}$-space.

A repeated index (Φ) means the integration:

$$V^{\mu(\Phi)} h_{\mu(\Phi)} = \int V^\mu(\Phi) h_\mu(\Phi) \mathrm{d}\Phi \qquad (6.194)$$

where

$$\mathrm{d}\Phi \equiv \prod_J \mathrm{d}\phi^J \qquad (6.195)$$

$$= \mathrm{d}\phi_0 \left(\prod_A \mathrm{d}\phi_A\right) \left(\prod_{A_1 < A_2} \mathrm{d}\phi^{A_1 A_2}\right) \ldots \left(\prod_{A_1 < A_2 < \ldots < A_n} \mathrm{d}\phi^{A_1\ldots A_n}\right).$$

We do not need to worry about the measure, since the measure is assumed to be included in the $\mathcal{M}_{(\Phi)}$-space metric which we define as the inner product

$$\rho_{\mu(\Phi)\nu(\Phi)} = h_{\mu(\Phi)} \cdot h_{\nu(\Phi')}. \qquad (6.196)$$

All expressions of the form $\rho_{\mu(\Phi)\nu(\Phi)} A^{\mu(\Phi)} B^{\nu(\Phi')}$ are then by definition invariant under reparametrizations of Φ. (See also the discussion in Sec. 4.1.)

The object of our study is a generalized world sheet. It is described by an $\mathcal{M}_{(\Phi)}$-space polyvector

$$X = X^{[Z]} h_{[Z]} = x_0 + X^{\mu(\Phi)} h_{\mu(\Phi)} + X^{\mu_1(\Phi_1)\mu_2(\Phi_2)} h_{\mu_1(\Phi_1)} h_{\mu_2(\Phi_2)} + \ldots, \qquad (6.197)$$

where $X^{\mu(\Phi)}$, $X^{\mu_1(\Phi_1)\mu_2(\Phi_2)}$, etc., are 1-vector, 2-vector, etc., coordinate functions. In a finite-dimensional sector V_N of $\mathcal{M}_{(\Phi)}$, i.e., at a fixed value of $\Phi = \Phi_1 = \Phi_2 = \ldots = \Phi_p$, the quantities $X^{\mu_1(\Phi)\mu_2(\Phi)\ldots\mu_p(\Phi)} \equiv X^{\mu_1\mu_2\ldots\mu_p}$,

$p = 1, 2, ..., N$, are holographic projections of points, 1-loops, 2-loops,..., N-loop onto the embedding space planes spanned by γ_μ, $\gamma_{\mu_1} \wedge \gamma_{\mu_2}$, ... , $\gamma_{\mu_1} \wedge ... \wedge \gamma_{\mu_N}$. Infinite loops, manifesting themselves as straight p-lines, are also included.

On the other hand, the polyvector parameters $\Phi = \phi^J e_J$, where ϕ^J are scalar parameters (coordinates) ϕ^A, $\phi^{A_1 A_2}$, ..., $\phi^{A_1...A_n}$, represent holographic projections of points, 1-loops, ..., p-loops, ... n-loops onto the planes, spanned by e_A, $e_{A_1} \wedge e_{A_2}$, ..., $e_{A_1} \wedge ... \wedge e_{A_n}$ in the *world sheet* V_n.

If we let the polyvector parameter Φ vary, then $X^{\mu_1(\Phi)\mu_2(\Phi)...\mu_p(\Phi)}$ become infinite sets.

For instance, if we let only ϕ^A vary, and keep ϕ^0, $\phi^{A_1 A_2}$, ...$\phi^{A_1...A_n}$ fixed, then:

$X^{\mu(\Phi)}$ becomes a set of 0-loops (i.e., points), constituting an ordinary n-dimensional world sheet;

$X^{\mu_1(\Phi)\mu_2(\Phi)}$ becomes a set of 1-loops, constituting a bivector type worldsheet;

$$\vdots$$

$X^{\mu_1(\Phi)\mu_2(\Phi)...\mu_N(\Phi)}$ becomes a set of $(N-1)$-loops, constituting an N-vector type worldsheet.

To an $(r-1)$-loop is associated an r-vector type worldsheet, since the differential of an $(r-1)$-loop, i.e., $\mathrm{d}X^{\mu_1(\Phi)\mu_2(\Phi)...\mu_{r-1}(\Phi)}$ is an r-vector. The multivector type of an ordinary worldsheet is 1.

If we let all compnents $\phi^J = \phi^0$, ϕ^A, $\phi^{A_1 A_2}$, ..., $\phi^{A_1...A_n}$ of Φ vary, then $X^{\mu_1(\Phi)\mu_2(\Phi)...\mu_r(\Phi)}$, $1 \leq r \leq N$, become even more involved infinite sets, whose precise properties will not be studied in this book.

In general, $\Phi_1, \Phi_2, ..., \Phi_n$ can be different, and $X^{\mu_1(\Phi_1)\mu_2(\Phi_2)...\mu_r(\Phi_r)}$ represent even more complicated sets of p-loops in $\mathcal{M}_{(\Phi)}$-space. In (6.197) all those objects of different multivector type are mixed together and X is a Clifford algebra valued position in the infinite-dimensional *Clifford manifold* which is just the Clifford algebra of $\mathcal{M}_{(\Phi)}$-space. Here we have an infinite-dimensional generalization of the concept of "*pandimensional continuum*" introduced by Pezzaglia [23] and used by Castro [75] under the name "*Clifford manifold*".

Let an action encoding the dynamics of such a generalized membrane be

$$\mathcal{I}[X] = \mathcal{I}[X^{[Z]}, e^A, h^{\mu(\Phi)}]$$

$$= \int \mathcal{D}X \sqrt{|\rho|} \left(I[X^{[Z]}] + I[e^A] + I[h^{\mu(\Phi)}] \right), \quad (6.198)$$

Extended objects and Clifford algebra

where
$$\mathcal{D}X \equiv \prod_{[Z]} \mathrm{d}X^{[Z]} \qquad (6.199)$$

and
$$\rho \equiv \det \rho_{[Z][Z']} \ , \quad \rho_{[Z][Z']} = h_{[Z]} * h_{[Z']} = \langle h_{[Z]} h_{[Z']} \rangle_0 \ , \qquad (6.200)$$

$$I[X^{[Z]}] = \tfrac{1}{2} V^2, \qquad (6.201)$$

$$I[e^A] = \int \mathrm{d}\Phi \left[\frac{1}{2}\sqrt{|\gamma|} + \frac{1}{16\pi G^{(n)}} \left([\partial_A, \partial_B] e^A\right) \cdot e^B \right], \qquad (6.202)$$

$$I[h^{\mu(\Phi)}] = \frac{1}{16\pi G^{(N)}} \left([\partial_{\mu(\Phi)}, \partial_{\nu(\Phi')}] h^{\mu(\Phi)}\right) \cdot h^{\nu(\Phi')}. \qquad (6.203)$$

If we consider a fixed background $\mathcal{M}_{(\phi)}$-space then the last term, as well as the integration over $X^{[Z]}$, can be omitted and we have

$$I[X] = I[X^{[Z]}] + I[e^A]. \qquad (6.204)$$

In eq. (6.198) we have a generalization of the action (6.187). The dynamical system described by (6.198) has no fixed background space with a fixed metric. Actually, instead of a metric, we have the basis vectors $h^{\mu(\Phi)}$ as dynamical variables (besides $\mathcal{M}_{(\Phi)}$-space coordinates $X^{[Z]}$ and V_n basis vectors e^A). The inner product $h_{\mu(\Phi)} \cdot h_{\nu(\Phi')} = \rho_{\mu(\Phi)\nu(\Phi')}$ gives the metric, but the latter is not prescribed. It comes out after a solution to the equations of motion for $h_{\mu(\Phi)}$ is found. The target space is an infinite-dimensional $\mathcal{M}_{(\Phi)}$-space. A finite-dimensional spacetime V_N, whose points can be identified, is associated with a particular worldsheet configuration $X^{\mu(\Phi)}$ (a system of many worldsheets).

The first term in (6.198), $I[X^{[Z]}]$, is the square of V, which is a worldsheet and a target space polyvector. We have seen in Sec. 2.5 that any polyvector can be written as a superposition of basis spinors which are elements of left and right ideals of the Clifford algebra. World sheet and target space spinors are automatically included in our action (6.198). There must certainly be a connection with descriptions which employ Grassmann odd variables. A step in this direction is indicated in Sec. 7.2, where it is shown that a polyvector can be written as a superposition of Grassmann numbers.

A theory of the generalized point-particle tracing a Clifford algebra-valued line $X(\Sigma)$ "living" in the Clifford manifold (C-space) has been convincingly outlined by Castro [75]. The polyvector

$$X = \Omega_{p+1} + \Lambda^p x_\mu \gamma^\mu + \Lambda^{p-1} \sigma_{\mu\nu} \gamma^\mu \gamma^\nu + \ldots$$

encodes at one single stroke the point history represented by the ordinary x_μ coordinates and the nested family of 1-loop, 2-loop, ..., p-loop histories. Castro [76] also investigated some important consequences for string theory and the conjectured M-theory [77]. Castro's work built on important contributions of Aurilia et al. [78], who formulated p-branes in terms of $(p+1)$-tangents to a p-brane's worldvolumes (which I call worldsheets),

$$X_{\nu_1\nu_2...\nu_{p+1}}(\sigma) = \epsilon^{\sigma^1\sigma^2...\sigma^{p+1}} \partial_{\sigma^1} X_{\nu_1}...\partial_{\sigma^{p+1}} X_{\nu_{p+1}}.$$

The action they considered contains the square of the latter expression. The same expression also enters the term of the highest degree in the actions (6.155) or (6.157). It also occurs in (6.187) if we identify the coefficients $v_{A_1...A_n}^{\mu_1...\mu_n}$ with $\partial_{A_1}X^{\mu_1}...\partial_{A_n}X^{\mu_n}$, and analogously for the coefficients $\partial X^{\mu_1(\Phi)...\mu_2(\Phi)}/\partial \phi^{A_1...A_n}$ in (6.192).

Here we have considered a natural generalization of the approaches investigated by Aurilia et al., and Castro. Our object of study is a mess which encompasses the set of points, 1-lines (1-loops), 2-lines (2-loops), ..., N-lines (N-loops). Spinors are automatically present because of the polyvector character of the basic quantities X and V. There is very probably a deep relationship between this approach and the ones known as loop quantum gravity [70], spin networks [71], and spin foams [72], pioneered by Smolin, Rovelli and Baez, to name just few. I beleive that in this book we have an outline of a possible general, unifying, background independent geometrical principle behind many of the approaches incorporating strings, p-branes, and quantum gravity.

Chapter 7

QUANTIZATION

Quantization of extended objects has turned out to be extremely difficult, and so far it has not yet been competely solved. In this Chapter I provide some ideas of how quantization might be carried out technically. First I demonstrate how quantization of the unconstrained membrane is more or less straightforward. Then I consider the subject from a much broader perspective, involving the Clifford algebra approach. Section 1 is largely based on a published paper [55] (and also [54]) with some additional material, whilst Section 2 is completely novel.

7.1. THE QUANTUM THEORY OF UNCONSTRAINED MEMBRANES

When studying the classical membrane we arrived in Sec. 4.3 at the *unconstrained action* (4.148) or (4.154). The latter action can be written in terms of the $\mathcal{M}$-space tensor notation

$$I[X^{\mu(\xi)}, p_{\mu(\xi)}] = \int d\tau \left[p_{\mu(\xi)} \dot{X}^{\mu(\xi)} - \tfrac{1}{2}(p^{\mu(\xi)} p_{\mu(\xi)} - K) - \Lambda^a \partial_a X^{\mu(\xi)} p_{\mu(\xi)} \right], \tag{7.1}$$

where the metric is

$$\rho_{\mu(\xi)\nu(\xi')} = \frac{1}{\Lambda} \eta_{\mu\nu} \delta(\xi - \xi') \tag{7.2}$$

and

$$K \equiv \int d^n \xi \, \Lambda \kappa^2 |f| = K[X] = \alpha^{(\xi)} \alpha_{(\xi)}, \tag{7.3}$$

where $\alpha^{(\xi)} \equiv \Lambda \kappa \sqrt{|f|}$ In principle the metric need not be restricted to the above form. In this section we shall often use, when providing specific examples, the metric

$$\rho_{\mu(\xi)\nu(\xi')} = \frac{\kappa\sqrt{|f|}}{\Lambda}\eta_{\mu\nu}\delta(\xi-\xi'), \tag{7.4}$$

which is just the metric (4.51) with fixed $\lambda = \Lambda$. eq. (7.1) holds for any fixed background metric of $\mathcal{M}$-space.

The Hamiltonian belonging to (7.1) is [53]-[55]

$$H = \tfrac{1}{2}(p^{\mu(\xi)}p_{\mu(\xi)} - K[X]) + \Lambda^a \partial_a X^{\mu(\xi)} p_{\mu(\xi)}. \tag{7.5}$$

In the following we are going to quantize the system described by the action (7.1) and the Hamiltonian (7.5).

THE COMMUTATION RELATIONS AND THE HEISENBERG EQUATIONS OF MOTION

Since there are no constraints on the dynamical variables $X^\mu(\xi)$ and the corresponding canonical momenta $p_\mu(\xi)$, quantization of the theory is straightforward. The classical variables become operators and the Poisson brackets are replaced by commutators, taken at equal τ:

$$[X^\mu(\xi), p_\nu(\xi')] = i\,\delta^\mu{}_\nu\,\delta(\xi-\xi'), \tag{7.6}$$

$$[X^\mu(\xi), X^\nu(\xi')] = 0 \ , \qquad [p_\mu(\xi), p_\nu(\xi')] = 0. \tag{7.7}$$

The Heisenberg equations of motion for an operator A read

$$\partial_a A = -i\,[A, H_a], \tag{7.8}$$

$$\dot{A} = -i\,[A, H]. \tag{7.9}$$

In particular, we have

$$\dot{p}_\mu(\xi) = -i\,[p_\mu(\xi), H], \tag{7.10}$$

$$\dot{X}^\mu(\xi) = -i\,[X^\mu(\xi), H]. \tag{7.11}$$

We may use the representation in which the operators $X^\mu(\xi)$ are diagonal, and

$$p_\mu(\xi) = -i\left(\frac{\delta}{\delta X^\mu(\xi)} + \frac{\delta F}{\delta X^\mu(\xi)}\right), \tag{7.12}$$

Quantization

where F is a suitable functional of X^μ (see eq. (7.26)). The extra term in $p_\mu(\xi)$ is introduced in order to take into account the metric (7.4) of the membrane space $\mathcal{M}$. In such a representation it is straightforward to calculate the useful commutators

$$[p_\mu(\xi), \sqrt{|f|}] = -i\frac{\delta\sqrt{|f(\xi')|}}{\delta X^\mu(\xi)} = i\partial_a\left(\sqrt{|f|}\,\partial^a X_\mu \delta(\xi-\xi')\right), \quad (7.13)$$

$$[p_\mu(\xi), \partial_a X^\nu(\xi')] = -i\frac{\delta \partial_a X^\nu(\xi')}{\delta X^\mu(\xi)} = -i\delta_\mu^{\ \nu}\,\partial_a \delta(\xi-\xi'). \quad (7.14)$$

For the Hamiltonian (7.5) the explicit form of eqs. (7.10) and (7.11) can be obtained straightforwardly by using the commutation relations (7.6), (7.7):

$$\dot{p}_\mu(\xi) = -\partial_a\left[\frac{\Lambda}{2\kappa}\,\partial^a X_\mu \sqrt{|f|}\left(\frac{p^2}{|f|}+\kappa^2\right)\right], \quad (7.15)$$

$$\dot{X}^\mu(\xi) = -i\int d\xi'\left[X^\mu(\xi), \frac{\Lambda}{2\kappa}\sqrt{|f|}\left(\frac{p^\alpha(\xi')p_\alpha(\xi')}{|f|}-\kappa^2\right)\right.$$
$$\left.-\Lambda^a\partial_a X^\alpha(\xi')p_\alpha(\xi')\right]$$
$$= \frac{\Lambda}{\sqrt{|f|}\kappa}p^\mu(\xi') - \Lambda^a\partial_a X^\mu. \quad (7.16)$$

We recognise that the operator equations (7.15), (7.16) have the same form as the classical equations of motion belonging to the action (7.1).

THE SCHRÖDINGER REPRESENTATION

The above relations (7.6)–(7.11), (7.15), (7.16) are valid regardless of representation. A possible representation is one in which the basic states $|X^\mu(\xi)\rangle$ have definite values $X^\mu(\xi)$ of the membrane's position operators[1] $\widehat{X}^\mu(\xi)$. An arbitrary state $|a\rangle$ can be expressed as

$$|a\rangle = \int |X^\mu(\xi)\rangle \mathcal{D}X^\mu(\xi)\langle X^\mu(\xi)|a\rangle \quad (7.17)$$

where the measure $\mathcal{D}X^\mu(\xi)$ is given in eq. (4.17) with $\alpha = \kappa/\Lambda$.

We shall now write the equation of motion for the wave functional $\psi \equiv \langle X^\mu(\xi)|a\rangle$. We adopt the requirement that, in the classical limit, the wave

---

[1] When necessary we use symbols with a hat in order to distinguish operators from their eigenvalues.

functional equation should reproduce the Hamilton–Jacobi equation (see ref. [55])

$$-\frac{\partial S}{\partial \tau} = H. \qquad (7.18)$$

The supplementary equation

$$\frac{\partial S}{\partial X^{\mu(\xi)}} = p_{\mu(\xi)} \qquad (7.19)$$

has to arise from the corresponding quantum equation. For this goal we admit that ψ evolves with the evolution parameter τ and is a functional of $X^\mu(\xi)$:

$$\psi = \psi[\tau, X^\mu(\xi)]. \qquad (7.20)$$

It is normalized according to

$$\int \mathcal{D}X \, \psi^* \psi = 1, \qquad (7.21)$$

which is a straightforward extension of the corresponding relation

$$\int \mathrm{d}^4 x \, \psi^* \psi = 1 \qquad (7.22)$$

for the unconstrained point particle in Minkowski spacetime [1]–[16], [53]–[55]. It is important to stress again [53]–[55] that, since (7.21) is satisfied at any τ, the evolution operator U which brings $\psi(\tau) \to \psi(\tau') = U\psi(\tau)$ is *unitary*.

The following equations are assumed to be satisfied ($\rho = \mathrm{Det}\, \rho_{\mu(\xi)\nu(\xi')}$):

$$-i\hbar \frac{1}{|\rho|^{1/4}} \partial_{\mu(\xi)} (|\rho|^{1/4} \psi) = \hat{p}_{\mu(\xi)} \psi, \qquad (7.23)$$

$$i\hbar \frac{1}{|\rho|^{1/4}} \frac{\partial (|\rho|^{1/4} \psi)}{\partial \tau} = H\psi, \qquad (7.24)$$

where[2]

$$H = \tfrac{1}{2}(\hat{p}^{\mu(\xi)} \hat{p}_{\mu(\xi)} - K) + \Lambda^a \partial_a X^{\mu(\xi)} \hat{p}_{\mu(\xi)}. \qquad (7.25)$$

When the metric $\rho_{\mu(\xi)\nu(\xi')}$ in $\mathcal{M}$ explicitly depends on τ (which is the case when $\dot{\Lambda} \neq 0$) such a modified τ-derivative in eq. (7.24) is required [33] in

---

[2] Using the commutator (7.14) we find that

$$\Lambda^a [p_{\mu(\xi)}, \partial_a X^{\mu(\xi)}] = \int \mathrm{d}\xi \, \mathrm{d}\xi' \, \Lambda^a \, [p_\mu(\xi), \partial_a X^\nu(\xi')] \, \delta_\nu^{\ \mu} \, \delta(\xi - \xi') = 0,$$

therefore the order of operators in the second term of eq. (7.25) does not matter.

Quantization

order to assure conservation of probability, as expressed by the τ-invariance of the integral (7.21).

The momentum operator given by

$$\hat{p}_{\mu(\xi)} = -i\hbar \left(\partial_{\mu(\xi)} + \tfrac{1}{2}\Gamma^{\nu(\xi')}_{\mu(\xi)\nu(\xi')} \right), \qquad (7.26)$$

where

$$\tfrac{1}{2}\Gamma^{\nu(\xi')}_{\mu(\xi)\nu(\xi')} = |\rho|^{-1/4} \partial_{\mu(\xi)} |\rho|^{1/4},$$

satisfies the commutation relations (7.6), (7.7) and is Hermitian with respect to the scalar product $\int \mathcal{D}X\, \psi^* \hat{p}_{\mu(\xi)} \psi$ in $\mathcal{M}$.

The expression (7.25) for the Hamilton operator H is obtained from the corresponding classical expression (7.5) in which the quantities $X^{\mu(\xi)}$, $p_{\mu(\xi)}$ are replaced by the operators $\hat{X}^{\mu(\xi)}$, $\hat{p}_{\mu(\xi)}$. There is an ordering ambiguity in the definition of $\hat{p}^{\mu(\xi)} \hat{p}_{\mu(\xi)}$. Following the convention in a finite-dimensional curved space [33], we use the identity

$$|\rho|^{1/4} \hat{p}_{\mu(\xi)} |\rho|^{-1/4} = -i\hbar\, \partial_{\mu(\xi)}$$

and define

$$\hat{p}^{\mu(\xi)} \hat{p}_{\mu(\xi)} \psi = |\rho|^{-1/2} |\rho|^{1/4} \hat{p}_{\mu(\xi)} |\rho|^{-1/4} |\rho|^{1/2} \rho^{\mu(\xi)\nu(\xi')} |\rho|^{1/4} \hat{p}_{\nu(\xi')} |\rho|^{-1/4} \psi$$

$$= -|\rho|^{-1/2} \partial_{\mu(\xi)} \left(|\rho|^{1/2} \rho^{\mu(\xi)\nu(\xi')} \partial_{\nu(\xi')} \psi \right)$$

$$= -\mathrm{D}_{\mu(\xi)} \mathrm{D}^{\mu(\xi)} \psi. \qquad (7.27)$$

Let us derive the classical limit of equations (7.23), (7.24). For this purpose we write

$$\psi = A[\tau, X^\mu(\xi)] \exp\left[\tfrac{i}{\hbar} S[\tau, X^\mu(\xi)]\right] \qquad (7.28)$$

with real A and S.

Assuming (7.28) and taking the limit $\hbar \to 0$ eq. (7.23) becomes

$$\hat{p}_{\mu(\xi)} \psi = \partial_{\mu(\xi)} S\, \psi. \qquad (7.29)$$

If we assume that in eq. (7.28) A is a slowly varying and S a quickly varying functional of $X^\mu(\xi)$ we find that $\partial_{\mu(\xi)} S$ is the expectation value of the momentum operator $\hat{p}_{\mu(\xi)}$.

Let us insert (7.28) into eq. (7.24). Taking the limit $\hbar \to 0$, and writing separately the real and imaginary part of the equation, we obtain

$$-\frac{\partial S}{\partial \tau} = \tfrac{1}{2}(\partial_{\mu(\xi)} S\, \partial^{\mu(\xi)} S - K) + \Lambda^a \partial_a X^{\mu(\xi)} \partial_{\mu(\xi)} S, \qquad (7.30)$$

$$\frac{1}{|\rho|^{1/2}}\frac{\partial}{\partial \tau}(|\rho|^{1/2}A^2) + D_{\mu(\xi)}[A^2(\partial^{\mu(\xi)}S + \Lambda^a \partial_a X^{\mu(\xi)})] = 0. \tag{7.31}$$

eq. (7.30) is just the functional Hamilton–Jacobi (7.18) equation of the classical theory. Eq. (7.31) is the continuity equation, where $\psi^*\psi = A^2$ is *the probability density* and

$$A^2\left(\partial^{\mu(\xi)}S - \Lambda^a \partial_a X^{\mu(\xi)}\right) = j^{\mu(\xi)} \tag{7.32}$$

is *the probability current*. Whilst the covariant components $\partial_{\mu(\xi)}S$ form a momentum vector p_μ, the contravariant components form a vector ∂X^μ (see also Sec. 4.2):

$$\begin{aligned}
\partial^{\mu(\xi)}S &= \rho^{\mu(\xi)\nu(\xi')}\partial_{\nu(\xi')}S \\
&= \int d\xi' \frac{\Lambda}{\kappa\sqrt{|f|}} \eta^{\mu\nu}\, \delta(\xi-\xi')\frac{\delta S}{\delta X^\nu(\xi')} \\
&= \frac{\Lambda}{\kappa\sqrt{|f|}}\eta^{\mu\nu}\frac{\delta S}{\delta X^\nu(\xi)} = \partial X^\mu, \tag{7.33}
\end{aligned}$$

where we have taken

$$\delta S/\delta X^\nu(\xi) = p_\nu(\xi) = \frac{\kappa\sqrt{|f|}}{\Lambda}\partial X_\nu(\xi),$$

and raised the index by $\eta^{\mu\nu}$, so that

$$\partial X^\mu(\xi) = \eta^{\mu\nu}\partial X_\nu(\xi). \tag{7.34}$$

So we have

$$\partial^{\mu(\xi)}S + \Lambda^a \partial_a X^{\mu(\xi)} = \dot X^{\mu(\xi)},$$

and the current (7.32) is proportional to the velocity, as it should be.

Since eq. (7.24) gives the correct classical limit it is consistent and can be taken as the equation of motion for the wave functional ψ. We shall call (7.24) *the (functional) Schrödinger equation*. In general, it admits the following *continuity equation*:

$$\frac{1}{|\rho|^{1/2}}\frac{\partial}{\partial\tau}(|\rho|^{1/2}\psi^*\psi) + D_{\mu(\xi)}j^{\mu(\xi)} = 0, \tag{7.35}$$

where

$$\begin{aligned}
j^{\mu(\xi)} &= \tfrac{1}{2}\psi^*(\hat p^{\mu(\xi)} + \Lambda^a\partial_a X^{\mu(\xi)})\psi + \text{h.c.} \\
&= -\frac{i}{2}(\psi^*\partial^{\mu(\xi)}\psi - \psi\,\partial^{\mu(\xi)}\psi^*) + \Lambda^a\partial_a X^{\mu(\xi)}\psi^*\psi. \tag{7.36}
\end{aligned}$$

Quantization

For exercise we prove below that the probability current (7.36) satisfies (7.35). First we observe that

$$\frac{\delta\, \partial'_a X^\mu(\xi')}{\delta X^\nu(\xi)} = \partial'_a \left(\frac{\delta X^\mu(\xi')}{\delta X^\nu(\xi)}\right)$$

$$= \delta^\mu{}_\nu \, \partial'_a \delta(\xi' - \xi) = -\delta^\mu{}_\nu \partial_a \delta(\xi' - \xi). \qquad (7.37)$$

Then we calculate

$$\partial_{\nu(\xi)}(\Lambda^a \partial_a X^{\mu(\xi')} \psi^* \psi) = \frac{\delta}{\delta X^\nu(\xi)}(\Lambda^a \partial_a X^\mu(\xi') \psi^* \psi)$$

$$= -\Lambda^a \delta^\mu{}_\nu \, \partial_a \delta(\xi' - \xi)\psi^* \psi + \Lambda^a \partial_a X^\mu(\xi') \left(\psi \frac{\delta\psi^*}{\delta X^\nu(\xi)} + \psi^* \frac{\delta\psi}{\delta X^\nu(\xi)}\right). \qquad (7.38)$$

Multiplying (7.38) by $\delta^\mu{}_\nu \delta(\xi' - \xi) \mathrm{d}\xi' \mathrm{d}\xi$, summing over μ, ν and integrating over ξ', ξ, we obtain

$$\partial_{\mu(\xi)}(\Lambda^a \partial_a X^{\mu(\xi)} \psi^* \psi) = -N \int \mathrm{d}\xi' \, \mathrm{d}\xi \, \Lambda^a \, \delta(\xi' - \xi) \, \partial_a \delta(\xi' - \xi) \psi^* \psi$$

$$+ \int \mathrm{d}\xi \, \Lambda^a \, \partial_a X^\mu(\xi) \left(\psi \frac{\delta \psi^*}{\delta X^\mu(\xi)} + \psi^* \frac{\delta\psi}{\delta X^\mu(\xi)}\right)$$

$$= \Lambda^a \, \partial_a X^{\mu(\xi)} (\psi \, \partial_{\mu(\xi)} \psi^* + \psi^* \, \partial_{\mu(\xi)} \psi). \qquad (7.39)$$

In eq. (7.39) we have taken $\delta^\mu{}_\nu \delta^\nu{}_\mu = N$ and $\int \mathrm{d}\xi \, \Lambda^a \delta(\xi' - \xi) \, \partial_a \delta(\xi' - \xi) = 0$. Next we take into account

$$\mathrm{D}_{\mu(\xi)}(\Lambda^a \partial_a X^{\mu(\xi)} \psi^* \psi) = (\mathrm{D}_{\mu(\xi)} \, \partial_a X^{\mu(\xi)}) \Lambda^a \psi^* \psi + \partial_a X^{\mu(\xi)} \, \partial_{\mu(\xi)}(\Lambda^a \psi^* \psi)$$

and

$$\mathrm{D}_{\mu(\xi)} \partial_a X^{\mu(\xi)} = \partial_{\mu(\xi)} \partial_a X^{\mu(\xi)} + \Gamma^{\mu(\xi)}_{\mu(\xi)\nu(\xi')} \partial'_a X^{\nu(\xi')}. \qquad (7.40)$$

From (7.36), (7.39) and (7.40) we have

$$\mathrm{D}_{\mu(\xi)} j^{\mu(\xi)} = -\frac{i}{2}(\psi^* \mathrm{D}_{\mu(\xi)} \mathrm{D}^{\mu(\xi)} \psi - \psi \mathrm{D}_{\mu(\xi)} \mathrm{D}^{\mu(\xi)} \psi^*) \qquad (7.41)$$

$$+ \Lambda^a \partial_a X^{\mu(\xi)} (\psi^* \partial_{\mu(\xi)} \psi + \psi \partial_{\mu(\xi)} \psi^* + \Gamma^{\nu(\xi')}_{\mu(\xi)\nu(\xi')} \psi^* \psi).$$

Using the Schrödinger equation

$$i|\rho|^{-1/4} \partial(|\rho|^{1/4} \psi)/\partial\tau = H\psi \qquad (7.42)$$

and the complex conjugate equation

$$-i|\rho|^{-1/4} \partial(|\rho|^{1/4} \psi^*)/\partial\tau = H^* \psi^*, \qquad (7.43)$$

210   *THE LANDSCAPE OF THEORETICAL PHYSICS: A GLOBAL VIEW*

where H is given in (7.25), we obtain that the continuity equation (7.35) is indeed satisfied by the probability density $\psi^*\psi$ and the current (7.36). For the Ansatz (7.28) the current (7.36) becomes equal to the expression (7.32), as it should.

We notice that the term with $\Gamma^{\nu(\xi')}_{\mu(\xi)\nu(\xi')}$ in eq. (7.41) is canceled by the same type of the term in H. The latter term in H comes from the definition (7.26) of the momentum operator in a (curved) membrane space $\mathcal{M}$, whilst the analogous term in eq. (7.41) results from the covariant differentiation. The definition (7.26) of $\hat{p}_{\mu(\xi)}$, which is an extension of the definition introduced by DeWitt [33] for a finite-dimensional curved space, is thus shown to be consistent also with the conservation of the current (7.36).

THE STATIONARY SCHRÖDINGER EQUATION FOR A MEMBRANE

The evolution of a generic membrane's state $\psi[\tau, X^\mu(\xi)]$ is given by the τ-dependent functional Schrödinger equation (7.24) and the Hamiltonian (7.25). We are now going to consider solutions which have the form

$$\psi[\tau, X^\mu(\xi)] = e^{-iE\tau}\phi[X^\mu(\xi)], \qquad (7.44)$$

where E is a constant. We shall call it *energy*, since it has a role analogous to energy in non-relativistic quantum mechanics. Considering the case $\dot\Lambda = 0$, $\dot\Lambda^a = 0$ and inserting the Ansatz (7.44) into eq. (7.24) we obtain

$$\left(-\tfrac{1}{2}\mathrm{D}^{\mu(\xi)}\mathrm{D}_{\mu(\xi)} + i\Lambda^a\partial_a X^{\mu(\xi)}(\partial_{\mu(\xi)} + \tfrac{1}{2}\Gamma^{\nu(\xi')}_{\mu(\xi)\nu(\xi')}) - \tfrac{1}{2}K\right)\phi = E\,\phi. \qquad (7.45)$$

So far the membrane's dimension and signature have not been specified. Let us now consider *Case 2* of Sec. 4.3. All the dimensions of our membrane have the same signature, and the index a of a membrane's coordinates assumes the values $a = 1, 2, ..., n = p$. Assuming a real ϕ eq. (7.45) becomes

$$\left(-\tfrac{1}{2}\mathrm{D}^{\mu(\xi)}\mathrm{D}_{\mu(\xi)} - \tfrac{1}{2}K - E\right)\phi = 0, \qquad (7.46)$$

$$\Lambda^a\partial_a X^{\mu(\xi)}(\partial_{\mu(\xi)} + \tfrac{1}{2}\Gamma^{\nu(\xi')}_{\mu(\xi)\nu(\xi')})\phi = 0. \qquad (7.47)$$

These are equations for a *stationary state*. They remind us of the well known p-brane equations [79].

In order to obtain from (7.46), (7.47) the conventional p-brane equations we have to assume that eqs. (7.46), (7.47) hold for any Λ and Λ^a, which is indeed the case. Then instead of eqs. (7.46), (7.47) in which we have the integration over ξ, we obtain the equations without the integration over ξ:

$$\left(-\frac{\Lambda}{2\kappa|f|}\eta^{\mu\nu}\frac{\mathrm{D}^2}{\mathrm{D}X^\mu(\xi)\mathrm{D}X^\nu(\xi)} - \frac{\Lambda}{2} - \mathcal{E}\right)\phi = 0, \qquad (7.48)$$

Quantization

$$\partial_a X^\mu(\xi) \left(\frac{\delta}{\delta X^\mu(\xi)} + \tfrac{1}{2} \Gamma^{\nu(\xi')}_{\mu(\xi)\nu(\xi')} \right) \phi = 0. \qquad (7.49)$$

The last equations are obtained from (7.46), (7.47) after writing the energy as the integral of the energy density $\mathcal{E}$ over the membrane, $E = \int d^n\xi \sqrt{|f|}\,\mathcal{E}$, taking into account that $K = \int d^n\xi \sqrt{|f|}\,\kappa\Lambda$, and omitting the integration over ξ.

Equations (7.48), (7.49), with $\mathcal{E} = 0$, are indeed the quantum analogs of the classical p-brane constraints used in the literature [79, 80] and their solution ϕ represents states of a conventional, constrained, p-brane with a tension κ. When $\mathcal{E} \neq 0$ the preceding statement still holds, provided that $\mathcal{E}(\xi)$ is proportional to $\Lambda(\xi)$, so that the quantity $\kappa(1 - 2\mathcal{E}/\Lambda)$ is a constant, identified with the effective tension. Only the particular stationary states (as indicated above) correspond to the conventional, Dirac–Nambu–Goto p-brane states, but in general they correspond to a sort of wiggly membranes [53, 81].

DIMENSIONAL REDUCTION OF THE SCHRÖDINGER EQUATION

Let us now consider the *Case 1*. Our membrane has signature $(+ - - - ...)$ and is actually an n-dimensional worldsheet. The index a of the worldsheet coordinates ξ^a assumes the values $a = 0, 1, 2, ..., p$, where $p = n - 1$.

Amongst all possible wave functional satisfying eqs. (7.24) there are also the special ones for which it holds (for an example see see eqs. (7.81), (7.84)–(7.86))

$$\frac{\delta\psi}{\delta X^\mu(\xi^0, \xi^i)} = \delta(\xi^0 - \xi^0_\Sigma)(\partial_0 X^\mu \partial_0 X_\mu)^{1/2} \frac{\delta\psi}{\delta X^\mu(\xi^0_\Sigma, \xi^i)}, \qquad (7.50)$$

$$i = 1, 2, ..., p = n - 1,$$

where ξ^0_Σ is a fixed value of the time-like coordinate ξ^0. In the compact tensorial notation in the membrane space $\mathcal{M}$ eq. (7.50) reads

$$\partial_{\mu(\xi^0,\xi^i)}\phi = \delta(\xi^0 - \xi^0_\Sigma)(\partial_0 X^\mu \partial_0 X_\mu)^{1/2} \partial_{\mu(\xi^0_\Sigma, \xi^i)}\psi. \qquad (7.51)$$

Using (7.51) we find that the dimension of the Laplace operator in

$$D^{\mu(\xi)} D_{\mu(\xi)}\psi = \int d^n\xi \frac{\Lambda}{\kappa\sqrt{|f|}} \eta^{\mu\nu} \frac{D^2\psi}{DX^\mu(\xi)DX^\nu(\xi)}$$

$$= \int d\xi^0\,d^p\xi \frac{\Lambda}{\kappa\sqrt{|f|}} (\partial_0 X^\mu \partial_0 X_\mu)^{1/2} \eta^{\mu\nu} \delta(\xi^0 - \xi^0_\Sigma) \times$$

$$\times \frac{\mathrm{D}^2\psi}{\mathrm{D}X^\mu(\xi^0,\xi^i)\mathrm{D}X^\nu(\xi^0_\Sigma,\xi^i)}$$

$$= \int \mathrm{d}^p\xi \frac{\Lambda}{\kappa\sqrt{|\bar f|}} \eta^{\mu\nu} \frac{\mathrm{D}^2\psi}{\mathrm{D}X^\mu(\xi^0_\Sigma,\xi^i)X^\nu(\xi^0_\Sigma,\xi^i)}$$

$$= \int \mathrm{d}^p\xi \frac{\Lambda}{\kappa\sqrt{|\bar f|}} \eta^{\mu\nu} \frac{\mathrm{D}^2\psi}{\mathrm{D}X^\mu(\xi^i)\mathrm{D}X^\nu(\xi^i)}. \tag{7.52}$$

Here $\bar f \equiv \det \bar f_{ij}$ is the determinant of the induced metric $\bar f_{ij} \equiv \partial_i X^\mu \partial_j X_\mu$ on V_p, and it is related to the determinant $f \equiv \det f_{ab}$ of the induced metric $f_{ab} = \partial_a X^\mu \partial_b X_\mu$ on V_{p+1} according to $f = \bar f \, \partial_0 X^\mu \partial_0 X_\mu$ (see refs. [61, 62]).

The differential operator in the last expression of eq. (7.52) (where we have identified $X^{\mu(\xi^0_\Sigma,\xi^i)} \equiv X^{\mu(\xi^i)}$) acts in the space of p-dimensional membranes, although the original operator we started from acted in the space of $(p+1)$-dimensional membranes ($n = p+1$). This comes from the fact that our special functional, satisfying (7.50), has vanishing functional derivative $\delta\psi/\delta X^\mu(\xi^0,\xi^i)$ for all values of ξ^0, except for $\xi^0 = \xi^0_\Sigma$. The xpression (7.50) has its finite-dimensional analog in the relation $\partial\phi/\partial x^A = \delta_A{}^\mu \partial\phi/\partial x^\mu$, $A = 0, 1, 2, 3, ..., 3+m$, $\mu = 0, 1, 2, 3$, which says that the field $\phi(x^A)$ is constant along the extra dimensions. For such a field the $(4+m)$-dimensional Laplace expression $\eta^{AB}\frac{\partial^2\phi}{\partial x^A \partial x^B}$ reduces to the 4-dimensional expression $\eta^{\mu\nu}\frac{\partial^2\phi}{\partial x^\mu \partial x^\nu}$.

The above procedure can be performed not only for ξ^0 but also for any of the coordinates ξ^a; it applies both to *Case 1* and *Case 2*.

Using (7.51), (7.52) we thus find that, for such a special wave functional ψ, the equation (7.24), which describes a state of a $(p+1)$-dimensional membrane reduces to the equation for a p-dimensional membrane. This is an important finding. Namely, at the beginning we may assume a certain dimension of a membrane and then consider lower-dimensional membranes as particular solutions of the higher-dimensional equation. This means that the point particle theory (0-brane), the string theory (1-brane), and in general a p-brane theory for arbitrary p, are all contained in the theory of a $(p+1)$-brane.

A PARTICULAR SOLUTION TO THE COVARIANT SCHRÖDINGER EQUATION

Let us now consider the covariant functional Schrödinger equation (7.24) with the Hamiltonian operator (7.25). The quantities Λ^a are arbitrary in principle. For simplicity we now take $\Lambda^a = 0$. Additionally we also take a

Quantization

τ-independent Λ, so that $\dot{\rho} = 0$. Then eq. (7.24) becomes simply ($\hbar = 1$)

$$i\frac{\partial \psi}{\partial \tau} = -\frac{1}{2}(D^{\mu(\xi)}D_{\mu(\xi)} + K)\psi. \tag{7.53}$$

The operator on the right hand side is the infinite-dimensional analog of the covariant Klein–Gordon operator. Using the definition of the covariant derivative (4.19) and the corresponding affinity we have [54, 55]

$$\begin{aligned}
D_{\mu(\xi)}D^{\mu(\xi)}\psi &= \rho^{\mu(\xi)\nu(\xi')}D_{\mu(\xi)}D_{\nu(\xi')}\psi \\
&= \rho^{\mu(\xi)\nu(\xi')}\left(\partial_{\mu(\xi)}\partial_{\nu(\xi')}\psi - \Gamma^{\alpha(\xi'')}_{\mu(\xi)\nu(\xi')}\partial_{\alpha(\xi'')}\psi\right). \tag{7.54}
\end{aligned}$$

The affinity is explicitly

$$\Gamma^{\alpha(\xi'')}_{\mu(\xi)\nu(\xi')} = \tfrac{1}{2}\rho^{\alpha(\xi'')\beta(\xi''')}\left(\rho_{\beta(\xi''')\mu(\xi),\nu(\xi')} + \rho_{\beta(\xi''')\nu(\xi'),\mu(\xi)} - \rho_{\mu(\xi)\nu(\xi'),\beta(\xi''')}\right), \tag{7.55}$$

where the metric is given by (7.4). Using

$$\begin{aligned}
\rho_{\beta(\xi'')\mu(\xi),\nu(\xi')} &= \eta_{\mu\nu}\,\alpha(\xi)\delta(\xi - \xi'')\frac{\delta\sqrt{|f(\xi)|}}{\delta X^\nu(\xi')} \tag{7.56} \\
&= \eta_{\mu\nu}\,\alpha(\xi)\delta(\xi - \xi'')\sqrt{|f(\xi)|}\partial^a X_\nu(\xi)\partial_a\delta(\xi - \xi')
\end{aligned}$$

the equation (7.54) becomes

$$D_{\mu(\xi)}D^{\mu(\xi)}\psi = \rho^{\mu(\xi)\nu(\xi')}\frac{\partial^2\psi}{\partial X^{\mu(\xi)}X^{\nu(\xi')}} \tag{7.57}$$

$$-\frac{\delta(0)}{\kappa}\int d^n\xi\,\frac{\delta\psi}{\delta X^\mu(\xi)}\left[\frac{N}{2}\frac{\Lambda}{\sqrt{|f|}}\frac{1}{\sqrt{|f|}}\partial_a(\sqrt{|f(\xi)|}\,\partial^a X^\mu)\right.$$

$$\left.+\left(\frac{N}{2}+1\right)\Lambda\partial^a X^\mu\,\partial_a\!\left(\frac{1}{\sqrt{|f(\xi)|}}\right) + \frac{\partial^a X^\mu \partial_a \Lambda}{\sqrt{|f(\xi)|}}\right],$$

where $N = \eta^{\mu\nu}\eta_{\mu\nu}$ is the dimension of spacetime. In deriving eq. (7.57) we encountered the expression $\delta^2(\xi - \xi')$ which we replaced by the corresponding approximate expression $F(a, \xi - \xi')\delta(\xi - \xi')$, where $F(a, \xi - \xi')$ is any finite function, e.g., $(1/\sqrt{\pi}a)\exp[-(\xi - \xi')^2/a^2]$, which in the limit $a \to 0$ becomes $\delta(\xi - \xi')$. The latter limit was taken after performing all the integrations, and $\delta(0)$ should be considered as an abbreviation for $\lim_{a\to 0} F(a, 0)$.

As already explained in footnote 5 of Chapter 5, the infinity $\delta(0)$ in an expression such as (7.57) can be regularized by taking into account the plausible assumption that a generic physical object is actually a fractal (i.e., an object with detailed structure on all scales). The objects $X^\mu(\xi)$ which we are using in our calculations are well behaved functions with definite derivatives and should be considered as approximations of the actual physical objects. This means that a description with a given $X^\mu(\xi)$ below a certain scale a has no physical meaning. In order to make physical sense of the expression (7.57), $\delta(0)$ should therefore be replaced by $F(a,0)$. Choice of the scale a is arbitrary and determines the precision of our description. (Cf., the length of a coast depends on the scale of the units with which it is measured.)

Expression (7.54) is analogous to the corresponding finite dimensional expression. In analogy with a finite-dimensional case, a metric tensor $\rho'_{\mu(\xi)\nu(\xi')}$ obtained from $\rho_{\mu(\xi)\nu(\xi')}$ by a coordinate transformation (4.5) belongs to the same space $\mathcal{M}$ and is equivalent to $\rho_{\mu(\xi)\nu(\xi')}$. Instead of a finite number of coordinate conditions which express a choice of coordinates, we now have infinite coordinate conditions. The second term in eq. (7.57) becomes zero if we take

$$\frac{N}{2}\frac{\Lambda}{\sqrt{|f|}}\frac{1}{\sqrt{|f|}}\partial_a\left(\sqrt{|f(\xi)|}\,\partial^a X_\mu\right) \tag{7.58}$$

$$+\left((\frac{N}{2}+1)\Lambda\partial_a(\frac{1}{\sqrt{|f(\xi)|}})+\frac{\partial_a\Lambda}{\sqrt{|f(\xi)|}}\right)\partial^a X_\mu = 0,$$

and these are just possible coordinate conditions in the membrane space $\mathcal{M}$. eq. (7.58), together with boundary conditions, determines a family of functions $X^\mu(\xi)$ for which the functional ψ is defined; in the operator theory such a family is called the domain of an operator. Choice of a family of functions $X^\mu(\xi)$ is, in fact, a choice of coordinates (a gauge) in $\mathcal{M}$.

If we contract eq. (7.58) by $\partial_b X_\mu$ and take into account the identity $\partial_c X^\mu D_a D_b X_\mu = 0$ we find

$$(\frac{N}{2}+1)\Lambda\partial_a(\frac{1}{\sqrt{|f(\xi)|}})+\frac{\partial_a\Lambda}{\sqrt{|f(\xi)|}} = 0. \tag{7.59}$$

From (7.59) and (7.58) we have

$$\frac{1}{\sqrt{|f(\xi)|}}\partial_a\left(\sqrt{|f(\xi)|}\,\partial^a X_\mu\right) = 0. \tag{7.60}$$

Interestingly, the gauge condition (7.58) in $\mathcal{M}$ automatically implies the gauge condition (7.59) in V_n. The latter condition is much simplified if we take $\Lambda \neq 0$ satisfying $\partial_a \Lambda = 0$; then for $|f| \neq 0$ eq. (7.59) becomes

$$\partial_a\sqrt{|f|} = 0 \tag{7.61}$$

Quantization

which is just the gauge condition considered by Schild [63] and Eguchi [82].

In the presence of the condition (7.58) (which is just a gauge, or coordinate, condition in the function space $\mathcal{M}$) the functional Schrödinger equation (7.53) can be written in the form

$$i\frac{\partial \psi}{\partial \tau} = -\frac{1}{2}\int d^n\xi \left(\frac{\Lambda}{\sqrt{|f|}\kappa}\frac{\delta^2}{\delta X^\mu(\xi)\delta X_\mu(\xi)} + \sqrt{|f|}\,\Lambda\kappa\right)\psi. \tag{7.62}$$

A particular solution to eq. (7.62) can be obtained by considering the following eigenfunctions of the momentum operator

$$\hat{p}_{\mu(\xi)}\psi_p[X^{\mu(\xi)}] = p_{\mu(\xi)}\psi_p[X^{\mu(\xi)}], \tag{7.63}$$

with

$$\psi_p[X^{\mu(\xi)}] = \mathcal{N}\exp\left[i\int_{X_0}^{X} p_{\mu(\xi)}\,dX'^{\mu(\xi)}\right]. \tag{7.64}$$

This last expression is invariant under reparametrizations of ξ^a (eq. (4.2)) and of $X^\mu(\xi^a)$ (eq. (4.5)). The momentum field $p_{\mu(\xi)}$, in general, functionally depends on $X^{\mu(\xi)}$ and satisfies (see [55])

$$\partial_{\mu(\xi)}p_{\nu(\xi')} - \partial_{\nu(\xi')}p_{\mu(\xi)} = 0\,, \qquad D_{\mu(\xi)}p^{\mu(\xi)} = 0. \tag{7.65}$$

In particular $p_{\mu(\xi)}$ may be just a constant field, such that $\partial_{\nu(\xi')}p_{\mu(\xi)} = 0$. Then (7.64) becomes

$$\psi_p[X^{\mu(\xi)}] = \mathcal{N}\exp\left[i\int d^n\xi\, p_\mu(\xi)(X^\mu(\xi) - X_0^\mu(\xi))\right]$$

$$= \mathcal{N}\exp\left[ip_{\mu(\xi)}(X^{\mu(\xi)} - X_0^{\mu(\xi)})\right]. \tag{7.66}$$

The latter expression holds only in a particular parametrization[3] (eq. (4.5)) of $\mathcal{M}$ space, but it is still invariant with respect to reparametrizations of ξ^a.

Let the τ-dependent wave functional be

$$\psi_p[\tau, X^\mu(\xi)]$$

$$= \mathcal{N}\exp\left[i\int d^n\xi\, p_\mu(\xi)\left(X^\mu(\xi) - X_0^\mu(\xi)\right)\right] \times$$

---

[3] *Solutions* to the equations of motion are always written in a particular parametrization (or gauge). For instance, a plane wave solution $\exp[ip_\mu x^\mu]$ holds in Cartesian coordinates, but not in spherical coordinates.

$$\times \exp\left[-\frac{i\tau}{2}\int \Lambda \mathrm{d}^n\xi \left(\frac{p^\mu(\xi)p_\mu(\xi)}{\sqrt{|f|}\kappa} - \sqrt{|f|}\kappa\right)\right]$$

$$\equiv \mathcal{N} \exp\left[i\int \mathrm{d}^n\xi\, \mathcal{S}\right]. \tag{7.67}$$

where $f \equiv \det \partial_a X^\mu(\xi)\partial_b X_\mu(\xi)$ should be considered as a functional of $X^\mu(\xi)$ and independent of τ. From (7.67) we find

$$-i\frac{\delta\psi_p[\tau, X^\mu]}{\delta X^\alpha} = -i\left(\frac{\partial \mathcal{S}}{\partial X^\alpha} - \partial_a \frac{\partial \mathcal{S}}{\partial \partial_a X^\alpha}\right)\psi_p[\tau, X^\mu]$$

$$= p_\alpha \psi_p - \tau \partial_a\left(\sqrt{|f|}\partial^a X_\mu\right)\frac{\Lambda}{2\kappa}\left(\frac{p^2}{|f|} + \kappa^2\right)\psi_p$$

$$- \tau\sqrt{|f|}\partial^a X_\mu \partial_a\left[\frac{\Lambda}{2\kappa}\left(\frac{p^2}{|f|} + \kappa^2\right)\right]\psi_p. \tag{7.68}$$

Let us now take the gauge condition (7.60). Additionally let us assume

$$\partial_a\left[\frac{\Lambda}{2\kappa}\left(\frac{p^2}{|f|} + \kappa^2\right)\right] \equiv \kappa\,\partial_a\mu = 0. \tag{7.69}$$

By inspecting the classical equations of motion with $\Lambda^a = 0$ we see [55] that eq. (7.69) is satisfied when the momentum of a classical membrane does not change with τ, i.e.,

$$\frac{\mathrm{d}p_\mu}{\mathrm{d}\tau} = 0. \tag{7.70}$$

Then the membrane satisfies the minimal surface equation

$$\mathrm{D}_a\mathrm{D}^a X_\mu = 0, \tag{7.71}$$

which is just our gauge condition[4] (7.60) in the membrane space $\mathcal{M}$. When $\dot\Lambda = 0$ the energy $E \equiv \int \Lambda \mathrm{d}^n\xi \left(\frac{p^\mu p_\mu}{\sqrt{|f|}\kappa} - \sqrt{|f|}\kappa\right)$ is a constant of motion. Energy conservation in the presence of eq. (7.70) implies

$$\frac{\mathrm{d}\sqrt{|f|}}{\mathrm{d}\tau} = 0. \tag{7.72}$$

---

[4]The reverse is not necessarily true: the imposition of the gauge condition (7.60) does not imply (7.69), (7.70). The latter are additional assumptions which fix a possible congruence of trajectories (i..e., of $X^{\mu(\xi)}(\tau)$) over which the wave functional is defined.

Quantization

Our stationary state (7.67) is thus defined over a congruence of classical trajectories satisfying (7.71) and (7.70), which imply also (7.69) and (7.72). eq. (7.68) then becomes simply

$$-i\frac{\delta\psi_p}{\delta X^\alpha} = p_\alpha \psi_p. \tag{7.73}$$

Using (7.73) it is straightforward to verify that (7.67) is a particular solution to the Schrödinger eqation (7.53). In ref. [54] we found the same relation (7.73), but by using a different, more involved procedure.

THE WAVE PACKET

From the particular solutions (7.67) we can compose a wave packet

$$\psi[\tau, X^\mu(\xi)] = \int \mathcal{D}p\, c[p]\, \psi_p[\tau, X^\mu(\xi)] \tag{7.74}$$

which is also a solution to the Schrödinger equation (7.53). However, since all $\psi_p[\tau, X^\mu(\xi)]$ entering (7.74) are assumed to belong to a restricted class of particular solutions with $p_\mu(\xi)$ which does not functionally depend on $X^\mu(\xi)$, a wave packet of the form (7.74) cannot represent every possible state of the membrane. This is just a particular kind of wave packet; a general wave packet can be formed from a complete set of particular solutions which are not restricted to momenta $p_{\mu(\xi)}$ satisfying $\partial_{\nu(\xi')} p_{\mu(\xi)} = 0$, but allow for $p_{\mu(\xi)}$ which do depend on $X^{\mu(\xi)}$. Treatment of such a general case is beyond the scope of the present book. Here we shall try to demonstrate some illustrative properties of the wave packet (7.74).

In the definition of the invariant measure in momentum space we use the metric (4.14) with $\alpha = \kappa/\Lambda$:

$$\mathcal{D}p \equiv \prod_{\xi,\mu} \left(\frac{\Lambda}{\sqrt{|f|}\kappa}\right)^{1/2} dp_\mu(\xi). \tag{7.75}$$

Let us take[5]

---

[5]This can be written compactly as

$$c[p] = \mathcal{B}\exp[-\frac{1}{2}\rho_{\mu(\xi)\nu(\xi'')}(p^{\mu(\xi)} - p_0^{\mu(\xi)})(p^{\nu(\xi')} - p_0^{\nu(\xi')})\sigma_{(\xi')}{}^{(\xi'')}],$$

where $\sigma_{(\xi')}{}^{(\xi'')} = \sigma(\xi')\delta(\xi', \xi'')$. Since the covariant derivative of the metric is zero, we have that $D_{\mu(\xi)} c[p] = 0$. Similarly, the measure $\mathcal{D}p = (\text{Det}\, \rho_{\mu(\xi)\nu(\xi')})^{1/2} \prod_{\mu,\xi} dp_\mu(\xi)$, and the covariant derivative of the determinant is zero. Therefore

$$D_{\mu(\xi)} \int \mathcal{D}p\, c[p]\, \psi_p[\tau, X^\mu(\xi)] = \int \mathcal{D}p\, c[p]\, D_{\mu(\xi)} \psi_p[\tau, X^\mu(\xi)].$$

This confirms that the superposition (7.74) is a solution if ψ_p is a solution of (7.24).

218   THE LANDSCAPE OF THEORETICAL PHYSICS: A GLOBAL VIEW

$$c[p] = \mathcal{B}\exp\left[-\frac{1}{2}\int d^n\xi \frac{\Lambda}{\sqrt{|f|}\kappa}(p^\mu - p_0^\mu)^2 \sigma(\xi)\right], \qquad (7.76)$$

where

$$\mathcal{B} = \lim_{\Delta\xi \to 0} \prod_{\xi,\mu} \left(\frac{\Delta\xi\,\sigma(\xi)}{\pi}\right)^{1/4} \qquad (7.77)$$

is the normalization constant, such that $\int \mathcal{D}p\, c^*[p]c[p] = 1$. For the normalization constant $\mathcal{N}$ occurring in (7.67) we take

$$\mathcal{N} = \lim_{\Delta\xi \to 0} \prod_{\xi,\mu} \left(\frac{\Delta\xi}{2\pi}\right)^{1/2}. \qquad (7.78)$$

From (7.74)–(7.76) and (7.67) we have

$$\psi[\tau, X(\xi)] =$$

$$\lim_{\Delta\xi\to 0}\prod_{\xi,\mu}\int \left(\frac{\Delta\xi}{2\pi}\right)^{1/2}\left(\frac{\Delta\xi\,\sigma(\xi)}{\pi}\right)^{1/4}\left(\frac{\Lambda}{\sqrt{|f|}\kappa}\right)^{1/2} dp_\mu(\xi)$$

$$\times \exp\left[-\frac{\Delta\xi}{2}\frac{\Lambda}{\sqrt{|f|}\kappa}\left((p^\mu - p_0^\mu)^2\sigma(\xi) - 2i\frac{\sqrt{|f|}\kappa}{\Lambda}p_\mu(X^\mu - X_0^\mu) + i\tau p_\mu p^\mu\right)\right]$$

$$\times \exp\left[\frac{i\tau}{2}\int d^n\xi\,\sqrt{|f|}\Lambda\kappa\right].$$

$$(7.79)$$

We assume no summation over μ in the exponent of the above expression and no integration (actually summation) over ξ, because these operations are now already included in the product which acts on the whole expression. Because of the factor $\left(\frac{\Delta\xi\Lambda}{\sqrt{|f|}\kappa}\right)^{1/2}$ occurring in the measure and the same factor in the exponent, the integration over p in eq. (7.79) can be performed straightforwardly. The result is

$$\psi[\tau, X] = \left[\lim_{\Delta\xi\to 0}\prod_{\xi,\mu}\left(\frac{\Delta\xi\,\sigma}{\pi}\right)^{1/4}\left(\frac{1}{\sigma+i\tau}\right)^{1/2}\right]$$

$$\times \exp\left[\int d^n\xi\,\frac{\Lambda}{\sqrt{|f|}\kappa}\left(\frac{\left(i\frac{\sqrt{|f|}\kappa}{\Lambda}(X^\mu - X_0^\mu) + p_0^\mu\sigma\right)^2}{2(\sigma+i\tau)} - \frac{p_0^2\sigma}{2}\right)\right] \times$$

Quantization

$$\times \exp\left[\frac{i\tau}{2}\int d^n\xi\sqrt{|f|}\Lambda\kappa\right]. \tag{7.80}$$

Eq. (7.80) is a generalization of the familiar Gaussian wave packet. At $\tau = 0$ eq. (7.80) becomes

$$\psi[0, X] = \left[\lim_{\Delta\xi\to 0}\prod_{\xi,\mu}\left(\frac{\Delta\xi}{\pi\sigma}\right)^{1/4}\right]\exp\left[-\int d^n\xi\frac{\sqrt{|f|}\kappa}{\Lambda}\frac{(X^\mu(\xi) - X_0^\mu(\xi))^2}{2\sigma(\xi)}\right]$$

$$\times \exp\left[i\int p_{0\mu}(X^\mu - X_0^\mu)d^n\xi\right]. \tag{7.81}$$

The probability density is given by

$$|\psi[\tau, X]|^2 = \left[\lim_{\Delta\xi\to 0}\prod_{\xi,\mu}\left(\frac{\Delta\xi\,\sigma}{\pi}\right)^{1/2}\left(\frac{1}{\sigma^2 + \tau^2}\right)^{1/2}\right]$$

$$\times \exp\left[-\int d^n\xi\frac{\sqrt{|f|}\kappa}{\Lambda}\frac{(X^\mu - X_0^\mu - \frac{\Lambda}{\sqrt{|f|}\kappa}p_0^\mu\tau)^2}{(\sigma^2 + \tau^2)/\sigma}\right], \tag{7.82}$$

and the normalization constant, although containing the infinitesimal $\Delta\xi$, gives precisely $\int |\psi|^2 \mathcal{D}X = 1$.

From (7.82) we find that the motion of the centroid membrane of our particular wave packet is determined by the equation

$$X_C^\mu(\tau, \xi) = X_0^\mu(\xi) + \frac{\Lambda}{\sqrt{|f|}\kappa}p_0^\mu(\xi)\tau. \tag{7.83}$$

From the classical equation of motion [55] derived from (7.1) (see also (7.70), (7.72)) we indeed obtain a solution of the form (7.83). At this point it is interesting to observe that the classical null strings considered, within different theoretical frameworks, by Schild [63] and Roshchupkin et al. [83] also move according to equation (7.83).

A special choice of wave packet. The function $\sigma(\xi)$ in eqs. (7.76)–(7.82) is arbitrary; the choice of $\sigma(\xi)$ determines how the wave packet is prepared. In particular, we may consider *Case 1* of Sec. 4.3 and take $\sigma(\xi)$ such that the wave packet of a $(p+1)$-dimensional membrane $\mathcal{V}_{p+1}$ is peaked around a space-like p-dimensional membrane $\mathcal{V}_p$. This means that the wave functional localizes $\mathcal{V}_{p+1}$ much more sharply around $\mathcal{V}_p$ than in other regions of spacetime. Effectively, such a wave packet describes the τ-evolution of $\mathcal{V}_p$ (although formally it describes the τ-evolution of $\mathcal{V}_{p+1}$). This can be clearly seen by taking the following limiting form of the wave packet (7.81), such that

$$\frac{1}{\sigma(\xi)} = \frac{\delta(\xi^0 - \xi_\Sigma^0)}{\sigma(\xi^i)(\partial_0 X^\mu \partial_0 X_\mu)^{1/2}}, \quad i = 1, 2, ..., p, \quad (7.84)$$

and choosing

$$p_{0\mu}(\xi^a) = \bar{p}_{0\mu}(\xi^i)\delta(\xi^0 - \xi_\Sigma^0). \quad (7.85)$$

Then the integration over the δ-function gives in the exponent of eq. (7.81) the expression

$$\int d^p\xi \frac{\sqrt{|\bar{f}|}\kappa}{\Lambda}(X^\mu(\xi^i) - X_0^\mu(\xi^i))^2 \frac{1}{2\sigma(\xi^i)} + i\int d^p\xi\, \bar{p}_{0\mu}(X^\mu(\xi^i)(\xi^i) - X_0^\mu(\xi^i)), \quad (7.86)$$

so that eq. (7.81) becomes a wave functional of a p-dimensional membrane $X^\mu(\xi^i)$. Here again $\bar{f}$ is the determinant of the induced metric on V_p, while f is the determinant of the induced metric on V_{p+1}. One can verify that such a wave functional (7.81), (7.86) satisfies the relation (7.50).

Considerations analogous to those described above hold for *Case 2* as well, so that a wave functional of a $(p-1)$-brane can be considered as a limiting case of a p-brane's wave functional.

THE EXPECTATION VALUES

The expectation value of an operator $\widehat{A}$ is defined by

$$\langle \widehat{A} \rangle = \int \psi^*[\tau, X(\xi)]\, \widehat{A}\, \psi[\tau, X(\xi)]\, \mathcal{D}X, \quad (7.87)$$

where the invariant measure (see (4.16)) in the membrane space $\mathcal{M}$ with the metric (7.4) is

$$\mathcal{D}X = \prod_{\xi,\mu}\left(\frac{\sqrt{|f(\xi)|}\kappa}{\Lambda}\right)^{1/2} dX^\mu(\xi). \quad (7.88)$$

Using (7.74) we can express eq. (7.87) in the momentum space

$$\langle \widehat{A} \rangle = \int \mathcal{D}X\, \mathcal{D}p\, \mathcal{D}p'\, c^*[p']c[p]\psi_{p'}^*\, \widehat{A}\, \psi_p, \quad (7.89)$$

where ψ_p is given in eq. (7.67). For the momentum operator[6] we have $\hat{p}_\mu \psi = p_\mu \psi$, and eq. (7.89) becomes

---

[6]In order to distinguish operators from their eigenvalues we use here the hatted notation.

Quantization

$$\langle \hat{p}_\mu(\xi) \rangle = \tag{7.90}$$

$$\lim_{\Delta\xi' \to 0} \prod_{\xi',\nu} \int \left(\frac{\Delta\xi'}{2\pi}\right) \left(\frac{\sqrt{|f|}\kappa}{\Lambda}\right)^{1/2} \mathrm{d}X^\nu(\xi')$$

$$\times \exp\left[i\Delta\xi(p_\nu - p_{0\nu})(X^\nu - X_0^\nu)\right]$$

$$\times \exp\left[-i\Delta\xi \left(\frac{\sqrt{|f|}\kappa}{\Lambda}\right)^{-1}(p_0'^\nu p'_{0\nu} - p_0^\nu p_{0\nu})\tau\right] p_\mu(\xi) c^*[p']c[p] \mathcal{D}p\, \mathcal{D}p'.$$

The following relations are satisfied:

$$\prod_{\xi,\mu} \int \left(\frac{\Delta\xi}{2\pi}\right) \left(\frac{\sqrt{|f|}\kappa}{\Lambda}\right)^{1/2} \mathrm{d}X^\mu(\xi) \exp\left[i\Delta\xi(p'_\mu - p_{0\mu})(X^\mu - X_0^\mu)\right]$$

$$= \prod_{\xi\mu} \frac{\delta\left(p'_\mu(\xi) - p_\mu(\xi)\right)}{\left(\frac{\sqrt{|f|}\kappa}{\Lambda}\right)^{-1/2}}, \tag{7.91}$$

$$\prod_{\xi,\mu} \int \left(\frac{\Delta\xi}{2\pi}\right) \left(\frac{\sqrt{|f|}\kappa}{\Lambda}\right)^{-1/2} \mathrm{d}p_\mu(\xi) \exp\left[i\Delta\xi\, p_\mu(X'^\mu - X^\mu)\right]$$

$$= \prod_{\xi,\mu} \frac{\delta\left(X'^\mu(\xi) - X^\mu(\xi)\right)}{\left(\frac{\sqrt{|f|}\kappa}{\Lambda}\right)^{1/2}}. \tag{7.92}$$

Using (7.92) in eq. (7.90) we have simply

$$\langle \hat{p}_\mu(\xi) \rangle = \int \mathcal{D}p\, p_\mu(\xi) c^*[p] c[p]. \tag{7.93}$$

Let us take for $c[p]$ the expression (7.76). From (7.93), using the measure (7.75), we obtain after the straightforward integration that the expectation value of the momentum is equal to the centroid momentum $p_{0\mu}$ of the Gaussian wave packet (7.76):

$$\langle \hat{p}_\mu(\xi) \rangle = p_\mu(\xi). \tag{7.94}$$

Another example is the expectation value of the membrane position $X^\mu(\xi)$ with respect to the evolving wave packet (7.80):

$$\langle \widehat{X}^\mu(\xi) \rangle = \int \mathcal{D}X \, \psi^*[\tau, X(\xi)] X^\mu(\xi) \psi[\tau, X(\xi)]. \tag{7.95}$$

Inserting the explicit expression (7.80), into (7.95) we find

$$\langle \widehat{X}^\mu \rangle = X_0^\mu(\xi) + \frac{\Lambda}{\sqrt{|f|}\kappa} p_0^\mu(\xi)\tau \equiv X_c^\mu(\tau,\xi), \tag{7.96}$$

which is the correct expectation value, in agreement with that obtained from the probability density (7.82).

Although the normalization constants in eqs. (7.76)-(7.82) contain the infinitesimal $\Delta\xi$, it turns out that, when calculating the expectation values, $\Delta\xi$ disappears from the expressions. Therefore our wave functionals (7.76) and (7.80) with such normalization constants are well defined operationally.

CONCLUSION

We have started to elaborate a theory of relativistic p-branes which is more general than the theory of conventional, constrained, p-branes. In the proposed generalized theory, p-branes are unconstrained, but amongst the solutions to the classical and quantum equations of motion there are also the usual, constrained, p-branes. A strong motivation for such a generalized approach is the elimination of the well known difficulties due to the presence of constraints. Since the p-brane theories are still at the stage of development and have not yet been fully confronted with observations, it makes sense to consider an enlarged set of classical and quantum p-brane states, such as, e.g., proposed in the present and some previous works [53]–[55]. What we gain is a theory without constraints, still fully relativistic, which is straightforward both at the classical and the quantum level, and is not in conflict with the conventional p-brane theory.

Our approach might shed more light on some very interesting developments concerning the duality [84], such as one of strings and 5-branes, and the interesting interlink between p-branes of various dimensions p. In this chapter we have demonstrated how a higher-dimensional p-brane equation naturally contains lower-dimensional p-branes as solutions.

The highly non-trivial concept of unconstrained membranes enables us to develop the elegant formulation of "point particle" dynamics in the infinite-dimensional space $\mathcal{M}$. It is fascinating that the action, canonical and Hamilton formalism, and, after quantization, the Schrödinger equation all look like nearly trivial extensions of the correspondings objects in the elegant Fock–Stueckelberg–Schwinger–DeWitt proper time formalism for a point particle in curved space. Just this "triviality", or better, simplicity, is

Quantization

a distinguished feature of our approach and we have reasons to expect that also the p-brane gauge field theory —not yet a completely solved problem— can be straightforwardly formulated along the lines indicated here.

7.2. CLIFFORD ALGEBRA AND QUANTIZATION

PHASE SPACE

Let us first consider the case of a 1-dimensional coordinate variable q and its conjugate momentum p. The two quantities can be considered as coordinates of a point in the 2-dimensional *phase space*. Let e_q and e_p be the basis vectors satisfying the Clifford algebra relations

$$e_q \cdot e_p \equiv \tfrac{1}{2}(e_q e_p + e_p e_q) = 0, \tag{7.97}$$

$$e_q^2 = 1, \qquad e_p^2 = 1. \tag{7.98}$$

An arbitrary vector in phase space is then

$$Q = q e_q + p e_p. \tag{7.99}$$

The product of two vectors e_p and e_q is the unit bivector in phase space and it behaves as the imaginary unit

$$i = e_p e_q, \qquad i^2 = -1. \tag{7.100}$$

The last relation immediately follows from (7.97), (7.98): $i^2 = e_p e_q e_p e_q = -e_p^2 e_q^2 = -1$.

Multiplying (7.99) respectively from the right and from the left by e_q we thus introduce the quantities Z and Z^*:

$$Q e_q = q + p e_p e_q = q + pi = Z, \tag{7.101}$$

$$e_q Q = q + p e_q e_p = q - pi = Z^*. \tag{7.102}$$

For the square we have

$$Q e_q e_q Q = Z Z^* = q^2 + p^2 + i(pq - qp), \tag{7.103}$$

$$e_q Q Q e_q = Z^* Z = q^2 + p^2 - i(pq - qp). \tag{7.104}$$

Upon quantization q, p do not commute, but satisfy

$$[q, p] = i, \tag{7.105}$$

therefore (7.103), (7.104) become

$$ZZ^* = q^2 + p^2 + 1, \tag{7.106}$$

$$Z^*Z = q^2 + p^2 - 1, \tag{7.107}$$

$$[Z, Z^*] = 1. \tag{7.108}$$

Even before quantization the natural variables for describing physics are the complex quantity Z and its conjugate Z^*. *The imaginary unit is the bivector of the phase space, which is 2-dimensional.*

Writing $q = \rho \cos \phi$ and $p = \rho \sin \phi$ we find

$$Z = \rho(\cos \phi + i \sin \phi) = \rho e^{i\phi}, \tag{7.109}$$

$$Z^* = \rho(\cos \phi - i \sin \phi) = \rho e^{-i\phi}, \tag{7.110}$$

where ρ and ϕ are real numbers. Hence taking into account that physics takes place in the phase space and that the latter can be described by complex numbers, we automatically introduce complex numbers into both the classical and quantum physics. And what is nice here is that *the complex numbers are nothing but the Clifford numbers of the 2-dimensional phase space.*

What if the configuration space has more than one dimension, say n? Then with each spatial coordinate is associated a 2-dimensional phase space. The dimension of the total phase space is then $2n$. A phase space vector then reads

$$Q = q^\mu e_{q\mu} + p^\mu e_{p\mu}. \tag{7.111}$$

The basis vectors have now two indices q, p (denoting the direction in the 2-dimensional phase space) and $\mu = 1, 2, ..., n$ (denoting the direction in the n-dimensional configuration space).

The basis vectors can be written as the product of the configuration space basis vectors e_μ and the 2-dimensional phase space basis vectors e_q, e_p:

$$e_{q\mu} = e_q e_\mu, \qquad e_{p\mu} = e_p e_\mu. \tag{7.112}$$

A vector Q is then

$$Q = (q^\mu e_q + p^\mu e_p)e_\mu = Q^\mu e_\mu, \tag{7.113}$$

Quantization

where
$$Q^\mu = q^\mu e_q + p^\mu e_p. \tag{7.114}$$

Eqs. (7.101), (7.102) generalize to

$$Q^\mu e_q = q^\mu + p^\mu e_p e_q = q^\mu + p^\mu i = Z^\mu, \tag{7.115}$$

$$e_q Q^\mu = q^\mu + p^\mu e_q e_p = q^\mu - p^\mu i = Z^{*\mu}. \tag{7.116}$$

Hence, even if configuration space has many dimensions, the imaginary unit i in the variables X^μ comes from the bivector $e_q e_p$ of the 2-dimensional phase space which is associated with every direction μ of the configuration space.

When passing to quantum mechanics it is then natural that in general the wave function is complex-valued. *The imaginary unit is related to the phase space which is the direct product of the configuration space and the 2-dimensional phase space.*

At this point let us mention that Hestenes was one of the first to point out clearly that imaginary and complex numbers need not be postulated separately, but they are automatically contained in the geometric calculus based on Clifford algebra. When discussing quantum mechanics Hestenes ascribes the occurrence of the imaginary unit i in the Schrödinger and especially in the Dirac equation to a chosen configuration space Clifford number which happens to have the square -1 and which commutes with all other Clifford numbers within the algebra. This brings an ambiguity as to which of several candidates should serve as the imaginary unit i. In this respect Hestenes had changed his point of view, since initially he proposed that one must have a 5-dimensional space time whose pseudoscalar unit $I = \gamma_0 \gamma_1 \gamma_2 \gamma_3 \gamma_4$ commutes with all the Clifford numbers of $\mathcal{C}_5$ and its square is $I^2 = -1$. Later he switched to 4-dimensional space time and chose the bivector $\gamma_1 \gamma_2$ to serve the role of i. I regard this as unsatisfactory, since $\gamma_2 \gamma_3$ or $\gamma_1 \gamma_3$ could be given such a role as well. In my opinion it is more natural to ascribe the role of i to the bivector of the 2-dimensional phase space sitting at every coordinate of the configuration space. A more detailed discussion about the relation between the geometric calculus in a generic 2-dimensional space (not necessarily interpreted as phase space) and complex number is to be found in Hestenes' books [22].

WAVE FUNCTION AS A POLYVECTOR

We have already seen in Sec. 2.5 that a wave function can in general be considered as a polyvector, i.e., as a Clifford number or Clifford aggregate generated by a countable set of basis vectors e_μ. Such a wave function contains spinors, vectors, tensors, etc., all at once. In particular, it may contain only spinors, or only vectors, etc. .

Let us now further generalize this important procedure. In Sec. 6.1 we have discussed vectors in an infinite-dimensional space V_∞ from the point of view of geometric calculus based on the Clifford algebra generated by the uncountable set of basis vectors $h(x)$ of V_∞. We now apply that procedure to the case of the wave function which, in general, is complex-valued.

For an arbitrary complex function we have

$$f(x) = \frac{1}{\sqrt{2}}(f_1(x) + if_2(x)), \qquad f^*(x) = \frac{1}{\sqrt{2}}(f_1(x) - if_2(x)), \quad (7.117)$$

where $f_1(x)$, $f_2(x)$ are real functions. From (7.117) we find

$$f_1(x) = \frac{1}{\sqrt{2}}(f(x) + f^*(x)), \qquad f_2(x) = \frac{1}{i\sqrt{2}}(f(x) - f^*(x)). \quad (7.118)$$

Hence, instead of a complex function we can consider a set of two independent real functions $f_1(x)$ and $f_2(x)$.

Introducing the basis vectors $h_1(x)$ and $h_2(x)$ satisfying the Clifford algebra relations

$$h_i(x) \cdot h_j(x') \equiv \tfrac{1}{2}(h_i(x)h_j(x') + h_j(x')h_i(x)) = \delta_{ij}\delta(x - x'), \quad i,j = 1,2, \quad (7.119)$$

we can expand an arbitrary vector F according to

$$F = \int \mathrm{d}x(f_1(x)h_1(x) + f_2(x)h_2(x)) = f^{i(x)}h_{i(x)}, \quad (7.120)$$

where $h_{i(x)} \equiv h_i(x)$, $f^{i(x)} \equiv f_i(x)$. Then

$$F \cdot h_1(x) = f_1(x), \qquad F \cdot h_2(x) = f_2(x) \quad (7.121)$$

are components of F.

Quantization

Introducing the imaginary unit i which commutes[7] with $h_i(x)$ we can form a new set of basis vectors

$$h(x) = \frac{h_1(x) + ih_2(x)}{\sqrt{2}}, \quad h^*(x) = \frac{h_1(x) - ih_2(x)}{\sqrt{2}}, \quad (7.122)$$

the inverse relations being

$$h_1(x) = \frac{h(x) + h^*(x)}{\sqrt{2}}, \quad h_2(x) = \frac{h(x) - h^*(x)}{i\sqrt{2}}. \quad (7.123)$$

Using (7.118), (7.123) and (7.120) we can re-express F as

$$F = \int dx (f(x)h^*(x) + f^*(x)h(x)) = f^{(x)} h_{(x)} + f^{*(x)} h^*_{(x)}, \quad (7.124)$$

where

$$f^{(x)} \equiv f^*(x), \quad f^{*(x)} \equiv f(x), \quad h_{(x)} \equiv h(x), \quad h^*_{(x)} \equiv h^*(x). \quad (7.125)$$

From (7.119) and (7.123) we have

$$h(x) \cdot h^*(x) \equiv \tfrac{1}{2}(h(x)h^*(x') + h^*(x')h(x)) = \delta(x - x'), \quad (7.126)$$

$$h(x) \cdot h(x') = 0, \quad h^*(x) \cdot h^*(x') = 0, \quad (7.127)$$

which are *the anticommutation relations for a fermionic field.*

A vector F can be straightforwardly generalized to a *polyvector*:

$$\begin{aligned}F &= f^{i(x)} h_{i(x)} + f^{i(x)j(x')} h_{i(x)} h_{j(x')} + f^{i(x)j(x')k(x'')} h_{i(x)} h_{j(x')} h_{k(x'')} + \cdots \\ &= f^{(x)} h_{(x)} + f^{(x)(x')} h_{(x)} h_{(x')} + f^{(x)(x')(x'')} h_{(x)} h_{(x')} h_{(x'')} + \cdots \\ &\quad + f^{*(x)} h^*_{(x)} + f^{*(x)(x')} h^*_{(x)} h^*_{(x')} + f^{*(x)(x')(x'')} h^*_{(x)} h^*_{(x')} h^*_{(x'')} + \cdots \end{aligned} \quad (7.128)$$

---

[7] Now, the easiest way to proceed is in forgetting how we have obtained the imaginary unit, namely as a bivector in 2-dimensional phase space, and define all the quantities i, $h_1(x)$, $h_2(x)$, etc., in such a way that i commutes with everything. If we nevertheless persisted in maintaining the geometric approach to i, we should then take $h_1(x) = e(x)$, $h_2(x) = e(x)e_p e_q$, satisfying

$$\begin{aligned} h_1(x) \cdot h_1(x') &= e(x) \cdot e(x') = \delta(x - x'), \\ h_2(x) \cdot h_2(x) &= -\delta(x - x'), \\ h_1(x) \cdot h_2(x') &= \delta(x - x')1 \cdot (e_p e_q) = 0, \end{aligned}$$

where according to Hestenes the inner product of a scalar with a multivector is zero. Introducing $h = (h_1 + h_2)/\sqrt{2}$ and $h^* = (h_1 - h_2)/\sqrt{2}$ one finds $h(x) \cdot h^*(x') = \delta(x - x')$, $h(x) \cdot h(x') = 0$, $h^*(x) \cdot h^*(x') = 0$.

where $f^{(x)(x')(x'')\cdots}$ are scalar coefficients, *antisymmetric* in $(x)(x')(x'')\ldots$

We have exactly the same expression (7.128) in the usual quantum field theory (QFT), where $f^{(x)}$, $f^{(x)(x')}$,..., are 1-particle, 2-particle,..., wave functions (wave packet profiles). Therefore a natural interpretation of the polyvector F is that it represents a superposition of multi-particle states.

In the usual formulation of QFT one introduces a vacuum state $|0\rangle$, and interprets $h(x)$, $h^*(x)$ as the operators which create or annihilate a particle or an antiparticle at x, so that (roughly speaking) e.g. $h^*(x)|0\rangle$ is a state with a particle at position x.

In the geometric calculus formulation (based on the Clifford algebra of an infinite-dimensional space) the Clifford numbers $h^*(x)$, $h(x)$ already represent vectors. At the same time $h^*(x)$, $h(x)$ also behave as operators, satisfying (7.126), (7.127). When we say that a state vector is expanded in terms of $h^*(x)$, $h(x)$ we mean that it is a superposition of states in which a particle has a definite position x. The latter states are just $h^*(x)$, $h(x)$. Hence the Clifford numbers (operators) $h^*(x)$, $h(x)$ need not act on a vacuum state in order to give the one-particle states. They are already the one-particle states. Similarly the products $h(x)h(x')$, $h(x)h(x')h(x'')$, $h^*(x)h^*(x')$, etc., already represent the multi-particle states.

When performing quantization of a classical system we arrived at *the wave function*. The latter can be considered as an uncountable (infinite) set of scalar components of a vector in an infinite-dimensional space, spanned by the basis vectors $h_1(x)$, $h_2(x)$. Once we have basis vectors we automatically have not only arbitrary vectors, but also arbitrary polyvectors which are *Clifford numbers* generated by $h_1(x)$, $h_2(x)$ (or equivalently by $h(x)$, $h^*(x)$. Hence *the procedure in which we replace infinite-dimensional vectors with polyvectors is equivalent to the second quantization.*

If one wants to consider *bosons* instead of *fermions* one needs to introduce a new type of fields $\xi_1(x)$, $\xi_2(x)$, satisfying *the commutation relations*

$$\tfrac{1}{2}[\xi_i(x), \xi_j(x')] \equiv \tfrac{1}{2}[\xi_i(x)\xi_j(x') - \xi_j(x')\xi_i(x)] = \epsilon_{ij}\Delta(x-x')\tfrac{1}{2}, \quad (7.129)$$

with $\epsilon_{ij} = -\epsilon_{ji}$, $\Delta(x-x') = -\Delta(x'-x)$, which stay instead of the anticommutation relations (7.119). Hence the numbers $\xi(x)$ are not Clifford numbers. By (7.129) the $\xi_i(x)$ generate a new type of algebra, which could be called an *anti-Clifford algebra*.

Instead of $\xi_i(x)$ we can introduce the basis vectors

$$\xi(x) = \frac{\xi_1(x) + i\xi_2}{\sqrt{2}}, \qquad \xi^*(x) = \frac{\xi_1(x) - i\xi_2}{\sqrt{2}} \quad (7.130)$$

which satisfy the commutation relations

$$[\xi(x), \xi^*(x')] = -i\Delta(x-x'), \quad (7.131)$$

Quantization

$$[\xi(x), \xi(x')] = 0, \qquad [\xi^*(x), \xi^*(x')] = 0. \qquad (7.132)$$

A polyvector representing a superposition of bosonic multi-particle states is then expanded as follows:

$$B = \phi^{i(x)} \xi_{i(x)} + \phi^{i(x)j(x')} \xi_{i(x)} \xi_{j(x')} + \cdots \qquad (7.133)$$

$$= \phi^{(x)} \xi_{(x)} + \phi^{(x)(x')} \xi_{(x)} \xi_{(x')} + \phi^{(x)(x')(x'')} \xi_{(x)} \xi_{(x')} \xi_{(x'')} + \cdots$$

$$+ \phi^{*(x)} \xi^*_{(x)} + \phi^{*(x)(x')} \xi^*_{(x)} \xi^*_{(x')} + \phi^{*(x)(x')(x'')} \xi^*_{(x)} \xi^*_{(x')} \xi^*_{(x'')} + \cdots,$$

where $\phi^{i(x)j(x')\cdots}$ and $\phi^{(x)(x')\cdots}$, $\phi^{*(x)(x')(x'')\cdots}$ are scalar coefficients, *symmetric* in $i(x)j(x')\ldots$ and $(x)(x')\ldots$, respectively. They can be interpreted as representing 1-particle, 2-particle,..., wave packet profiles. Because of (7.131) $\xi(x)$ and $\xi^*(x)$ can be interpreted as creation operators for *bosons*. Again, *a priori* we do not need to introduce a vacuum state. However, whenever convenient we may, of course, define a vacuum state and act on it by the operators $\xi(x)$, $\xi^*(x)$.

EQUATIONS OF MOTION FOR BASIS VECTORS

In the previous subsection we have seen how the geometric calculus naturally leads to the second quantization which incorporates superpositions of multi-particle states. We shall now investigate what are the equations of motion that the basis vectors satisfy.

For illustration let us consider the action for a *real scalar field* $\phi(x)$:

$$I[\phi] = \tfrac{1}{2} \int d^4 x \, (\partial_\mu \phi \partial^\mu \phi - m^2). \qquad (7.134)$$

Introducing the metric

$$\rho(x, x') = h(x) \cdot h(x') \equiv \tfrac{1}{2}(h(x)h(x') + h(x')h(x)) \qquad (7.135)$$

we have

$$I[\phi] = \tfrac{1}{2} \int dx \, dx' \left(\partial_\mu \phi(x) \partial'^\mu \phi(x') - m^2 \phi(x)\phi(x') \right) h(x)h(x'). \qquad (7.136)$$

If, in particular,

$$\rho(x, x') = h(x) \cdot h(x') = \delta(x - x') \qquad (7.137)$$

then the action (7.136) is equivalent to (7.134).

In general, $\rho(x, x')$ need not be equal to $\delta(x - x')$, and (7.136) is then a generalization of the usual action (7.134) for the scalar field. An action which is invariant under field redefinitions ('coordinate' transformations in the space of fields) has been considered by Vilkovisky [85]. Integrating (7.136) *per partes* over x and x' and omitting the surface terms we obtain

$$I[\phi] = \tfrac{1}{2} \int dx\, dx'\, \phi(x)\phi(x') \left(\partial_\mu h(x) \partial'^\mu h(x') - m^2 h(x) h(x') \right). \qquad (7.138)$$

Derivatives no longer act on $\phi(x)$, but on $h(x)$. If we fix $\phi(x)$ then instead of an action for $\phi(x)$ we obtain an action for $h(x)$.

For instance, if we take

$$\phi(x) = \delta(x-y) \qquad (7.139)$$

and integrate over y we obtain

$$I[h] = \frac{1}{2} \int dy \left(\frac{\partial h(y)}{\partial y^\mu} \frac{\partial h(y)}{\partial y_\mu} - m^2 h^2(y) \right). \qquad (7.140)$$

The same equation (7.140), of course, follows directly from (7.136) in which we fix $\phi(x)$ according to (7.139).

On the other hand, if instead of $\phi(x)$ we fix $h(x)$ according to (7.137), then we obtain the action (7.134) which governs the motion of $\phi(x)$.

Hence the same basic expression (7.136) can be considered either as an action for $\phi(x)$ or an action for $h(x)$, depending on which field we consider as fixed and which one as a variable. If we consider the basis vector field $h(x)$ as a variable and $\phi(x)$ as fixed according to (7.139), then we obtain the action (7.140) for $h(x)$. The latter field is actually an operator. The procedure from now on coincides with the one of quantum field theory.

Renaming y^μ as x^μ (7.140) becomes an action for a bosonic field:

$$I[h] = \tfrac{1}{2} \int dx\, (\partial_\mu h \partial^\mu h - m^2 h^2). \qquad (7.141)$$

The canonically conjugate variables are

$$h(t, \mathbf{x}) \qquad \text{and} \qquad \pi(t, \mathbf{x}) = \partial \mathcal{L}/\partial \dot{h} = \dot{h}(t, \mathbf{x}).$$

They satisfy the *commutation relations*

$$[h(t,\mathbf{x}), \pi(t,\mathbf{x}')] = i\delta^3(\mathbf{x}-\mathbf{x}'), \qquad [h(t,\mathbf{x}), h(t,\mathbf{x}')] = 0. \qquad (7.142)$$

At different times $t' \neq t$ we have

$$[h(x), h(x')] = i\Delta(x-x'), \qquad (7.143)$$

Quantization

where $\Delta(x-x')$ is the well known covariant function, antisymmetric under the exchange of x and x'.

The geometric product of two vectors can be decomposed as

$$h(x)h(x') = \tfrac{1}{2}\left(h(x)h(x') + h(x')h(x)\right) + \tfrac{1}{2}\left(h(x)h(x') - h(x')h(x)\right). \tag{7.144}$$

In view of (7.143) we have that *the role of the inner product is now given to the antisymmetric part, whilst the role of the outer product is given to the symmetric part*. This is characteristic for *bosonic vectors*; they generate what we shall call the *anti-Clifford algebra*. In other words, when the basis vector field $h(x)$ happens to satisfy the *commutation relation*

$$[h(x), h(x')] = f(x, x'), \tag{7.145}$$

where $f(x,x')$ is a scalar two point function (such as $i\Delta(x-x')$), it behaves as a *bosonic field*. On the contrary, when $h(x)$ happens to satisfy the *anticommutation relation*

$$\{h(x), h(x')\} = g(x, x'), \tag{7.146}$$

where $g(x,x')$ is also a scalar two point function, then it behaves as a *fermionic field*[8]

The latter case occurs when instead of (7.134) we take the action for the Dirac field:

$$I[\psi, \bar\psi] = \int d^4x\, \bar\psi(x)(i\gamma^\mu \partial_\mu - m)\psi(x). \tag{7.147}$$

Here we are using the usual spinor representation in which the spinor field $\psi(x) \equiv \psi_\alpha(x)$ bears the spinor index α. A generic vector is then

$$\begin{aligned}\Psi &= \int dx\, (\bar\psi_\alpha(x)h_\alpha(x) + \psi_\alpha(x)\bar h_\alpha(x)) \\ &\equiv \int dx\, (\bar\psi(x)h(x) + \psi(x)\bar h(x)).\end{aligned} \tag{7.148}$$

Eq. (7.147) is then equal to the scalar part of the action

$$\begin{aligned}I[\psi,\bar\psi] &= \int dx\, dx'\, \bar\psi(x')\bar h(x)h(x')(i\gamma^\mu\partial_\mu - m)\psi(x) \\ &= \int dx\, dx'\, \bar\psi(x')\left[\bar h(x)(i\gamma^\mu - m)h(x')\right]\psi(x),\end{aligned} \tag{7.149}$$

where $h(x), \bar h(x)$ are assumed to satisfy

$$\bar h(x)\cdot h(x') \equiv \tfrac{1}{2}\left(\bar h(x)h(x') + h(x')\bar h(x)\right) = \delta(x-x') \tag{7.150}$$

---

[8]In the previous section the bosonic basis vectors were given a separate name $\xi(x)$. Here we retain the same name $h(x)$ both for bosonic and fermionic basis vectors.

The latter relation follows from the Clifford algebra relations amongst the basis fields $h_i(x)$, $i = 1, 2$,

$$h_i(x) \cdot h_j(x') = \delta_{ij}(x - x') \qquad (7.151)$$

related to $\bar{h}(x)$, $h(x)$ accroding to

$$h_1(x) = \frac{h(x) + \bar{h}(x)}{\sqrt{2}}, \qquad h_2(x) = \frac{h(x) - \bar{h}(x)}{i\sqrt{2}}. \qquad (7.152)$$

Now we relax the condition (7.150) and (7.149) becomes a generalization of the action (7.147).

Moreover, if in (7.149) we fix the field ψ according to

$$\psi(x) = \delta(x - y), \qquad (7.153)$$

integrate over y, and rename y back into x, we find

$$I[h, \bar{h}] = \int dx\, \bar{h}(x)(i\gamma^\mu \partial_\mu - m)h(x). \qquad (7.154)$$

This is an action for the basis vector field $h(x)$, $\bar{h}(x)$, which are *operators*. The canonically conjugate variables are now

$$h(t, \mathbf{x}) \quad \text{and} \quad \pi(t, \mathbf{x}) = \partial \mathcal{L}/\partial \dot{h} = i\bar{h}\gamma^0 = ih^\dagger.$$

They satisfy the anticommutation relations

$$\{h(t, \mathbf{x}), h^\dagger(t, \mathbf{x}')\} = \delta^3(\mathbf{x} - \mathbf{x}'), \qquad (7.155)$$

$$\{h(t, \mathbf{x}), h(t, \mathbf{x}')\} = \{h^\dagger(t, \mathbf{x}), h^\dagger(t, \mathbf{x}')\} = 0. \qquad (7.156)$$

At different times $t' \neq t$ we have

$$\{h(x), \bar{h}(x')\} = (i\gamma^\mu + m)i\Delta(x - x'), \qquad (7.157)$$

$$\{h(x), h(x')\} = \{\bar{h}(x), \bar{h}(x')\} = 0. \qquad (7.158)$$

The basis vector fields $h_i(x)$, $i = 1, 2$, defined in (7.152) then satisfy

$$\{h_i(x), h_j(x')\} = \delta_{ij}(i\gamma^\mu + m)i\Delta(x - x'), \qquad (7.159)$$

which can be written as *the inner product*

$$h_i(x) \cdot h_j(x') = \tfrac{1}{2}\delta_{ij}(i\gamma^\mu + m)i\Delta(x - x') = \rho(x, x') \qquad (7.160)$$

Quantization

with the metric $\rho(x, x')$. We see that our procedure leads us to a metric which is different from the metric assumed in (7.151).

Once we have basis vectors we can form an arbitrary *vector* according to

$$\Psi = \int dx \, (\psi(x)\bar{h}(x) + \bar{\psi}(x)h(x)) = \psi^{(x)}h_{(x)} + \bar{\psi}^{(x)}\bar{h}_{(x)}. \tag{7.161}$$

Since the $h(x)$ generates a Clifford algebra we can form not only a vector but also an arbitrary multivector and a superposition of multivectors, i.e., a *polyvector* (or *Clifford aggregate*):

$$\begin{aligned}
\Psi &= \int dx \, (\psi(x)\bar{h}(x) + \bar{\psi}(x)h(x)) \\
&\quad + \int dx \, dx' \, (\psi(x,x')\bar{h}(x)\bar{h}(x') + \bar{\psi}(x,x')h(x)h(x')) + \ldots \\
&= \psi^{(x)}h_{(x)} + \psi^{(x)(x')}h_{(x)}h_{(x')} + \ldots \\
&\quad + \bar{\psi}^{(x)}\bar{h}_{(x)} + \bar{\psi}^{(x)(x')}\bar{h}_{(x)}\bar{h}_{(x')} + \ldots,
\end{aligned} \tag{7.162}$$

where $\psi(x, x', \ldots) \equiv \psi^{(x)(x')\ldots}$, $\bar{\psi}(x, x', \ldots) \equiv \bar{\psi}^{(x)(x')\ldots}$ are *antisymmetric* functions, interpreted as wave packet profiles for a system of free *fermions*.

Similarly we can form an arbitrary polyvector

$$\begin{aligned}
\Phi &= \int dx \, \phi(x)h(x) + \int dx \, dx' \, \phi(x,x')h(x)h(x') + \ldots \\
&\equiv \phi^{(x)}h_{(x)} + \phi^{(x)(x')}h_{(x)}h_{(x')} + \ldots
\end{aligned} \tag{7.163}$$

generated by *the basis vectors* which happen to satisfy the *commutation relations* (7.142). In such a case the uncountable set of basis vectors behaves as a *bosonic field*. The corresponding multi-particle wave packet profiles $\phi(x, x', \ldots)$ are *symmetric* functions of x, $x', \ldots$. If one considers a complex field, then the equations (7.141)–(7.142) and (7.163) are generalized in an obvious way.

As already mentioned, within the conceptual scheme of Clifford algebra and hence also of anti-Clifford algebra we do not need, if we wish so, to introduce a vacuum state[9], since the operators $h(x)$, $\bar{h}(x)$ already represent states. From the actions (7.141), (7.147) we can derive the corresponding Hamiltonian, and other relevant operators (e.g., the generators of spacetime translations, Lorentz transformations, etc.). In order to calculate their expectation values in a chosen multi-particle state one may simply sandwich those operators between the state and its Hermitian conjugate (or Dirac

---

[9]Later, when discussing the states of the quantized p-brane, we nevertheless introduce a vacuum state and the set of orthonormal basis states spanning the Fock space.

conjugate) and take the scalar part of the expression. For example, the expectation value of the Hamiltonian H in a bosonic 2-particle state is

$$\langle H \rangle = \langle \int dx\, dx'\, \phi(x,x') h(x) h(x') H \int dx''\, dx'''\, \phi(x'', x''') h(x'') h(x''') \rangle_0, \tag{7.164}$$

where, in the case of the real scalar field,

$$H = \tfrac{1}{2} \int d^3\mathbf{x}(\dot{h}^2(x) - \partial^i h \partial_i h + m^2 h^2). \tag{7.165}$$

Instead of performing the operation $\langle\ \rangle_0$ (which means taking the scalar part), in the conventional approach to quantum field theory one performs the operation $\langle 0|....|0\rangle$ (i.e., taking the vacuum expectation value). However, instead of writing, for instance,

$$\langle 0 | a(\mathbf{k}) a^*(\mathbf{k}') | 0 \rangle = \langle 0 | [a(\mathbf{k}), a^*(\mathbf{k}')] | 0 \rangle = \delta^3(\mathbf{k} - \mathbf{k}'), \tag{7.166}$$

we can write

$$\langle a(\mathbf{k}) a^*(\mathbf{k}') \rangle_0 = \tfrac{1}{2} \langle a(\mathbf{k}) a^*(\mathbf{k}') + a^*(\mathbf{k}') a(\mathbf{k}) \rangle_0$$

$$+ \tfrac{1}{2} \langle a(\mathbf{k}) a^*(\mathbf{k}') - a^*(\mathbf{k}') a(\mathbf{k}) \rangle_0$$

$$= \tfrac{1}{2} \delta^3(\mathbf{k} - \mathbf{k}'), \tag{7.167}$$

where we have taken into account that for a bosonic operator the symmetric part is not a scalar. Both expressions (7.166) and (7.167) give the same result, up to the factor $\tfrac{1}{2}$ which can be absorbed into the normalization of the states.

We leave to the interested reader to explore in full detail (either as an exercise or as a research project), for various operators and kinds of field, how much the results of the above procedure (7.164) deviate, if at all, from those of the conventional approach. Special attention should be paid to what happens with the vacuum energy (the cosmological constant problem) and what remains of the anomalies. According to a very perceptive explanation provided by Jackiw [86], anomalies are the true physical effects related to the choice of vacuum (see also Chapter 3). So they should be present, at least under certain circumstances, in the procedure like (7.164) which does not explicitly require a vacuum. I think that, e.g. for the Dirac field our procedure, in the language of QFT means dealing with bare vacuum. In other words, the momentum space Fourier transforms of the vectors $h(x)$, $\bar{h}(x)$ represent states which in QFT are created out of

QUANTIZATION OF THE STUECKELBERG FIELD

In Part I we have paid much attention to the unconstrained theory which involves a *Lorentz invariant evolution parameter* τ. We have also seen that such an unconstrained Lorentz invariant theory is embedded in a polyvector generalization of the theory. Upon quantization we obtain the Schrödinger equation for the wave function $\psi(\tau, x^\mu)$:

$$i\frac{\partial \psi}{\partial \tau} = \frac{1}{2\Lambda}(-\partial_\mu \partial^\mu - \kappa^2)\psi. \tag{7.168}$$

The latter equation follows from the action

$$I[\psi, \psi^*] = \int d\tau\, d^4x \left(i\psi^* \frac{\partial \psi}{\partial \tau} - \frac{\Lambda}{2}(\partial_\mu \psi^* \partial^\mu \psi - \kappa^2 \psi^* \psi) \right). \tag{7.169}$$

This is equal to the scalar part of

$$I[\psi, \psi^*] = \int d\tau\, d\tau'\, dx\, dx' \left[i\psi^*(\tau', x') \frac{\partial \psi(\tau, x)}{\partial \tau} - \frac{\Lambda}{2}(\partial'_\mu \psi^*(\tau', x') \partial^\mu \psi(\tau, x) \right.$$

$$\left. -\kappa^2 \psi^*(\tau', x')\psi(\tau, x) \right] h^*(\tau, x) h(\tau', x')$$

$$= \int d\tau\, d\tau'\, dx\, dx'\; \psi^*(\tau', x')\psi(\tau, x)$$

$$\times \left[-i\frac{\partial h^*(\tau, x)}{\partial \tau} h(\tau', x') \right.$$

$$\left. -\frac{\Lambda}{2}\left(\partial_\mu h^*(\tau, x) \partial'^\mu h(\tau', x') - \kappa^2 h^*(\tau, x) h(\tau', x') \right) \right] \tag{7.170}$$

where $h(\tau', x')$, $h^*(\tau, x)$ are assumed to satisfy

$$h(\tau', x') \cdot h^*(\tau, x) \equiv \tfrac{1}{2}\left(h(\tau', x')h^*(\tau, x) + h^*(\tau, x)h(\tau', x')\right),$$
$$= \delta(\tau - \tau')\delta^4(x - x')$$

$$h(\tau, x) \cdot h(\tau', x') = 0, \qquad h^*(\tau, x) \cdot h^*(\tau', x') = 0. \tag{7.171}$$

The relations above follow, as we have seen in previous subsection (eqs. (7.119)-(7.127)) from the Clifford algebra relations amongst the basis fields $h_i(x)$, $i = 1, 2$,
$$h_i(\tau, x) \cdot h_j(\tau', x') = \delta_{ij}\delta(\tau - \tau')\delta(x - x') \qquad (7.172)$$
related to $h(\tau, x)$, $h^*(\tau, x)$ according to
$$h_1 = \frac{h + h^*}{\sqrt{2}}, \qquad h_2 = \frac{h - h^*}{i\sqrt{2}}. \qquad (7.173)$$

Let us now relax the condition (7.171) so that (7.170) becomes a generalization of the original action (7.169).

Moreover, if in (7.170) we fix the field ψ according to
$$\psi(\tau, x) = \delta(\tau - \tau')\delta^4(x - x'), \qquad (7.174)$$
integrate over τ', x', and rename τ', x' back into τ, x, we find
$$I[h, h^*] = \int d\tau\, d^4x \left[ih^* \frac{\partial h}{\partial \tau} - \frac{\Lambda}{2}(\partial_\mu h^* \partial^\mu h - \kappa^2 h^* h) \right], \qquad (7.175)$$
which is an action for basis vector fields $h(\tau, x)$, $h^*(\tau, x)$. The latter fields are *operators*.

The usual canonical procedure then gives that the field $h(x)$ and its conjugate momentum $\pi = \partial \mathcal{L}/\partial \dot{h} = ih^*$, where $\dot{h} \equiv \partial h/\partial \tau$, satisfy *the commutation relations*
$$[h(\tau, x), \pi(\tau', x')]|_{\tau'=\tau} = i\delta(x - x'),$$
$$[h(\tau, x), h(\tau', x')]_{\tau'=\tau} = [h^*(\tau, x), h^*(\tau', x')]_{\tau'=\tau} = 0. \qquad (7.176)$$

From here on the procedure goes along the same lines as discussed in Chapter 1, Section 4.

QUANTIZATION OF THE PARAMETRIZED DIRAC FIELD

In analogy with the Stueckelberg field we can introduce an invariant evolution parameter for the Dirac field $\psi(\tau, x^\mu)$. Instead of the usual Dirac equation we have
$$i\frac{\partial \psi}{\partial \tau} = -i\gamma^\mu \partial_\mu \psi. \qquad (7.177)$$

The corresponding action is
$$I[\psi, \bar\psi] = \int d\tau\, d^4x \left(i\bar\psi \frac{\partial \psi}{\partial \tau} + i\bar\psi \gamma^\mu \partial_\mu \psi \right). \qquad (7.178)$$

Quantization

Introducing a basis $h(\tau,x)$ in function space so that a generic vector can be expanded according to

$$\begin{aligned}\Psi &= \int d\tau\, d\tau'\, dx\, dx'\, (\bar\psi_\alpha(\tau,x)h_\alpha(\tau,x)+\psi_\alpha(\tau,x)\bar h_\alpha(\tau,x))\\ &\equiv \int d\tau\, d\tau'\, dx\, dx'\, (\bar\psi(\tau,x)h(\tau,x)+\psi(\tau,x)\bar h(\tau,x))\,. \end{aligned} \quad (7.179)$$

We can write (7.178) as the scalar part of

$$I[\psi,\bar\psi] = \int d\tau\, d\tau'\, dx\, dx'\, i\bar\psi(\tau',x')\bar h(\tau,x)h(\tau',x')\left(\frac{\partial\psi(\tau,x)}{\partial\tau}+i\gamma^\mu\partial_\mu\psi(\tau,x)\right), \quad (7.180)$$

where we assume

$$\begin{aligned}\bar h(\tau,x)\cdot h(\tau',x') &\equiv \tfrac{1}{2}\left(\bar h(\tau,x)h(\tau',x')+h(\tau',x')\bar h(\tau,x)\right)\\ &= \delta(\tau-\tau')\delta(x-x') \end{aligned} \quad (7.181)$$

For simplicity, in the relations above we have suppressed the spinor indices.

Performing partial integrations in (7.180) we can switch the derivatives from ψ to h, as in (7.149):

$$I = \int d\tau\, d\tau'\, dx\, dx'\, \bar\psi(\tau,x) \quad (7.182)$$
$$\times\left[-i\frac{\bar h(\tau,x)}{\partial\tau}h(\tau',x')-i\gamma^\mu\partial_\mu\bar h(\tau,x)h(\tau',x')\right]\psi(\tau',x').$$

We now relax the condition (7.181). Then eq. (7.182) is no longer equivalent to the action (7.178). Actually we shall no more consider (7.182) as an action for ψ. Instead we shall fix[10] ψ according to

$$\psi(\tau,x)=\delta(\tau-\tau')\delta^4(x-x'). \quad (7.183)$$

Integrating (7.182) over τ'', x'' and renaming τ'', x'' back into τ, x, we obtain an action for basis vector fields $h(\tau,x)$, $\bar h(\tau,x)$:

$$I[h,\bar h] = \int d\tau\, d^4x\left[i\bar h\frac{\partial h}{\partial\tau}+i\bar h\gamma^\mu\partial_\mu h\right]. \quad (7.184)$$

Derivatives now act on h, since we have performed additional partial integrations and have omitted the surface terms.

---

[10] Taking also the spinor indices into account, instead of (7.183) we have
$$\psi_\alpha(\tau,x)=\delta_{\alpha,\alpha'}\delta(\tau-\tau'')\delta^4(x-x'').$$

Again we have arrived at an action for field operators $h, \bar{h}$. The equations of motion (the field equations) are

$$i\frac{\partial h}{\partial \tau} = -i\gamma^\mu \partial_\mu h. \tag{7.185}$$

The canonically conjugate variables are h and $\pi = \partial \mathcal{L}/\partial \dot{h} = i\bar{h}$, and they satisfy *the anticommutation relations*

$$\{h(\tau,x), \pi(\tau',x')\}|_{\tau'=\tau} = i\delta(x-x'), \tag{7.186}$$

or

$$\{h(\tau,x), \bar{h}(\tau',x')\}|_{\tau'=\tau} = \delta(x-x'), \tag{7.187}$$

and

$$\{h(\tau,x), h(\tau',x')\}|_{\tau'=\tau} = \{\bar{h}(\tau,x), \bar{h}(\tau',x')\}|_{\tau'=\tau} = 0. \tag{7.188}$$

The anticommutation relations above being satisfied, the Heisenberg equation

$$\frac{\partial h}{\partial \tau} = i[h, H], \qquad H = \int dx\, i\bar{h}\gamma^\mu \partial_\mu h, \tag{7.189}$$

is equivalent to the field equation (7.185).

QUANTIZATION OF THE p-BRANE: A GEOMETRIC APPROACH

We have seen that a field can be considered as an uncountable set of components of an infinite-dimensional vector. Instead of considering the action which governs the dynamics of components, we have considered the action which governs the dynamics of the basis vectors. The latter behave as operators satisfying the Clifford algebra. The quantization consisted of the crucial step in which we abolished the requirement that the basis vectors satisfy the Clifford algebra relations for a "flat" metric in function space (which is proportional to the δ-function). We admitted an arbitrary metric in principle. The action itself suggested which are the (commutation or anti commutation) relations the basis vectors (operators) should satisfy. Thus we arrived at the conventional procedure of the field quantization.

Our geometric approach brings a new insight about the nature of field quantization. In the conventional approach classical fields are replaced by operators which satisfy the canonical commutation or anti-commutation relations. In the proposed geometric approach we observe that *the field operators are, in fact, the basis vectors* $h(x)$. By its very definition a basis vector $h(x)$, for a given x, "creates" a particle at the position x. Namely, an arbitrary vector Φ is written as a superposition of basis vectors

$$\Phi = \int dx' \, \phi(x') h(x') \tag{7.190}$$

and $\phi(x)$ is "the wave packet" profile. If in particular $\phi(x') = \delta(x' - x)$, i.e., if the "particle" is located at x, then

$$\Phi = h(x). \tag{7.191}$$

We shall now explore further the possibilities brought by such a geometric approach to quantization. Our main interest is to find out how it could be applied to the quantization of strings and p-branes in general. In Sec. 4.2, we have found out that a conventional p-brane can be described by the following action

$$I[X^{\alpha(\xi)}(\tau)] = \int d\tau' \, \rho_{\alpha(\xi')\beta(\xi'')} \dot{X}^{\alpha(\xi')} \dot{X}^{\beta(\xi'')} = \rho_{\alpha(\phi')\beta(\phi'')} \dot{X}^{\alpha(\phi')} \dot{X}^{\beta(\phi'')}, \tag{7.192}$$

where

$$\rho_{\alpha(\phi')\beta(\phi'')} = \frac{\kappa \sqrt{|f|}}{\sqrt{\dot{X}^2}} \delta(\tau' - \tau'') \delta(\xi' - \xi'') g_{\alpha\beta}. \tag{7.193}$$

Here $\dot{X}^{\alpha(\phi)} \equiv \dot{X}^{\alpha(\tau,\xi)} \equiv \dot{X}^{\alpha(\xi)}(\tau)$, where $\xi \equiv \xi^a$ are the p-brane coordinates, and $\phi \equiv \phi^A = (\tau, \xi^a)$ are coordinates of the world surface which I call worldsheet.

If the $\mathcal{M}$-space metric $\rho_{\alpha(\phi')\beta(\phi'')}$ is different from (7.193), then we have a deviation from the usual Dirac–Nambu–Goto p-brane theory. Therefore in the classical theory $\rho_{\alpha(\phi')\beta(\phi'')}$ was made dynamical by adding a suitable kinetic term to the action.

Introducing the basis vectors $h_{\alpha(\phi)}$ satisfying

$$h_{\alpha(\phi')} \cdot h_{\beta(\phi'')} = \rho_{\alpha(\phi')\beta(\phi'')} \tag{7.194}$$

we have

$$I[X^{\alpha(\phi)}] = h_{\alpha(\phi')} h_{\beta(\phi'')} \dot{X}^{\alpha(\phi')} \dot{X}^{\beta(\phi'')}. \tag{7.195}$$

Here $h_{\alpha(\phi')}$ are fixed while $X^{\alpha(\phi')}$ are variables. If we now admit that $h_{\alpha(\phi')}$ also change with τ, we can perform the partial integrations over τ' and τ'' so that eq. (7.195) becomes

$$I = \dot{h}_{\alpha(\phi')} \dot{h}_{\beta(\phi'')} X^{\alpha(\phi')} X^{\beta(\phi'')}. \tag{7.196}$$

We now assume that $X^{\alpha(\phi')}$ is an arbitrary configuration, not necessarily the one that solves the variational principle (7.192). In particular, let us take

$$X^{\alpha(\phi')} = \delta^{\alpha(\phi')}{}_{\mu(\phi)}, \qquad X^{\beta(\phi'')} = \delta^{\beta(\phi'')}{}_{\mu(\phi)}, \tag{7.197}$$

which means that our p-brane is actually a *point* at the values of the parameters $\phi \equiv (\tau, \xi^a)$ and the value of the index μ. So we have

$$I_0 = \dot{h}_{\mu(\phi)} \dot{h}_{\mu(\phi)} \quad \text{no sum and no integration.} \tag{7.198}$$

By taking (7.197) we have in a sense "quantized" the classical action. The above expression is a "quantum" of (7.192).

Integrating (7.198) over ϕ and summing over μ we obtain

$$I[h_{\mu(\phi)}] = \int d\phi \sum_\mu \dot{h}_\mu(\phi) \dot{h}_\mu(\phi). \tag{7.199}$$

The latter expression can be written as[11]

$$\begin{aligned} I[h_{\mu(\phi)}] &= \int d\phi\, d\phi'\, \delta(\phi - \phi') \eta^{\mu\nu} \dot{h}(\phi) \dot{h}_\nu(\phi) \\ &\equiv \eta^{\mu(\phi)\nu(\phi')} \dot{h}_{\mu(\phi)} \dot{h}_{\nu(\phi')}, \end{aligned} \tag{7.200}$$

where

$$\eta^{\mu(\phi)\nu(\phi')} = \eta^{\mu\nu} \delta(\phi - \phi') \tag{7.201}$$

is the flat $\mathcal{M}$-space metric. In general, of course, $\mathcal{M}$-space is not flat, and we have to use arbitrary metric. Hence (7.200) generalizes to

$$I[h_{\mu(\phi)}] = \rho^{\mu(\phi)\nu(\phi')} \dot{h}_{\mu(\phi)} \dot{h}_{\nu(\phi')}, \tag{7.202}$$

where

$$\rho^{\mu(\phi)\nu(\phi')} = h^{\mu(\phi)} \cdot h^{\nu(\phi')} = \tfrac{1}{2}(h^{\mu(\phi)} h^{\nu(\phi')} + h^{\nu(\phi')} h^{\mu(\phi)}). \tag{7.203}$$

Using the expression (7.203), the action becomes

$$I[h_{\mu(\phi)}] = h^{\mu(\phi)} h^{\nu(\phi')} \dot{h}_{\mu(\phi)} \dot{h}_{\nu(\phi')}. \tag{7.204}$$

---

[11] Again summation and integration convention is assumed.

Quantization

METRIC IN THE SPACE OF OPERATORS

In the definition of the action (7.202), or (7.204), we used the relation (7.203) in which the $\mathcal{M}$-space metric is expressed in terms of the basis vectors $h_{\mu(\phi)}$. In order to allow for a more general case, we shall introduce the *metric* $Z^{\mu(\phi)\nu(\phi')}$ in the "space" of operators. In particular it can be

$$\tfrac{1}{2} Z^{\mu(\phi)\nu(\phi')} = h^{\mu(\phi)} h^{\nu(\phi')}, \qquad (7.205)$$

or

$$\tfrac{1}{2} Z^{\mu(\phi)\nu(\phi')} = \tfrac{1}{2}(h^{\mu(\phi)} h^{\nu(\phi')} + h^{\nu(\phi')} h^{\mu(\phi)}), \qquad (7.206)$$

but in general, $Z^{\mu(\phi)\nu(\phi')}$ is expressed arbitrarily in terms of $h^{\mu(\phi)}$. Then instead of (7.204), we have

$$I[h] = \tfrac{1}{2} Z^{\mu(\phi)\nu(\phi')} \dot{h}_{\mu(\phi)} \dot{h}_{\nu(\phi')} = \tfrac{1}{2} \int \mathrm{d}\tau\, Z^{\mu(\xi)\nu(\xi')} \dot{h}_{\mu(\xi)} \dot{h}_{\nu(\xi')}. \qquad (7.207)$$

The factor $\tfrac{1}{2}$ is just for convenience; it does not influence the equations of motion.

Assuming $Z^{\mu(\phi)\nu(\phi')} = Z^{\mu(\xi)\nu(\xi')} \delta(\tau - \tau')$ we have

$$I[h] = \tfrac{1}{2} \int \mathrm{d}\tau\, Z^{\mu(\xi)\nu(\xi')} \dot{h}_{\mu(\xi)} \dot{h}_{\nu(\xi')}. \qquad (7.208)$$

Now we could continue by assuming the validity of the scalar product relations (7.194) and explore the equations of motion derived from (7.207) for a chosen $Z^{\mu(\phi)\nu(\phi')}$. This is perhaps a possible approach to geometric quantization, but we shall not pursue it here.

Rather we shall forget about (7.194) and start directly from the action (7.208), considered as an action for the operator field $h_{\mu(\xi)}$ where the commutation relations should now be determined. The canonically conjugate variables are

$$h_{\mu(\xi)}, \qquad \pi^{\mu(\xi)} = \partial L / \partial \dot{h}_{\mu(\xi)} = Z^{\mu(\xi)\nu(\xi')} \dot{h}_{\nu(\xi')}. \qquad (7.209)$$

They are assumed to satisfy the equal τ commutation relations

$$[h_{\mu(\xi)}, h_{\nu(\xi')}] = 0, \qquad [\pi^{\mu(\xi)}, \pi^{\nu(\xi')}] = 0, \qquad (7.210)$$

$$[h_{\mu(\xi)}, \pi^{\nu(\xi')}] = i \delta_{\mu(\xi)}{}^{\nu(\xi')}. \qquad (7.211)$$

By imposing (7.210) and (7.211) we have abolished the Clifford algebra relation (7.194) in which the inner product (defined as the symmetrized Clifford product) is equal to a *scalar valued* metric.

The Heisenberg equations of motion are

$$\dot{h}_{\mu(\xi)} = -i[h_{\mu(\xi)}, H], \tag{7.212}$$

$$\dot{\pi}^{\mu(\xi)} = -i[\pi^{\mu(\xi)}, H], \tag{7.213}$$

where the Hamiltonian is

$$H = \tfrac{1}{2} Z_{\mu(\xi)\nu(\xi')} \pi^{\mu(\xi)} \pi^{\nu(\xi')}. \tag{7.214}$$

In particular, we may take a trivial metric which does not contain $h_{\mu(\xi)}$, e.g.,

$$Z^{\mu(\xi)\nu(\xi')} = \eta^{\mu\nu} \delta(\xi - \xi'). \tag{7.215}$$

Then the equation of motion resulting from (7.213) or directly from the action (7.208) is

$$\dot{\pi}_{\mu(\xi)} = 0. \tag{7.216}$$

Such a dynamical system cannot describe the usual p-brane, since the equations of motion is too simple. It serves here for the purpose of demonstrating the procedure. In fact, in the quantization procedures for the Klein–Gordon, Dirac, Stueckelberg field, etc., we have in fact used a fixed prescribed metric which was proportional to the δ-function.

In general, the metric $Z^{\mu(\xi)\nu(\xi')}$ is an expression containing $h_{\mu(\xi)}$. The variation of the action (7.207) with respect to $h_{\mu(\xi)}$ gives

$$\frac{\mathrm{d}}{\mathrm{d}\tau}(Z^{\mu(\phi)\nu(\phi')} \dot{h}_{\nu(\phi)}) - \frac{1}{2} \frac{\delta Z^{\alpha(\phi')\beta(\phi'')}}{\delta h_{\mu(\phi)}} \dot{h}_{\alpha(\phi')} \dot{h}_{\beta(\phi'')} = 0. \tag{7.217}$$

Using (7.210),(7.211) one finds that the Heisenberg equation (7.213) is equivalent to (7.217).

THE STATES OF THE QUANTIZED BRANE

According to the traditional approach to QFT one would now introduce a vacuum *state vector* $|0\rangle$ and define

$$h_{\alpha(\xi)}|0\rangle \ , \quad h_{\alpha(\xi)} h_{\beta(\xi)}|0\rangle \ , \ldots \tag{7.218}$$

as *vectors* in Fock space. Within our geometric approach we can do something quite analogous. First we realize that because of the commutation relations (7.208) $h_{\mu(\xi)}$ are in fact not elements of the Clifford algebra. Therefore they are not vectors in the usual sense. In order to obtain vectors we introduce an object v_0 which, by definition, is a Clifford number satisfying[12]

$$v_0 v_0 = 1 \tag{7.219}$$

---

[12] The procedure here is an alternative to the one considered when discussing quantization of the Klein–Gordon and other fields.

Quantization

and has the property that the products

$$h_{\mu(\xi)}v_0 , \quad h_{\mu(\xi)}h_{\nu(\xi)}v_0 \ldots \tag{7.220}$$

are also Clifford numbers. Thus $h_{\mu(\xi)}v_0$ behaves as a vector. The inner product between such vectors is defined as usually in Clifford algebra:

$$\begin{aligned}(h_{\mu(\xi)}v_0) \cdot (h_{\nu(\xi')}v_0) &\equiv \tfrac{1}{2}\left[(h_{\mu(\xi)})v_0)(h_{\nu(\xi')}v_0) + (h_{\nu(\xi')}v_0)(h_{\mu(\xi)})v_0)\right] \\ &= \rho_{0\mu(\xi)\nu(\xi')},\end{aligned} \tag{7.221}$$

where $\rho_{0\mu(\xi)\nu(\xi')}$ is a scalar-valued metric. The choice of $\rho_{0\mu(\xi)\nu(\xi')}$ is determined by the choice of v_0. We see that the vector v_0 corresponds to the vacuum state vector of QFT. The vectors (7.220) correspond to the other basis vectors of Fock space. And we see here that choice of the vacuum vector v_0 determines the metric in Fock space. Usually basis vector of Fock space are orthonormal, hence we take

$$\rho_{0\mu(\xi)\nu(\xi')} = \eta_{\mu\nu}\delta(\xi - \xi'). \tag{7.222}$$

In the conventional field-theoretic notation the relation (7.221) reads

$$\langle 0|\tfrac{1}{2}(h_{\mu(\xi)}h_{\nu(\xi')} + h_{\nu(\xi')}h_{\mu(\xi)})|0\rangle = \rho_{0\mu(\xi)\nu(\xi')}. \tag{7.223}$$

This is the vacuum expectation value of the operator

$$\hat{\rho}_{\mu(\xi)\nu(\xi')} = \tfrac{1}{2}(h_{\mu(\xi)}h_{\nu(\xi')} + h_{\nu(\xi')}h_{\mu(\xi)}), \tag{7.224}$$

which has the role of the $\mathcal{M}$-space metric operator.

In a generic state $|\Psi\rangle$ of Fock space the expectation value of the operator $\hat{\rho}_{\mu(\xi)\nu(\xi')}$ is

$$\langle\Psi|\hat{\rho}_{\mu(\xi)\nu(\xi')}|\Psi\rangle = \rho_{\mu(\xi)\nu(\xi')}. \tag{7.225}$$

Hence, in a given state, for the expectation value of the metric operator we obtain a certain scalar valued $\mathcal{M}$-space metric $\rho_{\mu(\xi)\nu(\xi')}$.

In the geometric notation (7.225) reads

$$\langle V h_{\mu(\xi)} h_{\nu(\xi')} V \rangle_0 = \rho_{\mu(\xi)\nu(\xi')}. \tag{7.226}$$

This means that we choose a Clifford number (Clifford aggregate) V formed from (7.220)

$$V = (\phi^{\mu(\xi)} h_{\mu(\xi)} + \phi^{\mu(\xi)\nu(\xi')} h_{\mu(\xi)} h_{\nu(\xi')} + \ldots)v_0, \tag{7.227}$$

where $\phi^{\mu(\xi)}$, $\phi^{\mu(\xi)\nu(\xi')}$, ... , are the wave packet profiles, then we write the expression $V h_{\mu(\xi)} h_{\nu(\xi')} V$ and take its *scalar part*.

Conceptually our procedure appears to be very clear. We have an action (7.208) for operators $h_{\mu(\xi)}$ satysfying the commutation relations (7.210), (7.211) and the Heisenberg equation of motion (7.213). With the aid of those operators we form a Fock space of states, and then we calculate the expectation value of the metric operator $\hat{\rho}_{\mu(\xi)\nu(\xi')}$ in a chosen state. *We interpret this expectation value as the classical metric of $\mathcal{M}$-space.* This is justified because there is a correspondence between the operators $h_{\mu(\xi)}$ and the $\mathcal{M}$-space basis vectors (also denoted $h_{\mu(\xi)}$, but obeying the Clifford algebra relations (7.198)). The operators create, when acting on $|0\rangle$ or v_0, the many brane states and it is natural to interpret the expectation value of $\hat{\rho}_{\mu(\xi)\nu(\xi')}$ as the classical $\mathcal{M}$-space metric for such a many brane configuration.

WHICH CHOICE FOR THE OPERATOR METRIC $Z^{\mu(\xi)\nu(\xi')}$?

A question now arise of how to choose $Z^{\mu(\xi)\nu(\xi')}$. In principle any combination of operators $h_{\mu(\xi)}$ is good, provided that $Z^{\mu(\xi)\nu(\xi')}$ has its inverse defined according to

$$Z^{\mu(\xi)\alpha(\xi'')} Z_{\alpha(\xi'')\nu(\xi')} = \delta^{\mu(\xi)}{}_{\nu(\xi')} \equiv \delta^{\mu}{}_{\nu}\delta(\xi - \xi'). \qquad (7.228)$$

Different choices of $Z^{\mu(\xi)\nu(\xi')}$ mean different membrane theories, and hence different expectation values $\rho_{\mu(\xi)\nu(\xi')} = \langle\hat{\rho}_{\mu(\xi)\nu(\xi')}\rangle$ of the $\mathcal{M}$-space metric operator $\hat{\rho}_{\mu(\xi)\nu(\xi')}$. We have already observed that different choices of $\rho_{\mu(\xi)\nu(\xi')}$ correspond to different classical membrane theories. In order to get rid of a fixed background we have given $\rho_{\mu(\xi)\nu(\xi')}$ the status of a dynamical variable and included a kinetic term for $\rho_{\mu(\xi)\nu(\xi')}$ in the action (or equivalently for $h_{\mu(\xi)}$, which is the "square root" of $\rho_{\mu(\xi)\nu(\xi')}$, since classically $\rho_{\mu(\xi)\nu(\xi')} = h_{\mu(\xi)} \cdot h_{\nu(\xi')}$). In performing the quantization we have seen that to the classical vectors $h_{\mu(\xi)}$ there correspond quantum operators[13] $\hat{h}_{\mu(\xi)}$ which obey the equations of motion determined by the action (7.208). Hence in the quantized theory we do need a separate kinetic term for $\hat{h}_{\mu(\xi)}$. But now we have something new, namely, $Z^{\mu(\xi)\nu(\xi')}$ which is a background metric in the space of operators $\hat{h}_{\mu(\xi)}$. In order to obtain a background independent theory we need a kinetic term for $Z^{\mu(\xi)\nu(\xi')}$. The search for such a kinetic term will remain a subject of future investigations. It has its parallel in the attempts to find a background independent string or p-brane theory. However, it may turn out that we do not need a kinetic term for $Z^{\mu(\xi)\nu(\xi')}$ and that it is actually given by the expression (7.205) or (7.206), so that (7.204) is already the "final" action for the quantum p-brane.

---

[13] Now we use hats to make a clear distinction between the classical vectors $h_{\mu(\xi)}$, satisfying the Clifford algebra relations (7.194), and the quantum operators $\hat{h}_{\mu(\xi)}$ satisfying (7.210), (7.211).

Quantization

In the field theory discussed previously we also have a fixed metric $Z^{\mu(\xi)\nu(\xi')}$, namely, $Z^{(\xi)(\xi')} = \delta(\xi - \xi')$ for the Klein–Gordon and similarly for the Dirac field. Why such a choice and not some other choice? This clearly points to the plausible possibility that the usual QFT is not complete. That QFT is not yet a finished story is clear from the occurrence of infinities and the need for "renormalization"[14].

---

[14] An alternative approach to the quantization of field theories, also based on Clifford algebra, has been pursued by Kanatchikov [87].

BRANE WORLD

Chapter 8

SPACETIME AS A MEMBRANE IN A HIGHER-DIMENSIONAL SPACE

When studying dynamics of a system of membranes, as seen from the $\mathcal{M}$-space point of view, we have arrived in Chapter 5 at a fascinating conclusion that all that exists in such a world model is a membrane configuration. *The membrane configuration itself is a 'spacetime'*. Without membranes there is no spacetime. According to our basic assumption, at the fundamental level we have an $\mathcal{M}$-space — the space of all possible membrane configurations — and nothing else. If the membrane configuration consists of the membranes of various dimensions n, lower and higher than the dimension of our observed word ($n = 4$), then we are left with a model in which our 4-dimensional spacetime is one of those (4-dimensional) membranes (which I call *worldsheets*).

What is the space our worldsheet is embedded in? It is just the space formed by the other n-dimensional ($n = 0, 1, 2, ...$) extended objects (say membranes) entering the membrane configuration. If all those other membranes are sufficiently densely packed together, then as an approximation a concept of a continuous embedding space can be used. Our spacetime can then be considered as a 4-dimensional worldsheet embedded into a higher-dimensional space.

8.1. THE BRANE IN A CURVED EMBEDDING SPACE

We are now going to explore a brane moving in a curved background embedding space V_N. Such a brane sweeps an n-dimensional surface which

I call *worldsheet*[1] The dynamical principle governing motion of the brane requires that its worldsheet is a minimal surface. Hence the action is

$$I[\eta^a] = \int \sqrt{|\tilde{f}|}\, \mathrm{d}^n x, \qquad (8.1)$$

where

$$\tilde{f} = \det \tilde{f}_{\mu\nu}, \qquad \tilde{f}_{\mu\nu} = \partial_\mu \eta^a \partial_\nu \eta^b \gamma_{ab}. \qquad (8.2)$$

Here x^μ, $\mu = 0, 1, 2, ..., n-1$, are coordinates on the worldsheet V_n, whilst $\eta^a(x)$ are the embedding functions. The metric of the embedding space (from now on also called *bulk*) is γ_{ab}, and the induced metric on the worldsheet is $\tilde{f}_{\mu\nu}$.

In this part of the book I shall use the notation which is adapted to the idea that our world is a *brane*. Position coordinates in our world are commonly denoted as x^μ, $\mu = 0, 1, 2, ..., n-1$, and usually it is assumed that $n = 4$ (for good reasons, of course, unless one considers Kaluza–Klein theories). The notation in (8.1) (8.2) is "the reverse video" of the notation used so far. The correspondence between the two notations is the following

worldsheet coordinates	$\xi^a,\ \xi^A, \phi^A$	x^μ
embedding space coordinates	x^μ	η^a
embedding functions	$X^\mu(\xi^a),\ X^\mu(\phi^A)$	$\eta^a(x^\mu)$
worldsheet metric	$\gamma_{ab},\ \gamma_{AB}$	$g_{\mu\nu}$
embedding space metric	$g_{\mu\nu}$	γ_{ab}

Such a reverse notation reflects the change of role given to spacetime. So far 'spacetime' has been associated with the embedding space, whilst the brane has been an object in spacetime. Now spacetime is associated with a brane, so spacetime itself is an object in the embedding space[2].

For the extended object described by the minimal surface action (8.1) I use the common name *brane*. For a more general extended object described by a Clifford algebra generalization of the action (8.1) I use the name *membrane* (and occasionally also *worldsheet*, when I wish to stress that the object of investigation is a direct generalization of the object V_n described by (8.1) which is now understood as a special kind of (generalized) worldsheet).

---

[1] Usually, when $n > 2$ such a surface is called a *world volume*. Here I prefer to retain the name *worldsheet*, by which we can vividly imagine a surface in an embedding space.

[2] Such a distinction is only manifest in the picture in which we already have an effective embedding space. In a more fundamental picture the embedding space is inseparable from the membrane configuration, and in general is not a manifold at all.

Suppose now that the metric of V_N is conformally flat (with η_{ab} being the Minkowski metric tensor in N-dimensions):

$$\gamma_{ab} = \phi\,\eta_{ab}. \tag{8.3}$$

Then from (8.2) we have

$$\tilde{f}_{\mu\nu} = \phi\,\partial_\mu \eta^a \partial_\nu \eta^b \eta_{ab} \equiv \phi f_{\mu\nu}, \tag{8.4}$$

$$\tilde{f} \equiv \det \tilde{f}_{\mu\nu} = \phi^n \det f_{\mu\nu} \equiv \phi^n f \tag{8.5}$$

$$\sqrt{|\tilde{f}|} = \omega |f|, \quad \omega \equiv \phi^{n/2}. \tag{8.6}$$

Hence the action (8.1) reads

$$I[\eta^a] = \int \omega(\eta)\sqrt{|f|}\mathrm{d}^n x, \tag{8.7}$$

which looks like an action for a brane in a flat embedding space, except for a function $\omega(\eta)$ which depends on the position[3] η^a in the embedding space V_N.

Function $\omega(\eta)$ is related to the fixed background metric which is arbitrary in principle. Let us now assume [88] that $\omega(\eta)$ consists of a constant part ω_0 and a singular part with support on another brane's worldsheet $\widehat{V}_m$:

$$\omega(\eta) = \omega_0 + \kappa \int \mathrm{d}^m \hat{x}\, \sqrt{|\hat{f}|}\, \frac{\delta^N(\eta - \hat{\eta})}{\sqrt{|\gamma|}}. \tag{8.8}$$

Here $\hat{\eta}^a(\hat{x})$ are the embedding functions of the m-dimensional worldsheet $\widehat{V}_m$, $\hat{f}$ is the determinant of the induced metric on $\widehat{V}_m$, and $\sqrt{|\gamma|}$ allows for taking curved coordinates in otherwise flat V_N.

The action for the brane which sweeps a worldsheet V_n is then given by (8.7) in which we replace $\omega(\eta)$ with the specific expression (8.8):

$$I[\eta] = \int \omega_0\,\mathrm{d}^n x\,\sqrt{|f|} + \kappa \int \mathrm{d}^n x\,\mathrm{d}^m \hat{x}\,\sqrt{|f|}\sqrt{|\hat{f}|}\,\frac{\delta^N(\eta - \hat{\eta})}{\sqrt{|\gamma|}}. \tag{8.9}$$

If we take the second brane as dynamical too, then the kinetic term for $\hat{\eta}^a$ should be added to (8.9). Hence the total action for both branes is

$$I[\eta,\hat{\eta}] = \int \omega_0\,\mathrm{d}^n x\,\sqrt{|f|} + \int \omega_0\,\mathrm{d}^m \hat{x}\,\sqrt{|\hat{f}|} + \kappa \int \mathrm{d}^n x\,\mathrm{d}^m \hat{x}\,\sqrt{|f|}\sqrt{|\hat{f}|}\,\frac{\delta^N(\eta - \hat{\eta})}{\sqrt{|\gamma|}}. \tag{8.10}$$

---

[3] We use here the same symbol η^a either for position coordinates in V_N or for the embedding functions $\eta^a(x)$.

The first two terms are the actions for free branes, whilst the last term represents the interaction between the two branes. The interaction occurs when the branes intersect. If we take $m = N - n + 1$ then the *intersection* of V_n and $\hat{V}_m$ can be a (one-dimensional) line, i.e., a *worldline* V_1. In general, when $m = N - n + (p+1)$, the intersection can be a $(p+1)$-dimensional worldsheet representing the motion of a p-brane.

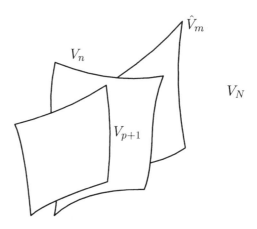

Figure 8.1. The intersection between two different branes V_n and $\hat{V}_m$ can be a p-brane V_{p+1}.

In eq. (8.10) we assume contact interaction between the branes (i.e., the interaction at the intersection). This could be understood by imagining that gravity decreases so quickly in the transverse direction from the brane that it can be approximated by a δ-function. More about this will be said in Section 4.

The equations of motion derived from the (8.10) by varying respectively η^a and $\hat{\eta}$ are:

$$\partial_\mu \left[\sqrt{|f|} \partial^\mu \eta_a \left(\omega_0 + \kappa \int d^m \hat{x} \sqrt{|\hat{f}|}\, \frac{\delta^N(\eta - \hat{\eta})}{\sqrt{|\gamma|}} \right) \right] = 0 \qquad (8.11)$$

$$\hat{\partial}_{\hat{\mu}} \left[\sqrt{|\hat{f}|} \hat{\partial}^{\hat{\mu}} \hat{\eta}_a \left(\omega_0 + \kappa \int d^n x \sqrt{|f|}\, \frac{\delta^N(\eta - \hat{\eta})}{\sqrt{|\gamma|}} \right) \right] = 0 \qquad (8.12)$$

where $\partial_\mu \equiv \partial/\partial x^\mu$ and $\hat{\partial}_{\hat{\mu}} \equiv \partial/\partial \hat{x}^{\hat{\mu}}$. When deriving eq. (8.11) we have taken into account that

$$\frac{\partial}{\partial \eta^a} \int \kappa \, \mathrm{d}^m \hat{x} \sqrt{|\hat{f}|} \, \delta^N(\eta - \hat{\eta}) = -\int \kappa \, \mathrm{d}^m \hat{x} \sqrt{|\hat{f}|} \frac{\partial}{\partial \hat{\eta}^a} \delta^N(\eta - \hat{\eta}) \quad (8.13)$$

$$= \kappa \int \mathrm{d}^m \hat{x} \frac{\partial \sqrt{|\hat{f}|}}{\partial \hat{\eta}^a} \delta^N(\eta - \hat{\eta}) = 0,$$

since

$$\frac{\partial \sqrt{|\hat{f}|}}{\partial \hat{\eta}^a} = \frac{\partial \sqrt{|\hat{f}|}}{\partial \hat{f}} \frac{\partial \hat{f}_{\mu\nu}}{\partial \hat{\eta}^a} = 0, \quad (8.14)$$

because

$$\hat{f}_{\mu\nu} = \hat{\partial}_{\hat{\mu}} \hat{\eta}^a \, \hat{\partial}_{\hat{\nu}} \hat{\eta}_a, \qquad \frac{\partial \hat{f}_{\mu\nu}}{\partial \hat{\eta}^a} = 0. \quad (8.15)$$

Analogous holds for eq. (8.12).

Assuming that the intersection $V_{p+1} = V_n \cap \hat{V}_m$ does exist, and, in particular, that it is a worldline (i.e., $p = 0$), then we can write

$$\int \mathrm{d}^m \hat{x} \sqrt{|\hat{f}|} \frac{\delta^N(\eta - \hat{\eta})}{\sqrt{|\gamma|}} = \int \mathrm{d}\tau \frac{\delta^n(x - X(\tau))}{\sqrt{|f|}} (\dot{X}^\mu \dot{X}_\mu)^{1/2}. \quad (8.16)$$

The result above was obtained by writing

$$\mathrm{d}^m \hat{x} = \mathrm{d}^{m-1} \hat{x} \, \mathrm{d}\tau, \qquad \sqrt{|\hat{f}|} = \sqrt{|\hat{f}^{(m-1)}|} (\dot{X}^\mu \dot{X}_\mu)^{1/2}$$

and taking the coordinates η^a such that $\eta^a = (x^\mu, \eta^n, \eta^{n+1}, ..., \eta^{N-1})$, where x^μ are (curved) coordinates on V_n. The determinant of the metric of the embedding space V_N in such a curvilinear coordinates is then $\gamma = \det \partial_\mu \eta^a \partial_\nu \eta_a = f$.

In general, for arbitrary intersection we have

$$\int \mathrm{d}^m \hat{x} \sqrt{|\hat{f}|} \frac{\delta^N(\eta - \hat{\eta})}{\sqrt{|\gamma|}} = \int \mathrm{d}^{p+1}\xi \frac{\delta^n(x - X\xi)}{\sqrt{|f|}} (\det \partial_A X^\mu \partial_B X_\mu)^{1/2}, \quad (8.17)$$

where $X^\mu(\xi^A)$, $\mu = 0, 1, 2, ..., n-1$, $A = 1, 2, ..., p$, are the embedding functions of the p-brane's worlsdsheet V_{p+1} in V_n.

Using (8.16) the equations of motion become

$$\partial_\mu \left[\sqrt{|f|} (\omega_0 f^{\mu\nu} + T^{\mu\nu}) \partial_\nu \eta_a \right] = 0, \quad (8.18)$$

$$\hat{\partial}_{\hat{\mu}} \left[\sqrt{|\hat{f}|} (\omega_0 \hat{f}^{\hat{\mu}\hat{\nu}} + \hat{T}^{\hat{\mu}\hat{\nu}}) \hat{\partial}_{\hat{\nu}} \hat{\eta}_a \right] = 0, \quad (8.19)$$

where
$$T^{\mu\nu} = \int \frac{\kappa}{\sqrt{|f|}} \delta^n(x - X(\tau)) \frac{\dot{X}^\mu \dot{X}^\nu}{(\dot{X}^\alpha \dot{X}_\alpha)^{1/2}} \, d\tau \tag{8.20}$$

and

$$\widehat{T}^{\hat{\mu}\hat{\nu}} = \int \frac{\kappa}{\sqrt{|\hat{f}|}} \delta^n(\hat{x} - \hat{X}(\tau)) \frac{\dot{\hat{X}}^{\hat{\mu}} \dot{\hat{X}}^{\hat{\nu}}}{(\dot{\hat{X}}^{\hat{\alpha}} \dot{\hat{X}}_{\hat{\alpha}})^{1/2}} \, d\tau \tag{8.21}$$

are the stress–energy tensors of the point particle on V_n and $\widehat{V}_m$, respectively.

If dimensions m and n are such that the intersection V_{p+1} is a worldsheet with a dimension $p \geq 1$, then using (8.17) we obtain the equations of motion of the same form (8.18),(8.19), but with the stress–energy tensor

$$T^{\mu\nu} = \int \frac{\kappa}{\sqrt{|f|}} \delta^n(x - X(\xi)) \partial_A X^\mu \partial^A X^\nu (\det \partial_C X^\alpha \partial_D X_\alpha)^{1/2} \, d^{p+1}\xi, \tag{8.22}$$

$$\widehat{T}^{\hat{\mu}\hat{\nu}} = \int \frac{\kappa}{\sqrt{|\hat{f}|}} \delta^n(\hat{x} - \hat{X}(\xi)) \partial_A \hat{X}^{\hat{\mu}} \partial^A \hat{X}^{\hat{\nu}} (\det \partial_C \hat{X}^{\hat{\alpha}} \partial_D \hat{X}_{\hat{\alpha}})^{1/2} \, d^{p+1}\xi. \tag{8.23}$$

This can also be seen directly from the action (8.9) in which we substitute eq. (8.16)

$$I[\eta^a, X^\mu] = \omega_0 \int d^n x \sqrt{|f|} + \kappa \int d^n x \, d\tau \, \delta^n(x - X(\tau))(f_{\mu\nu} \dot{X}^\mu \dot{X}^\nu)^{1/2} \tag{8.24}$$

or if we substitute (8.17)

$$I[\eta^a, X^\mu] = \omega_0 \int d^n x \sqrt{|f|} \tag{8.25}$$
$$+ \kappa \int d^n x \, d^{p+1}\xi \, \delta^n(x - X(\xi)) (\det \partial_A X^\mu \partial_B X^\nu f_{\mu\nu})^{1/2}.$$

Remembering that

$$f_{\mu\nu} = \partial_\mu \eta^a \partial_\nu \eta^b \eta_{ab} \tag{8.26}$$

we can vary (8.24) or (8.25) with respect to $\eta^a(x)$ and we obtain (8.18).
Eq.(8.18) can be written as

$$\omega_0 D_\mu D^\mu \eta_a + D_\mu(T^{\mu\nu} \partial_\nu \eta_a) = 0. \tag{8.27}$$

where D_μ denotes covariant derivative in V_n. If we multiply the latter equation by $\partial^\alpha \eta^a$, sum over a, and take into account the identity

$$\partial^\alpha \eta^a D_\mu D_\nu \eta_a = 0, \tag{8.28}$$

Spacetime as a membrane in a higher-dimensional space

which follows from $D_\mu(\partial_\rho\eta^a\partial_\sigma\eta_a) = D_\mu f_{\rho\sigma} = 0$, we obtain

$$D_\mu T^{\mu\nu} = 0. \tag{8.29}$$

Equation (8.29) implies that $X^\mu(\tau)$ is a geodesic equation in a space with metric $f_{\mu\nu}$, i.e., $X^\mu(\tau)$ is a geodesic on V_n. This can be easily shown by using the relation

$$DT^{\mu\nu} = \frac{1}{\sqrt{|f|}} \partial_\mu(\sqrt{|f|}\, T^{\mu\nu}) + \Gamma^\nu_{\rho\mu} T^{\rho\mu} = 0. \tag{8.30}$$

Taking (8.20) we have

$$\int d\tau \frac{\partial}{\partial x^\mu} \delta^n(x - X(\tau)) \frac{\dot{X}^\mu \dot{X}^\nu}{(\dot{X}^\alpha \dot{X}_\alpha)^{1/2}} d^n x$$

$$+ \Gamma^\nu_{\rho\mu} \int d\tau \delta^n(x - X(\tau)) \frac{\dot{X}^\mu \dot{X}^\nu}{(\dot{X}^\alpha \dot{X}_\alpha)^{1/2}} d^n x = 0. \tag{8.31}$$

The first term in the latter equation gives

$$-\int d\tau \frac{\partial}{\partial X^\mu(\tau)} \delta^n(x - X(\tau)) \frac{\dot{X}^\mu \dot{X}^\nu}{(\dot{X}^\alpha \dot{X}_\alpha)^{1/2}} d^n x$$

$$= -\int d\tau \frac{d}{d\tau} \delta^n(x - X(\tau)) \frac{\dot{X}^\nu}{(\dot{X}^\alpha \dot{X}_\alpha)^{1/2}} d^n x$$

$$= \int d\tau \frac{d}{d\tau} \left(\frac{\dot{X}^\nu}{(\dot{X}^\alpha \dot{X}_\alpha)^{1/2}} \right). \tag{8.32}$$

Differentiating eq. (8.31) with respect to τ we indeed obtain the geodesic equation.

In a similar way we find for $T^{\mu\nu}$, as given in eq. (8.22), that (8.30) implies

$$\frac{1}{\sqrt{|\det \partial_C X^\alpha \partial_D X_\alpha|}} \partial_A(\sqrt{|\det \partial_C X^\alpha \partial_D X_\alpha|}\, \partial^A X^\nu) + \Gamma^\nu_{\rho\mu} \partial_A X^\rho \partial^A X^\mu = 0, \tag{8.33}$$

which is the equation of motion for a p-brane in a background metric $f_{\mu\nu} = \partial_\mu\eta^a\partial_\nu\eta_a$. Do not forget that the latter p-brane is the intersection between two branes:

$$V_{p+1} = V_n \cap \widehat{V}_m. \tag{8.34}$$

It is instructive to integrate (8.27) over $d^n x$. We find

$$\omega_0 \oint \sqrt{|f|}\, d\Sigma_\mu\, \partial^\mu \eta_a \qquad (8.35)$$

$$= -\kappa \int d^{p+1}\xi (|\det \partial_C X^\alpha \partial_D X_\alpha|)^{1/2}\, \partial_A X^\mu \partial^A X^\nu\, D_\mu D_\nu \eta_a \Big|_{x=X(\xi)}$$

where $d\Sigma_\mu$ is an element of an $(n-1)$-dimensional hypersurface Σ on V_n. Assuming that the integral over the time-like part of Σ vanishes (either because $\partial^\mu \eta_a \to 0$ at the infinity or because V_n is closed) we have

$$\omega_0 \int \sqrt{|f|}\, d\Sigma_\mu\, \partial^\mu \eta_a \Big|_{\tau_2} - \omega_0 \int \sqrt{|f|}\, d\Sigma_\mu\, \partial^\mu \eta_a \Big|_{\tau_1} \qquad (8.36)$$

$$= -\kappa \int d\tau\, d^p \xi (|\det \partial_C X^\alpha \partial_D X_\alpha|)^{1/2}\, \partial_A X^\mu \partial^A X^\nu\, D_\mu D_\nu \eta_a \Big|_{x=X(\xi)}$$

or

$$\frac{dP_a}{d\tau} = -\kappa \int d^p \xi (|\det \partial_C X^\alpha \partial_D X_\alpha|)^{1/2}\, \partial_A X^\mu \partial^A X^\nu\, D_\mu D_\nu \eta_a \Big|_{x=X(\xi)} \qquad (8.37)$$

where

$$P_a \equiv \omega_0 \int \sqrt{|f|}\, d\Sigma_\mu\, \partial^\mu \eta_a . \qquad (8.38)$$

When $p = 0$, i.e., when the intersection is a worldline, eq. (8.37) reads

$$\frac{dP_a}{d\tau} = -\kappa \frac{\dot X^\mu \dot X^\nu}{(\dot X^\alpha \dot X_\alpha)^{1/2}}\, D_\mu D_\nu \eta_a \Big|_{x=X(\xi)} . \qquad (8.39)$$

8.2. A SYSTEM OF MANY INTERSECTING BRANES

Suppose we have a sytem of branes of various dimensionalities. They may intersect and their intersections are the branes of lower dimensionality. The action governing the dynamics of such a system is a generalization of (8.10) and consists of the free part plus the interactive part ($i, j = 1, 2, \dots$):

$$I[\eta_i] = \sum_i \int \omega_0 \sqrt{|f_i|}\, dx_i + \frac{1}{2} \sum_{i \neq j} \int \omega_{ij} \frac{\delta^N(\eta_i - \eta_j)}{\sqrt{|\gamma|}} \sqrt{|f_i|}\sqrt{|f_j|}\, dx_i dx_j .$$

$$(8.40)$$

The equations of motion for the i-th brane are

$$\partial_\mu \left[\sqrt{|f_i|} \partial^\mu \eta_i^a \left(\omega_0 + \sum_{i \neq j} \int \omega_{ij} \frac{\delta^N(\eta_i - \eta_j)}{\sqrt{|\gamma|}} \sqrt{|f_j|} dx_j \right) \right] = 0. \quad (8.41)$$

Neglecting the kinetic term for all other branes the action leading to (8.41) is (for a fixed i)

$$I[\eta_i] = \int \omega_0 \sqrt{|f_i|} \, dx_i + \sum_{i \neq j} \int \omega_{ij} \frac{\delta^N(\eta_i - \eta_j)}{\sqrt{|\gamma|}} \sqrt{|f_i|} \sqrt{|f_j|} \, dx_i dx_j \quad (8.42)$$

or

$$I[\eta_i] = \int \omega_i(\eta) \sqrt{|f_i|} \, dx_i, \quad (8.43)$$

with

$$\omega_i(\eta) = \omega_0 + \sum_j \kappa_j \frac{\delta^N(\eta - \eta_j)}{\sqrt{|\gamma|}} \sqrt{|f_j|} dx_j, \quad (8.44)$$

where $\kappa_j \equiv \omega_{ij}$.

Returning now to eqs. (8.3)–(8.6) we see that $\omega_i(\eta)$ is related to the conformally flat background metric as experienced by the i-th brane. The action (8.43) is thus the action for a brane in a background metric γ_{ab}, which is conformally flat:

$$I[\eta_i] = \int \sqrt{|\tilde{f}|} \, dx_i. \quad (8.45)$$

Hence the interactive term in (8.40) can be interpreted as a contribution to the background metric in which the i-th brane moves. Without the interactive term the metric is simply a flat metric (multiplied by ω); with the interactive term the background metric is singular on all the branes within our system.

The total action (8.40), which contains the kinetic terms for all the other branes, renders the metric of the embedding space V_N dynamical. The way in which other branes move depends on the dynamics of the whole system. It may happen that for a system of many branes, densely packed together, the effective (average) metric could no longer be conformally flat. We have already seen in Sec 6.2 that the effective metric for a system of generalized branes (which I call *membranes*) indeed satisfies the Einstein equations.

Returning now to the action (8.42) as experienced by one of the branes whose worldsheet V_n is represented by $\eta_i^a(x_i) \equiv \eta^a(x)$ we find, after integrating out x_j, $j \neq i$, that

$$I[\eta^a, X^\mu] = \omega_0 \int d^n x \sqrt{|f|} \qquad (8.46)$$

$$+ \sum_j \kappa_j \int d^n x\, d^{p_j+1}\xi\, (\det \partial_A X_j^\mu \partial_B X_j^\nu f_{\mu\nu})^{1/2} \delta^n(x - X_j(\xi)).$$

For various p_j the latter expression is an action for a system of point particles ($p_j = 0$), strings ($p_j = 1$), and higher-dimensional branes ($p_j = 2, 3, ...$) moving in the background metric $f_{\mu\nu}$, which is the induced metric on our brane V_n represented by $\eta^a(x)$. Variation of (8.46) with respect to X_k^μ gives the equations of motion (8.33) for a p-brane with $p = p_k$. Variation of (8.46) with respect to $\eta^a(x)$ gives the equations of motion (8.18 for the $(n-1)$-brane. If we vary (8.46) with respect to $\eta^a(x)$ then we obtain the equation of motion (III1.18) for an $(n-1)$-brane. The action (8.46) thus describes the dynamics of the $(n-1)$-brane (world sheet V_n) and the dynamics of the p-branes living on V_n.

We see that the interactive term in (8.40) manifests itself in various ways, depending on how we look at it. It is a manifestation of the metric of the embedding space being curved (in particular, the metric is singular on the system of branes). From the point of view of a chosen brane V_n the interactive term becomes the action for a system of p-branes (including point particles) moving on V_n. If we now adopt the *brane world view*, where V_n is our spacetime, we see that *matter on V_n comes from other branes' worldsheets which happen to intersect our worldsheet V_n*. Those other branes are responsible for the non trivial metric of the embedding space, also called the *bulk*.

THE BRANE INTERACTING WITH ITSELF

In (8.42) or (8.46) we have a description of a brane interacting with other branes. What about self-interaction? In the second term of the action (8.40) (8.42) we have excluded self-interaction. In principle we should not exclude self-interaction, since there is no reason why a brane could not interact with itself.

Let us return to the action (8.9) and let us calculate $\omega(\eta)$, this time assuming that $\hat{V}_m$ coincides with our brane V_n. Hence the intersection is the brane V_n itself, and according to (8.17) we have

$$\omega(\eta) = \omega_o + \kappa \int d^n \hat{x}\, \sqrt{|\hat{f}|}\, \frac{\delta^N(\eta - \hat{\eta}(\hat{x}))}{\sqrt{|\gamma|}}$$

$$= \omega_0 + \kappa \int d^n \xi\, \frac{\delta^n(x - X(\xi))}{\sqrt{|f|}} \sqrt{|\hat{f}|}$$

$$= \omega_0 + \kappa \int d^n x\, \delta^n(x - X(x)) = \omega_0 + \kappa. \qquad (8.47)$$

Spacetime as a membrane in a higher-dimensional space 259

Here the coordinates ξ^A, $A = 0, 1, 2, ..., n-1$, cover the manifold V_n, and $\hat{f}_{AB}$ is the metric of V_n in coordinates ξ^A. The other coordinates are x^μ, $\mu = 0, 1, 2, ..., n-1$. In the last step in (8.47) we have used the property that the measure is invariant, $d^n\xi\sqrt{|\hat{f}|} = d^n x\sqrt{|f|}$.

The result (8.47) demonstrates that we do not need to separate a constant term ω_0 from the function $\omega(\eta)$. For a brane moving in a background of many branes we can replace (8.44) with

$$\omega(\eta) = \sum_j \kappa_j \frac{\delta^N(\eta - \eta_j)}{\sqrt{|\gamma|}} \sqrt{|f_j|} dx_j, \tag{8.48}$$

where j runs over *all* the branes within the system. Any brane feels the same background, and its action for a fixed i is

$$I[\eta_i] = \int \omega(\eta_i) \sqrt{|f|} dx = \sum_j \int \kappa_j \frac{\delta^N(\eta_i - \eta_j)}{\sqrt{|\gamma|}} \sqrt{|f_i|} \sqrt{|f_j|} dx_i\, dx_j. \tag{8.49}$$

However the background is self-consistent: it is a solution to the variational principle given by the action

$$I[\eta_i] = \sum_{i \geq j} \omega_{ij}\, \delta^N(\eta_i - \eta_j) \sqrt{|f_i|} \sqrt{|f_j|} dx_i\, dx_j, \tag{8.50}$$

where now also i runs over *all* the branes within the system; the case $i = j$ is also allowed.

In (8.50) the self-interaction or self coupling occurs whenever $i = j$. The self coupling term of the action is

$$\begin{aligned}
I_{\text{self}}[\eta_i] &= \sum_i \kappa_i \int \delta^N(\eta_i(x_i) - \eta_i(x_i'))\sqrt{|f_i(x_i)|}\sqrt{|f_i(x_i')|}dx_i dx_i' \\
&= \sum_i \kappa_i \int \delta^N(\eta - \eta_i(x_i))\delta^N(\eta - \eta_i(x_i')) \\
&\qquad\times \sqrt{|f_i(x_i)|}\sqrt{|f_i(x_i')|}dx_i dx_i' d^N\eta \\
&= \sum_i \kappa_i \int \delta^N(\eta - \eta_i(x_i))\delta^{n_i}(x_i' - x_i'')\sqrt{|f_i(x_i)|}dx_i dx_i' d^N\eta \\
&= \sum_i \kappa_i \sqrt{|f_i(x_i)|} d^{n_i} x_i, \tag{8.51}
\end{aligned}$$

where we have used the same procedure which led us to eq. (8.17) or (8.47). We see that the interactive action (8.50) automatically contains the minimal surface terms as well, so they do not need to be postulated separately.

A SYSTEM OF MANY BRANES CREATES THE BULK AND ITS METRIC

We have seen several times in this book (Chapters 5, 6) that a system of membranes (a membrane configuration) can be identified with the embedding space in which a single membrane moves. Here we have a concrete realization of that idea. We have a system of branes which intersect. The only interaction between the branes is owed to intersection ('contact' interaction). The interaction at the intersection influences the motion of a (test) brane: it feels a potential because of the presence of other branes. If there are many branes and a test brane moves in the midst of them, then on average it feels a metric field which is approximately continuous. Our test brane moves in the effective metric of the embedding space.

A single brane or several branes give the singular conformal metric. Many branes give, on average, an arbitrary metric.

There is a close inter-relationship between the presence of branes and the bulk metric. In the model we discuss here the bulk metric is singular on the branes, and zero elsewhere. Without the branes there is no metric and no bulk. Actually the bulk consists of the branes which determine its metric.

Figure 8.2. A system of many intersecting branes creates the bulk metric. In the absence of the branes there is no bulk (no embedding space).

8.3. THE ORIGIN OF MATTER IN THE BRANE WORLD

Our principal idea is that we have a system of branes (a brane configuration). With all the branes in the system we associate the embedding space (bulk). One of the branes (more precisely, its worldsheet) represents our spacetime. Interactions between the branes (occurring at the intersections) represent *matter* in spacetime.

MATTER FROM THE INTERSECTION OF OUR BRANE WITH OTHER BRANES

We have seen that matter in V_n naturally occurs as a result of the intersection of our worldsheet V_n with other worldsheets. We obtain exactly the stress–energy tensor for a dust of point particles, or p-branes in general. Namely, varying the action (8.46) with respect to $\eta^a(x)$ we obtain

$$\omega_0 D_\mu D^\mu \eta_a + D_\mu(T^{\mu\nu} \partial_\nu \eta_a) = 0, \qquad (8.52)$$

with

$$T^{\mu\nu} = \sum_j \kappa_j \int d^{p_j+1}\xi \, (\det \partial_A X_j^\mu \partial_B X_j^\nu f_{\mu\nu})^{1/2} \frac{\delta^n(x - X_j(\xi))}{\sqrt{|f|}} \qquad (8.53)$$

being the stress–energy tensor for a system of p-branes (which are the intersections of V_n with the other worldsheets). The above expression for $T^{\mu\nu}$ holds if the extended objects have any dimensions p_j. In particular, when *all* objects have $p_j = 0$ (point particles) eq. (8.53) becomes

$$T^{\mu\nu} = \sum_j \kappa_j \int d\tau \, \frac{\dot{X}^\mu \dot{X}^\nu}{\sqrt{\dot{X}^2}} \frac{\delta(x - X(\tau))}{\sqrt{|f|}}. \qquad (8.54)$$

From the equations of motion (8.53) we obtain (see eqs. (8.27)–(8.29))

$$D_\mu T^{\mu\nu} = 0, \qquad (8.55)$$

which implies (see (8.30)–(8.33)) that any of the objects follows a geodesic in V_n.

MATTER FROM THE INTERSECTION OF OUR BRANE WITH ITSELF

Our model of intersecting branes allows for the possibility that a brane intersects with itself, as is schematically illustrated in Fig. 8.3. The analysis used so far is also valid for situations like that in Fig. 8.3, if we divide the

262   THE LANDSCAPE OF THEORETICAL PHYSICS: A GLOBAL VIEW

worldsheet V_n into two pieces which are glued together at a submanifold C (see Fig. 8.4).

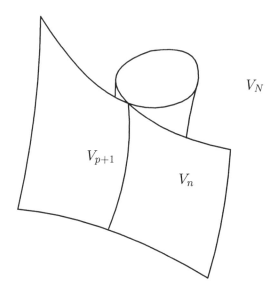

Figure 8.3. Illustration of a self-intersecting brane. At the intersection V_{p+1}, because of the contact interaction the stress–energy tensor on the brane V_n is singular and it manifests itself as matter on V_n. The manifold V_{p+1} is a worldsheet swept by a p-brane and it is is minimal surface (e.g., a geodesic, when = 0) in V_n.

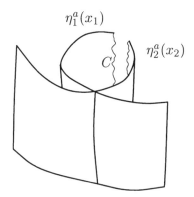

Figure 8.4. A self-intersecting worldsheet is cut into two pieces, described by $\eta_1^a(x_1)$ and $\eta_2^a(x_2)$, which are glued together at a submanifold C where the boundary condition $\eta_1^a(x_1)|_C = \eta_2^a(x_2)|_C$ is imposed.

There is a variety of ways a worldsheet can self-intersect. Some of them are sketched in Fig. 8.5.

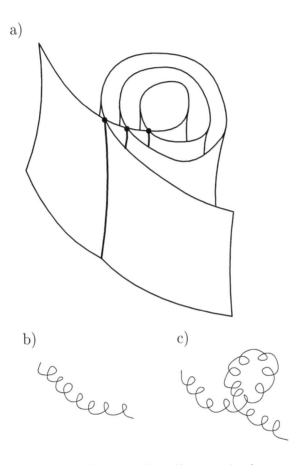

Figure 8.5. Some possible self-intersecting branes.

In this respect some interesting new possibilities occur, waiting to be explored in detail. For instance, it is difficult to imagine how the three particles entangled in the topology of the situation (a) in Fig. 8.5 could be separated to become asymptotically free. Hence this might be a possible classical model for hadrons composed of quarks; the extra dimensions of V_n would bring, via the Kaluza–Klein mechanism, the chromodynamic force into the action.

To sum up, it is obvious that a self-intersecting brane can provide a variety of matter configurations on the brane. This is a fascinating and intuitively clear mechanism for the origin of matter in a brane world.

8.4. COMPARISON WITH THE RANDALL–SUNDRUM MODEL

In our brane world model, which starts from the $\mathcal{M}$-space Einstein equations, we have assumed that gravity is localized on the brane. This was formally represented by the δ-function. In a more conventional approach the starting point is Einstein's equations in the ordinary space, not in $\mathcal{M}$-space. Let us therefore explore a little what such an approach has to say about gravity around a brane embedded in a "bulk".

Randall and Sundrum [95] have considered a model in which a 3-brane with tension κ is coupled to gravity, the cosmological constant Λ being different from zero. After solving the Einstein equations they found that the metric tensor decreases exponentially with the distance from the brane. Hence gravity is localized on the brane.

More precisely, the starting point is the action

$$I = \kappa \int d^n x \sqrt{|f|}\, \delta^N(\eta - \eta(x))\, d^N \eta + \frac{1}{16\pi G^{(N)}} \int d^N \eta \sqrt{|f|}(2\Lambda + R), \quad (8.56)$$

which gives the Einstein equations

$$G_{ab} \equiv R_{ab} - \tfrac{1}{2} R \gamma_{ab} = -\Lambda \gamma_{ab} - 8\pi G^{(N)} T_{ab}, \quad (8.57)$$

$$T_{ab} = \int \kappa\, d^n x \sqrt{|f|}\, f^{\mu\nu}\, \partial_\mu \eta_a \partial_\nu \eta_b\, \delta^N(\eta - \eta(x)). \quad (8.58)$$

Let us consider a 3-brane ($n = 4$) embedded in a 5-dimensional bulk ($N = 5$). In a particular gauge the worldsheet embedding functions are $\eta^\mu = x^\mu$, $\eta^5 = \eta^5(x^\mu)$. For a flat worldsheet $\eta^5(x^\mu) = y_0$, where y_0 is independent of x^μ, it is convenient to take $y_0 = 0$. For such a brane located at $\eta^5 \equiv y = 0$ the appropriate Ansatz for the bulk metric respecting the symmetry of the brane configuration is

$$ds^2 = a^2(y)\eta_{\mu\nu}\, dx^\mu\, dx^\nu - dy^2. \quad (8.59)$$

The Einstein equations read

$$G^0{}_0 = G^1{}_1 = G^2{}_2 = G^3{}_3$$
$$= \frac{3a''}{a} + \frac{3a'^2}{a^2} = -\Lambda - 8\pi G^{(N)} T^0{}_0, \quad (8.60)$$

$$G^5{}_5 = \frac{6a'^2}{a^2} = -\Lambda, \quad (8.61)$$

where

$$T_{\alpha\beta} = \int \kappa\, d^4 x \sqrt{|f|} f_{\alpha\beta}\, \delta^4(\eta^\mu - x^\mu)\delta(y) = \kappa \sqrt{|f|} f_{\alpha\beta} \delta(y), \quad (8.62)$$

whilst $T_{\alpha 5} = 0$, $T_{55} = 0$. The induced metric is

$$f_{\alpha\beta} = \partial_\alpha \eta^a \partial_\beta \eta_a = \eta_{\alpha\beta} a^2(y).$$

Hence

$$T^0{}_0 = T^1{}_1 = T^2{}_2 = T^3{}_3 = \kappa a^4 \delta(y). \tag{8.63}$$

From eq. (8.61), which can be easily integrated, we obtain

$$a = a_0 \, e^{-|y|\sqrt{-\Lambda/6}}. \tag{8.64}$$

Such a solution makes sense if $\Lambda < 0$ and it respects the symmetry $a(y) = a(-y)$, so that the bulk metric is the same on both sides of the brane.

Introducing $\alpha' = a'/a$ (where $a' \equiv da/dy$) eq. (8.60) can be written as

$$3\alpha'' = -8\pi G^{(N)} \kappa \, a^4 \delta(y). \tag{8.65}$$

Integrating both sides of the latter equation over y we find

$$3(\alpha'(0^+) - \alpha'(0^-)) = -8\pi G^{(N)} \kappa a^4(0). \tag{8.66}$$

Using (8.64) we have

$$\frac{a'(0^+)}{a} = \alpha'(0^+) = -\sqrt{\frac{-\Lambda}{6}},$$

$$\frac{a'(0^-)}{a} = \alpha'(0^-) = \sqrt{\frac{-\Lambda}{6}}, \tag{8.67}$$

$$a(0) = a_0 = 1.$$

Hence (8.66) gives

$$6\sqrt{\frac{-\Lambda}{6}} = 8\pi G^{(N)} \kappa \tag{8.68}$$

which is a relation between the cosmological constant Λ and the brane tension κ.

From (8.59) and (8.64) it is clear that the metric tensor is localized on the brane's worldsheet and falls quickly when the transverse coordinates y goes off the brane.

An alternative Ansatz. We shall now consider an alternative Ansatz in which the metric is conformally flat:

$$ds^2 = b^2(z)(\eta_{\mu\nu} dx^\mu dx^\nu - dz^2). \tag{8.69}$$

The Einstein equations read

$$G^0{}_0 = \frac{3b''}{b^3} = -\Lambda - 8\pi G^{(N)}\kappa b^4(z) \qquad (8.70)$$

$$G^5{}_5 = \frac{6b'^2}{b^4} = -\Lambda \qquad (8.71)$$

The solution of (8.71) is

$$b = -\frac{1}{C + \sqrt{\frac{-\Lambda}{6}}|z|}. \qquad (8.72)$$

From (8.70) (8.71) we have

$$3\left(\frac{b''}{b^2} - \frac{b'^2}{b^3}\right) = -8\pi G^{(N)}\kappa b^5 \delta(z). \qquad (8.73)$$

Introducing $\beta' = b'/b^2$ the latter equation becomes

$$3\beta'' = -8\pi G^{(N)}\kappa b^5 \delta(z). \qquad (8.74)$$

After integrating over z we have

$$3(\beta'(0^+) - \beta'(0^-)) = -8\pi G^{(N)}\kappa b^5(0), \qquad (8.75)$$

where

$$\beta'(0^+) = C^{-4}\sqrt{\frac{-\Lambda}{6}}, \quad \beta'(0^-) = -C^{-4}\sqrt{\frac{-\Lambda}{6}}, \quad b(0) = -C^{-1}. \qquad (8.76)$$

Hence

$$6\sqrt{\frac{-\Lambda}{6}} = 8\pi G^{(N)}\kappa C^{-1}. \qquad (8.77)$$

If we take $C = 1$ then the last relation coincides with (8.68). The metric in the Ansatz (8.69) is of course obtained from that in (8.59) by a coordinate transformation.

THE METRIC AROUND A BRANE IN A HIGHER-DIMENSIONAL BULK

It would be very interesting to explore what happens to the gravitational field around a brane embedded in more than five dimensions. One could set an appropriate Ansatz for the metric, rewrite the Einstein equations and attempt to solve them. My aim is to find out whether in a space of sufficiently high dimension the metric — which is a solution to the Einstein equations — can be approximated with the metric (8.3), the conformal factor being localized on the brane.

Let us therefore take the Ansatz

$$\gamma_{ab} = \Omega^2 \bar{\gamma}_{ab} . \tag{8.78}$$

We then find

$$R_a{}^b = \Omega^{-2}\bar{R}_a^b + (N-2)\Omega^{-3}\Omega_{;a}{}^{;b} - 2(N-2)\Omega^{-4}\Omega_{,a}\Omega^{,b}$$
$$+ \Omega^{-3}\delta_a{}^b\Omega_{;c}{}^{;c} + (N-3)\Omega^{-4}\delta_a{}^b\Omega_{,c}\Omega^{,c} , \tag{8.79}$$

$$R = \Omega^{-2}\bar{R} + 2(N-1)\Omega^{-3}\Omega_{;c}{}^{;c} + (N-1)(N-4)\Omega^{-4}\Omega_{,c}\Omega^{,c}. \tag{8.80}$$

Splitting the coordinates according to

$$\eta^a = (x^\mu, y^{\bar{\mu}}), \tag{8.81}$$

where $y^{\bar{\mu}}$ are the transverse coordinates and assuming that Ω depends on $y^{\bar{\mu}}$ only, the Einstein equations become

$$G_\mu{}^\nu = \Omega^{-2}\bar{G}_\mu^\nu + \Omega^{-3}\delta_\mu{}^\nu \left[(2-N)\Omega_{;\bar{\mu}}{}^{;\bar{\mu}} + (N-3)\Omega^{-1}\Omega_{,\bar{\mu}}\Omega^{,\bar{\mu}} \right]$$
$$= -8\pi G^{(N)} T_\mu{}^\nu - \Lambda\delta_\mu{}^\nu , \tag{8.82}$$

$$G_{\bar{\mu}}{}^{\bar{\nu}} = \Omega^{-2}\bar{G}_{\bar{\mu}}^{\bar{\nu}} + \Omega^{-3}\left[(N-2)\Omega_{;\bar{\mu}}{}^{;\bar{\nu}} - 2(N-2)\Omega^{-1}\Omega_{,\bar{\mu}}\Omega^{,\bar{\nu}}\right.$$
$$\left. + \delta_{\bar{\mu}}{}^{\bar{\nu}} \left((2-N)\Omega_{;\bar{\alpha}}{}^{;\bar{\alpha}} + \Omega^{-1}\Omega_{,\bar{\alpha}}\Omega^{,\bar{\alpha}}((N-3) + (N-1)(N-4))\right)\right]$$
$$= -8\pi G^{(N)} T_{\bar{\mu}}{}^{\bar{\nu}} - \Lambda\delta_{\bar{\mu}}{}^{\bar{\nu}} . \tag{8.83}$$

Let $T_a{}^b$ be the stress–energy tensor of the brane itself. Then $T_{\bar{\mu}}{}^{\bar{\nu}} = 0$ (see eq. (8.62). Using (8.83) we can express $\Omega^{-1}\Omega_{,\bar{\alpha}}\Omega^{,\bar{\alpha}}$ in terms of $\Omega_{;\bar{\alpha}}{}^{;\bar{\alpha}}$ and insert it into (8.82). Taking[4] $N > 5$, $\Lambda = 0$ and assuming that close to the

---

[4]If dimension $N = 5$ then Λ must be different from zero, otherwise eq. (8.83) gives $\Omega_{,5}\Omega^{,5} = 0$, which is inconsistent with eq. (8.82).

brane the term $\Omega^{-2}\bar{G}_a^b$ can be neglected we obtain the Laplace equation for Ω

$$\Omega_{;\bar{\mu}}{}^{;\bar{\mu}} = 16\pi G^{(N)} T \Omega^3 A \tag{8.84}$$

where

$$A = (N-2)(N-1) \tag{8.85}$$

$$\times \left[2 - (N-4) \frac{(N-2) + \bar{n}(2-N)}{-2(N-2) + \bar{n}(N-3) - \frac{\bar{n}}{2}(N-1)(N-4)} \right],$$

$\bar{n}$ being the dimension of the transverse space, $\bar{n} = \delta_{\bar{\mu}}{}^{\bar{\mu}}$, and $T \equiv T_a{}^a = T_\mu{}^\mu$.

The above procedure has to be taken with reserve. Neglect of the term $\Omega^{-2}\bar{G}_a^b$ in general is not expected to be consistent with the Bianchi identities. Therefore equation (8.85) is merely an approximation to the exact equation. Nevertheless it gives an idea about the behavior of the function $\Omega(y^{\bar{\mu}})$.

The solution of eq. (8.84) has the form

$$\Omega = -\frac{k}{r^{\bar{n}-2}}, \tag{8.86}$$

where r is the radial coordinate in the transverse space. For a large transverse dimension $\bar{n}$ the function Ω falls very quickly with r. The gravitational field around the brane is very strong close to the brane, and negligible anywhere else. The interaction is practically a *contact interaction* and can be approximated by the δ-function. Taking a cutoff r_c determined by the thickness of the brane we can normalize Ω according to

$$\int_{r_c}^{\infty} \Omega(r)\, dr = \frac{k}{(\bar{n}-1)r_c^{\bar{n}-1}} = 1 \tag{8.87}$$

and take $\Omega(r) \approx \delta(r)$.

The analysis above is approximate and requires a more rigorous study. But intuitively it is clear that in higher dimensions gravitational interaction falls very quickly. For a *point particle* the gravitational potential has the asymptotic behavior $\gamma_{00} - 1 \propto r^{-(N-3)}$, and for a sufficiently high spacetime dimension N the interaction is practically contact (like the Van der Waals force). Particles then either do not feel each other, or they form bound states upon contact. Network-like configurations are expected to be formed, as shown in Fig.8.6. Such configurations mimic very well the intersecting branes considered in Secs. 8.1–8.3.

In this section we have started from the conventional theory of gravitation and found strong arguments that in a space of very high dimension the

Spacetime as a membrane in a higher-dimensional space 269

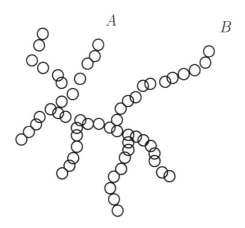

Figure 8.6. In a space of very large dimension separated point particles do not feel gravitational interaction, since it is negligible. When two particles meet they form a bound system which grows when it encounters other particles. There is (practically) no force between the 'tails' (e.g., between the points A and B). However, there is tension within the tail. (The tail, of course, need not be 1-dimensional; it could be a 2, 3 or higher-dimensional brane.)

gravitational force is a contact force. Various *network-like configurations* are then possible and they are stable. Effectively there is no gravity outside such a network configuration. Such a picture matches very well the one we postulated in the previous three sections of this chapter, and also the picture we considered when studying the $\mathcal{M}$-space formulation of the membrane theory.

Chapter 9

THE EINSTEIN–HILBERT ACTION ON THE BRANE AS THE EFFECTIVE ACTION

After so many years of intensive research the quantization of gravity is still an unfinished project. Amongst many approaches followed, there is the one which seems to be especially promising. This is the so called induced gravity proposed by Sakharov [96]. His idea was to treat the metric not as a fundamental field but as one induced from more basic fields. The idea has been pursued by numerous authors [97]; especially illuminating are works by Akama, Terazawa and Naka [98]. Their basic action contains N scalar fields and it is formally just a slight generalization of the well-known Dirac–Nambu–Goto action for an n-dimensional world sheet swept by an $(n-1)$-dimensional membrane.

Here we pursue such an approach and give a concrete physical interpretation of the N scalar fields which we denote $\eta^a(x)$. We assume that the spacetime is a surface V_4, called the spacetime sheet, embedded in a higher-dimensional space V_N, and $\eta^a(x)$ are the embedding functions. An embedding model has been first proposed by Regge and Teitelboim [99] and investigated by others [100]. In that model the action contains the Ricci scalar expressed in terms of the embedding functions. In our present model [88, 90, 89], on the contrary, we start from an action which is essentially the minimal surface action[1] weighted with a function $\omega(\eta)$ in V_N. For a suitably chosen ω, such that it is singular (δ-function like) on certain surfaces $\widehat{V}_m$, also embedded in V_N, we obtain on V_4 a set of world lines. In the previous chapter it was shown that these worldlines are geodesics of V_4, provided that V_4 described by $\eta^a(x)$ is a solution to our variational

---

[1] Recently Bandos [91] has considered a string-like description of gravity by considering bosonic p-branes coupled to an antisymmetric tensor field.

procedure. I will show that after performing functional integrations over $\eta^a(x)$ we obtain two contributions to the path integral. One contribution comes from all possible $\eta^a(x)$ not intersecting $\widehat{V}_m$, and the other from those $\eta^a(x)$ which do intersect the surfaces $\widehat{V}_m$. In the effective action so obtained the first contribution gives the Einstein–Hilbert term R plus higher-order terms like R^2. The second contribution can be cast into the form of a path integral over all possible worldlines $X^\mu(\tau)$. Thus we obtain an action which contains matter sources and a kinetic term for the metric field (plus higher orders in R). So in the proposed approach both the metric field and the matter field are induced from more basic fields $\eta^a(x)$.

9.1. THE CLASSICAL MODEL

Let us briefly summarize the discussion of Chapter 8. We assume that the arena where physics takes place is an N-dimensional space V_N with $N \geq 10$. Next we assume that an n-dimensional surface V_n living in V_N represents a possible spacetime. The parametric equation of such a 'spacetime sheet' V_n is given by the embedding functions $\eta^a(x^\mu)$, $a = 0, 1, 2, ..., N$, where x^μ, $\mu = 0, 1, 2, \ldots, n-1$, are coordinates (parameters) on V_n. We assume that the action is just that for a minimal surface V_n

$$I[\eta^a(x)] = \int (\det \partial_\mu \eta^a \, \partial_\nu \eta^b \gamma_{ab})^{1/2} \mathrm{d}^n x, \qquad (9.1)$$

where γ_{ab} is the metric tensor of V_N. The dimension of the spacetime sheet V_n is taken here to be arbitrary, in order to allow for the Kaluza–Klein approach. In particular, we may take $n = 4$. We admit that the embedding space is curved in general. In particular, let us consider the case of a conformally flat V_N, such that $\gamma_{ab} = \omega^{2/n} \eta_{ab}$, where η_{ab} is the N-dimensional Minkowski tensor. Then Eq. (9.1) becomes

$$I[\eta^a(x)] = \int \omega(\eta) (\det \partial_\mu \eta^a \, \partial_\nu \eta^b \eta_{ab})^{1/2} \mathrm{d}^n x. \qquad (9.2)$$

From now on we shall forget about the origin of $\omega(\eta)$ and consider it as a function of position in a *flat* embedding space. The indices a, b, c will be raised and lowered by η^{ab} and η_{ab}, respectively.

In principle $\omega(\eta)$ is arbitrary. But it is very instructive to choose the following function:

$$\omega(\eta) = \omega_0 + \sum_i \int m_i \frac{\delta^N(\eta - \hat{\eta}_i)}{\sqrt{|\gamma|}} \mathrm{d}^m \hat{x} \sqrt{|\hat{f}|}, \qquad (9.3)$$

The Einstein–Hilbert action on the brane as the effective action 273

where $\eta^a = \hat\eta_i^a(\hat x)$ is the parametric equation of an m-dimensional surface $\hat V_m^{(i)}$, called the *matter sheet*, also embedded in V_N, $\hat f$ is the determinant of the induced metric on $\hat V_m^{(i)}$, and $\sqrt{|\gamma|}$ allows for taking curved coordinates in otherwise flat V_N. If we take $m = N - n + 1$ then the intersection of V_n and $\hat V_m^{(i)}$ can be a (one-dimensional) line, i.e., a worldline C_i on V_n. In general, when $m = N - n + (p+1)$ the intersection can be a $(p+1)$-dimensional world sheet representing the motion of a p-dimensional membrane (also called p-brane). In this chapter we confine our consideration to the case $p = 0$, that is, to the motion of a point particle.

Inserting (9.3) into (9.2) and writing $f_{\mu\nu} \equiv \partial_\mu \eta^a \partial_\nu \eta_a$, $f \equiv \det f_{\mu\nu}$ we obtain

$$I[\eta] = \omega_0 \int d^n x \sqrt{|f|} + \int d^n x \sum_i m_i \, \delta^n(x - X_i)(f_{\mu\nu} \dot X_i^\mu \dot X_i^\nu)^{1/2} \, d\tau. \qquad (9.4)$$

As already explained in Sec. 8.1, the result above was obtained by writing

$$d^m \hat x = d^{m-1} \hat x \, d\tau, \qquad \hat f = \hat f^{(m-1)}(\dot X_i^\mu \dot X_{i\mu})^{1/2}$$

and taking the coordinates η^a such that $\eta^a = (x^\mu, \eta^n, ..., \eta^{N-1})$, where x^μ are (curved) coordinates on V_n. The determinant of the metric of the embedding space V_N in such curvilinear coordinates is then $\gamma = \det \partial_\mu \eta^a \partial_\nu \eta_a = f$.

If we vary the action (9.4) with respect to $\eta^a(x)$ we obtain

$$\partial_\mu \left[\sqrt{|f|} (\omega_0 f^{\mu\nu} + T^{\mu\nu}) \partial_\nu \eta_a \right] = 0, \qquad (9.5)$$

where

$$T^{\mu\nu} = \frac{1}{\sqrt{|f|}} \sum_i \int d^n x \, m_i \, \delta^n(x - X_i) \frac{\dot X_i^\mu \dot X_i^\nu}{(\dot X_i^\alpha \dot X_{i\alpha})^{1/2}} \, d\tau \qquad (9.6)$$

is the stress–energy tensor of dust. Eq. (9.5) can be rewritten in terms of the covariant derivative D_μ on V_n:

$$D_\mu \left[(\omega_0 f^{\mu\nu} + T^{\mu\nu}) \partial_\nu \eta_a \right] = 0. \qquad (9.7)$$

The latter equation gives

$$\partial_\nu \eta_a D_\mu T^{\mu\nu} + (\omega_0 f^{\mu\nu} + T^{\mu\nu}) D_\mu D_\nu \eta_a = 0, \qquad (9.8)$$

where we have taken into account that a covariant derivative of metric is zero, i.e., $D_\alpha f_{\mu\nu} = 0$ and $D_\alpha f^{\mu\nu} = 0$, which implies also $\partial_\alpha \eta^c D_\mu D_\nu \eta_c = 0$, since $f_{\mu\nu} \equiv \partial_\mu \eta^a \partial_\nu \eta_a$. Contracting Eq.(9.8) by $\partial^\alpha \eta^a$ we have

$$D_\mu T^{\mu\nu} = 0. \qquad (9.9)$$

The latter are the well known equations of motion for the sources. In the case of dust (9.9) implies that dust particles move along geodesics of the spacetime V_n. We have thus obtained the very interesting result that the worldlines C_i which are obtained as intersections $V_n \cap \widehat{V}_m^{(i)}$ are geodesics of the spacetime sheet V_n. The same result is also obtained directly by varying the action (9.4) with respect to the variables $X_i^\mu(\tau)$.

A solution to the equations of motion (9.5) (or (9.7)) gives both: the spacetime sheet $\eta^a(x)$ and the worldlines $X_i^\mu(\tau)$. Once $\eta^a(x)$ is determined the induced metric $g_{\mu\nu} = \partial_\mu \eta^a \partial_\nu \eta_a$ is determined as well. But such a metric in general does not satisfy the Einstein equations. In the next section we shall see that quantum effects induce the necessary Einstein–Hilbert term $(-g)^{1/2} R$.

9.2. THE QUANTUM MODEL

For the purpose of quantization we shall use the classical action [88], that is, a generalization of the well known Howe–Tucker action [31], which is equivalent to (9.2) (see also Sec. 4.2):

$$I[\eta^a, g^{\mu\nu}] = \tfrac{1}{2} \int d^n x \sqrt{|g|} \omega(\eta) (g^{\mu\nu} \partial_\mu \eta^a \partial_\nu \eta_a + 2 - n). \tag{9.10}$$

It is a functional of the embedding functions $\eta^a(x)$ and the Lagrange multipliers $g^{\mu\nu}$. Varying (9.10) with respect to $g^{\mu\nu}$ gives the constraints

$$-\frac{\omega}{4} \sqrt{|g|} \, g_{\alpha\beta} (g^{\mu\nu} \partial_\mu \eta^a \partial_\nu \eta_a + 2 - n) + \frac{\omega}{2} \sqrt{|g|} \, \partial_\alpha \eta^a \partial_\beta \eta_a = 0. \tag{9.11}$$

Contracting (9.11) with $g^{\alpha\beta}$, we find $g^{\mu\nu} \partial_\mu \eta^a \partial_\nu \eta_a = n$, and after inserting the latter relation back into (9.11) we find

$$g_{\alpha\beta} = \partial_\alpha \eta^a \partial_\beta \eta_a, \tag{9.12}$$

which is the expression for an induced metric on the surface V_n. In the following paragraphs we shall specify $n = 4$; however, whenever necessary, we shall switch to the generic case of arbitrary n.

In the classical theory we may say that a 4-dimensional spacetime sheet is swept by a 3-dimensional space-like hypersurface Σ which moves forward in time. The latter surface is specified by initial conditions, and the equations of motion then determine Σ at every value of a time-like coordinate $x^0 = t$. Knowledge of a particular hypersurface Σ implies knowledge of the corresponding intrinsic 3-geometry specified by the 3-metric $g_{ij} = \partial_i \eta^a \partial_j \eta_a$ induced on Σ ($i, j = 1, 2, 3$). However, knowledge of the data $\eta^a(t, x^i)$ on an entire infinite Σ is just a mathematical idealization which cannot be realized in a practical situation by an observer because of the finite speed of light.

In quantum theory a state of a surface Σ is not specified by the coordinates $\eta^a(t, x^i)$, but by a wave functional $\psi[t, \eta^a(x^i)]$. The latter represents the probability amplitude that at time t an observer would obtain, as a result of measurement, a particular surface Σ.

The probability amplitude for the transition from a state with definite Σ_1 at time t_1 to a state Σ_2 at time t_2 is given by the Feynman path integral

$$K(2,1) = \langle \Sigma_2, t_2 | \Sigma_1, t_1 \rangle = \int e^{i\,I[\eta,g]} \mathcal{D}\eta \mathcal{D}g. \tag{9.13}$$

Now, if in Eq.(9.13) we perform integration only over the embedding functions $\eta^a(x^\mu)$, then we obtain the so called *effective action* I_{eff}

$$e^{iI_{\text{eff}}[g]} \equiv \int e^{i\,I[\eta,g]} \mathcal{D}\eta, \tag{9.14}$$

which is a functional of solely the metric $g^{\mu\nu}$. From eq. (9.14) we obtain by functional differentiation

$$\frac{\delta I_{\text{eff}}[g]}{\delta g^{\mu\nu}} = \frac{\int \frac{\delta I[\eta,g]}{\delta g^{\mu\nu}} e^{i\,I[\eta,g]} \mathcal{D}\eta}{\int e^{i\,I[\eta,g]} \mathcal{D}\eta} \equiv \left\langle \frac{\delta I[\eta, g]}{\delta g^{\mu\nu}} \right\rangle = 0. \tag{9.15}$$

On the left hand side of eq. (9.15) we have taken into account the constraints $\delta I[\eta, g]/\delta g^{\mu\nu} = 0$ (explicitly given in eq. (9.11)).

The expression $\delta I_{\text{eff}}[\eta,g]/\delta g^{\mu\nu} = 0$ gives the classical equations for the metric $g_{\mu\nu}$, derived from the effective action.

Let us now consider a specific case in which we take for $w(\eta)$ the expression (9.3). Then our action (9.10) splits into two terms

$$I[\eta, g] = I_0[\eta, g] + I_m[\eta, g] \tag{9.16}$$

with

$$I_0[\eta, g] = \frac{w_0}{2} \int d^n x \sqrt{|g|} \, (g^{\mu\nu} \partial_\mu \eta^a \partial_\nu \eta_a + 2 - n), \tag{9.17}$$

$$I_m[\eta, g] = \frac{1}{2} \int d^n x \sqrt{|g|} \sum_i m_i \frac{\delta^N(\eta - \hat{\eta}_i)}{\sqrt{|\gamma|}} d^m \hat{x}$$

$$\times \sqrt{|\hat{f}|} \, (g^{\mu\nu} \partial_\mu \eta^a \partial_\nu \eta_a + 2 - n). \tag{9.18}$$

The last expression can be integrated over $m-1$ coordinates $\hat{x}^\mu$, while $\hat{x}^0$ is chosen so to coincide with the parameter τ of a worldline C_i. We also split the metric as $g^{\mu\nu} = n^\mu n^\nu/n^2 + \bar{g}^{\mu\nu}$, where n^μ is a time-like vector and $\bar{g}^{\mu\nu}$ is the projection tensor, giving $\bar{g}^{\mu\nu} \partial_\mu \eta^a \partial_\nu \eta_a = n - 1$. So we obtain

$$I_m[\eta, g] = \frac{1}{2} \int d^n x \sqrt{|g|} \sum_i \frac{\delta^n(x - X_i(\tau))}{\sqrt{|g|}}$$
$$\times \left(\frac{g_{\mu\nu} \dot{X}_i^\mu \dot{X}_i^\nu}{\mu_i} + \mu_i \right) d\tau = I_m[X_i, g]. \quad (9.19)$$

Here $\mu_i \equiv 1/\sqrt{n^2}|_{C_i}$ are the Lagrange multipliers giving, after variation, the worldline constraints $\mu_i^2 = \dot{X}_i^2$. Eq. (9.19) is the well known Howe-Tucker action [31] for point particles.

Now let us substitute our specific action (9.16)–(9.19) into the expression (9.14) for the effective action. The functional integration now runs over two distinct classes of spacetime sheets V_n [represented by $\eta^a(x)$]:

(a) those V_n which either *do not intersect* the matter sheets $\widehat{V}_m^{(i)}$ [represented by $\hat{\eta}_i^a(\hat{x})$] and do not self-intersect, or, if they do, the intersections are just single points, and

(b) those V_n which *do intersect* $\widehat{V}_m^{(i)}$ and/or self-intersect, the intersections being worldlines C_i.

The sheets V_n which correspond to the case (b) have two distinct classes of points (events):

(b1) *the points outside the intersection*, i.e., outside the worldlines C_i, and

(b2) *the points on the intersection*, i.e., the events belonging to C_i.

The measure $\mathcal{D}\eta^a(x)$ can be factorized into the contribution which corresponds to the case (a) or (b1) ($x \notin C_i$), and the contribution which corresponds to the case (b2) ($x \in C_i$):

$$\begin{aligned} \mathcal{D}\eta &= \prod_{a,x} (|g(x)|)^{1/4} d\eta^a(x) \\ &= \prod_{a, x \notin C_i} (|g(x)|)^{1/4} d\eta^a(x) \prod_{a, x \in C_i} (|g(x)|)^{1/4} d\eta^a(x) \\ &\equiv \mathcal{D}_0 \eta \, \mathcal{D} m. \end{aligned} \quad (9.20)$$

The additional factor $(|g(x)|)^{1/4}$ comes from the requirement that the measure be invariant under reparametrizations of x^μ (see Ref. [102] and Sec. 4.1 for details). From the very definition of $\prod_{a, x \in C_i}(|g(x)|)^{1/4} d\eta^a(x)$ as the measure of the set of points on the worldlines C_i (each C_i being represented by an equation $x = X_i^\mu(\tau)$) we conclude that

$$\mathcal{D}_m \eta^a(x) = \mathcal{D} X_i^\mu(\tau). \quad (9.21)$$

The effective action then satisfies [owing to (9.16)–(9.21)]

$$e^{iI_{\text{eff}}[g]} = \int e^{iI_0[\eta,g]} \mathcal{D}_0\eta \, e^{iI_m[X_i,g]} \mathcal{D}X_i \equiv e^{iW_0} e^{iW_m},$$
(9.22)

$$I_{\text{eff}} = W_0 + W_m.$$
(9.23)

The measure $\mathcal{D}_0\eta$ includes all those sheets V_n that do not intersect the matter sheet and do not self-intersect [case (a)], and also all those sheets which do intersect and/or self-intersect [case (b1)], apart from the points on the intersections.

The first factor in the product (9.22) contains the action (9.17). The latter has the same form as the action for N scalar fields in a curved background spacetime with the metric $g_{\mu\nu}$. The corresponding effective action has been studied and derived in Refs. [103]. Using the same procedure and substituting our specific constants $w_0/2$ and $(n-1)$ occurring in eq. (9.17), we find the following expression for the effective action:

$$I_{\text{eff}} = \lim_{\mu^2 \to 0} \int d^n x \sqrt{|g|} \left(N w_0^{-1} (4\pi)^{-n/2} \sum_{j=0}^{\infty} (\mu^2)^{n/2-j} a_j(x) \Gamma(j - \frac{n}{2}) \right.$$

$$\left. + \frac{w_0}{2}(2-n) \right)$$
(9.24)

with

$$a_0(x) = 1$$
(9.25)

$$a_1(x) = R/6$$
(9.26)

$$a_2(x) = \frac{1}{12}R^2 + \frac{1}{180}(R_{\alpha\beta\gamma\delta}R^{\alpha\beta\gamma\delta} - R_{\alpha\beta}R^{\alpha\beta}) - \frac{1}{30}D_\mu D^\mu R$$
(9.27)

where R, $R_{\alpha\beta}$ and $R_{\alpha\beta\gamma\delta}$ are the Ricci scalar, the Ricci tensor and the Riemann tensor, respectively. The function $\Gamma(y) = \int_0^\infty e^{-t} t^{y-1} dt$; it is divergent at negative integers y and finite at $y = \frac{3}{2}, \frac{1}{2}, -\frac{1}{2}, -\frac{3}{2}, -\frac{5}{2}, \ldots$. The effective action (9.24) is thus divergent in even-dimensional spaces V_n. For instance, when $n = 4$, the argument in eq. (9.24) is $j - 2$, which for $j = 0, 1, 2, \ldots$ is indeed a negative integer. Therefore, in order to obtain a finite effective action one needs to introduce a suitable cut-off parameter Λ and replace

$$(\mu^2)^{n/2-j} \Gamma(j - \frac{n}{2}) = \int_0^\infty t^{j-1-n/2} e^{-\mu^2 t} dt$$

with

$$\int_{1/\Lambda^2}^\infty t^{j-1-n/2} e^{-\mu^2 t} dt$$

Then we find

$$I_{\text{eff}} = \int d^n x \sqrt{|g|} \left(\lambda_0 + \lambda_1 R + \lambda_2 R^2 + \lambda_3 (R_{\alpha\beta\gamma\delta} R^{\alpha\beta\gamma\delta} - R_{\alpha\beta} R^{\alpha\beta}) \right.$$
$$\left. + \lambda_4 D_\mu D^\mu R + O(\Lambda^{n-6}) \right), \qquad (9.28)$$

where[2], for $n > 4$,

$$\lambda_0 = \frac{N\omega_0}{4(4\pi)^{n/2}} \frac{2\Lambda^n}{n} + \frac{\omega_0}{2}(2-n),$$

$$\lambda_1 = \frac{N\omega_0}{4(4\pi)^{n/2}} \frac{2\Lambda^{n-2}}{6(n-2)},$$

$$\lambda_2 = \frac{N\omega_0}{4(4\pi)^{n/2}} \frac{2\Lambda^{n-4}}{12(n-4)},$$

$$\lambda_3 = \frac{N\omega_0}{4(4\pi)^{n/2}} \frac{2\Lambda^{n-4}}{180(n-4)},$$

$$\lambda_4 = -\frac{N\omega_0}{4(4\pi)^{n/2}} \frac{2\Lambda^{n-4}}{30(n-4)}. \qquad (9.29)$$

Here λ_0 is the cosmological constant, whilst λ_1 is related to the gravitational constant G in n-dimensions according to

$$\lambda_1 \equiv (16\pi G)^{-1}. \qquad (9.30)$$

This last relation shows how the induced gravitational constant is calculated in terms of ω_0, which is a free parameter of our embedding model, and the cutoff parameter Λ. According to Akama [98], et al., and Sugamoto [92], we consider the cutoff Λ to be a physical quantity, the inverse thickness of a membrane, because the original action (9.17) describes an idealized theory of extended objects with vanishing thickness, but the real extended objects have non-vanishing thickness playing a role of the ultraviolet cutoff. In the case of thin extended objects, we can ignore the $O(\Lambda^{n-4})$ terms in eq. (9.28).

In the above calculation of the effective action we have treated all functions $\eta^a(x)$ entering the path integral (9.14) as those representing distinct spacetime sheets V_n. However, owing to the reparametrization invariance, there exist equivalence classes of functions representing the same V_n. This complication must be taken into account when calculating the entire amplitude (9.13). The conventional approach is to introduce ghost fields which

---

[2]In the case $n = 4$ it is convenient to use Akama's regularization [98]. Similarly for the problematic coefficients in higher dimensions.

cancel the non-physical degrees of freedom. An alternative approach, first explored in ref. [53]–[55] and much discussed in this book, is to assume that all possible embedding functions $\eta^a(x)$ can nevertheless be interpreted as describing physically distinct spacetime sheets $\mathcal{V}_n$. This is possible if the extra degrees of freedom in $\eta^a(x)$ describe tangent deformations of $\mathcal{V}_n$. Such a deformable surface $\mathcal{V}_n$ is then a different concept from a non-deformable surface $\mathcal{V}_n$. The path integral can be performed in a straightforward way in the case of $\mathcal{V}_n$, as was done in arriving at the result (9.24). However, even from the standard point of view [93], gauge fixing is not required for the calculation of the effective action, since in the η^a integration, $g_{\mu\nu}$ is treated as a fixed background.

Let us now return to eq. (9.22). In the second factor of eq. (9.22) the functional integration runs over all possible worldlines $X^\mu(\tau)$. Though they are obtained as intersections of *various* $\mathcal{V}_n$ with $\widehat{\mathcal{V}}_m^{(i)}$, we may consider all those worldlines to be lying in *the same* effective spacetime $\mathcal{V}_n^{(\text{eff})}$ with the intrinsic metric $g_{\mu\nu}$. In other words, in the effective theory we identify all those various $\mathcal{V}_n$'s, having the same induced (intrinsic) metric $g_{\mu\nu}$, as one and the same spacetime. If one considers the embedding space $\mathcal{V}_N$ of sufficiently high dimension N, then there is enough freedom to obtain as an intersection any possible worldline in the effective spacetime $\mathcal{V}_n^{(\text{eff})}$.

This is even more transparent when the spacetime sheet self-intersects. Then we can have a situation in which various sheets coincide in all the points, apart from the points in the vicinity of the intersections

When the condition for the classical approximation is satisfied, i.e., when $I_m \gg \hbar = 1$, then only those trajectories X_i^μ which are close to the classically allowed ones effectively contribute:

$$e^{iW_{\text{m}}} = e^{iI_m[X_i,g]}, \qquad W_{\text{m}} = I_m \qquad (9.31)$$

The effective action is then a sum of the gravitational kinetic term W_0 given in eq. (9.28) and the source term I_m given in (9.19). Variation of I_{eff} with respect to $g^{\mu\nu}$ then gives the gravitational field equation in the presence of point particle sources with the stress–energy tensor $T^{\mu\nu}$ as given in Eq. (9.6):

$$R^{\mu\nu} - \frac{1}{2}g^{\mu\nu} + \lambda_0 g^{\mu\nu} + \text{(higher order terms)} = -8\pi G\, T^{\mu\nu}. \qquad (9.32)$$

However, in general the classical approximation is not satisfied, and in the evaluation of the matter part W_{m} of the effective action one must take into account the contributions of all possible paths $X_i^\mu(\tau)$. So we have

(confining ourselves to the case of only one particle, omitting the subscript i, and taking $\mu = 1$)

$$e^{iW_m} = \int_{x_a}^{x_b} \mathcal{D}X \exp\left(\frac{i}{2}\int_{\tau_a}^{\tau_b} d\tau\, m\left(g_{\mu\nu}\dot{X}^\mu \dot{X}^\nu + 1\right)\right)$$

$$= \mathcal{K}(x_b,\tau_b;x_a,\tau_a) \equiv \mathcal{K}(b,a) \quad (9.33)$$

which is a propagator or a Green's function satisfying (for $\tau_b \geq \tau_a$) the equation

$$\left(i\frac{\partial}{\partial \tau_b} - H\right)\mathcal{K}(x_b,\tau_b;x_a,\tau_a) = -\frac{1}{\sqrt{|g|}}\delta^n(x_b - x_a)\delta(\tau_b - \tau_a), \quad (9.34)$$

where $H = (|g|)^{-1/2}\partial_\mu((|g|)^{1/2}\partial^\mu)$. From (9.34) it follows that

$$\mathcal{K}(b,a) = -\left[i\frac{\partial}{\partial \tau} - H\right]^{-1}_{x_b,\tau_b;x_a,\tau_a} \quad (9.35)$$

where the inverse Green's function is treated as a matrix in the (x,τ) space.

Using the following relation [32] for Gaussian integration

$$\int y_m y_n \prod_{i=1}^N dy_i\, e^{-\sum_{ij} y_i A_{ij} y_j} \propto \frac{(A^{-1})_{mn}}{(\det |A_{ij}|)^{1/2}}, \quad (9.36)$$

we can rewrite the Green's function in terms of the second quantized field:

$$\mathcal{K}(a,b) = \int \psi^*(x_b,\tau_b)\,\psi(x_b,\tau_b)\,\mathcal{D}\psi^*\,\mathcal{D}\psi$$

$$\times \exp\left[-i\int d\tau\, d^n x\, \sqrt{|g|}\,\psi^*(i\partial_\tau - H)\psi\right]. \quad (9.37)$$

If the conditions for the "classical" approximation are satisfied, so that the phase in (9.37) is much greater than $\hbar = 1$, then only those paths $\psi(\tau,x)$, $\psi^*(\tau,x)$ which are close to the extremal path, along which the phase is zero, effectively contribute to $\mathcal{K}(a,b)$. Then the propagator is simply

$$\mathcal{K}(b,a) \propto \exp\left[-i\int d\tau\, d^n x\sqrt{|g|}\,\psi^*(i\partial_\tau - H)\psi\right]. \quad (9.38)$$

The effective one-particle "matter" action W_m is then

$$W_m = -\int d\tau\, d^n x\,\sqrt{|g|}\,\psi^*(\tau,x)(i\partial_\tau - H)\psi(\tau,x). \quad (9.39)$$

If we assume that the τ-dependence of the field $\psi(\tau,x)$ is specified[3] by $\psi(\tau,x) = e^{im\tau}\phi(x)$, then eq. (9.39) simplifies to the usual well known expression for a scalar field:

$$W_m = \int d^n x \sqrt{|g|} \phi^*(x) \left[\frac{1}{\sqrt{|g|}} \partial_\mu(\sqrt{|g|}\partial^\mu) + m^2 \right] \phi(x)$$

$$= -\tfrac{1}{2} \int d^n x \sqrt{|g|} \left(g^{\mu\nu} \partial_\mu \phi^* \partial_\nu \phi - m^2 \right), \qquad (9.40)$$

where the surface term has been omitted.

Thus, starting from our basic fields $\eta^a(x)$, which are the embedding functions for a spacetime sheet V_n, we have arrived at the effective action I_{eff} which contains the kinetic term W_0 for the metric field $g^{\mu\nu}$ (see Eq. (9.28)) and the source term W_m (see eq. (9.32) and (9.18), or (9.40)). Both the metric field $g_{\mu\nu}$ and the bosonic matter field ϕ are induced from the basic fields $\eta^a(x)$.

9.3. CONCLUSION

We have investigated a model which seems to be very promising in attempts to find a consistent relation between quantum theory and gravity. Our model exploits the approach of induced gravity and the concept of spacetime embedding in a higher-dimensional space and has the following property of interest: what appear as worldlines in, e.g., a 4-dimensional spacetime, are just the intersections of a spacetime sheet V_4 with "matter" sheets $\widehat{V}_m^{(i)}$, or self-intersections of V_4. Various choices of spacetime sheets then give various configurations of worldlines. Instead of V_4 it is convenient to consider a spacetime sheet V_n of arbitrary dimension n. When passing to the quantized theory a spacetime sheet is no longer definite. All possible alternative spacetime sheets are taken into account in the expression for a wave functional or a Feynman path integral. The intersection points of V_n with itself or with a matter sheet $\widehat{V}_m^{(i)}$ are treated specially, and it is found that their contribution to the path integral is identical to the contribution of a point particle path. We have paid special attention to the effective action which results from functionally integrating out all possible embeddings with the same induced metric tensor. We have found that the effective ac-

---

[3] By doing so we in fact project out the so called physical states from the set of all possible states. Such a procedure, which employs a "fictitious" evolution parameter τ is often used (see Chapter 1). When gauge fixing the action (9.33) one pretends that such an action actually represents an "evolution" in parameter τ. Only later, when all the calculations (e.g., path integral) are performed, one integrates over τ and thus projects out the physical quantities.

tion, besides the Einstein–Hilbert term and the corresponding higher-order terms, also contains a source term. The expression for the latter is equal to that of a classical (when such an approximation can be used) or quantum point particle source described by a scalar (bosonic) field.

In other words, we have found that the (n-dimensional) Einstein equations (including the R^2 and higher derivative terms) with classical or quantum point-particle sources are effective equations resulting from performing a quantum average over all possible embeddings of the spacetime. Gravity —as described by Einstein's general relativity— is thus considered not as a fundamental phenomenon, but as something induced quantum-mechanically from more fundamental phenomena.

In our embedding model of gravity with bosonic sources, new and interesting possibilities are open. For instance, instead of a 4-dimensional spacetime sheet we can consider a sheet which possesses additional dimensions, parametrized either with usual or Grassmann coordinates. In such a way we expect to include, on the one hand, via the Kaluza–Klein mechanism, other interactions as well besides the gravitational one, and, on the other hand, fermionic sources. The latter are expected also to result from the polyvector generalization of the action, as discussed in Sec. 2.5.

It is well known that the quantum field theory based on the action (9.40) implies the infinite vacuum energy density and consequently the infinite (or, more precisely, the Planck scale cutoff) cosmological constant. This is only a part of the total cosmological constant predicted by the complete theory, as formulated here. The other part comes from eqs. (9.28), (9.29). The parameters N, ω_0, Λ and n could in principle be adjusted so to give a small or vanishing total cosmological constant. If carried out successfully this would be then an alternative (or perhaps a complement) to the solution of the cosmological constant problem as suggested in Chapter 3.

Finally, let us observe that within the "brane world" model of Randall and Sundrum [95], in which matter fields are localized on a 3-brane, whilst gravity propagates in the bulk, it was recently proposed [104] to treat the Einstein–Hilbert term on the brane as being induced in the quantum theory of the brane. It was found that the localized matter on a brane can induce via loop correction a 4D kinetic term for gravitons. This also happens in our quantum model if we consider the effective action obtained after integrating out the second quantized field in (9.37). (The procedure expounded in refs. [21] is then directly applicable.) In addition, in our model we obtain, as discussed above, a kinetic term for gravity on the brane after functionally integrating out the embedding functions.

Chapter 10

ON THE RESOLUTION OF TIME PROBLEM IN QUANTUM GRAVITY

Since the pioneering works of Sakharov [96] and Adler [97] there has been increasing interest in various models of induced gravity [98]. A particularly interesting and promising model seems to be the one in which spacetime is a 4-dimensional manifold (a 'spacetime sheet') V_4 embedded in an N-dimensional space V_N [101, 88], [53]–[55]. The dynamical variables are the embedding functions $\eta^a(x)$ which determine positions (coordinates) of points on V_4 with respect to V_N. The action is a straightforward generalization of the Dirac–Nambu–Goto action. The latter can be written in an equivalent form in which there appears the induced metric $g_{\mu\nu}(x)$ and $\eta^a(x)$ as variables which have to be varied independently. Quantization of such an action enables one to express *an effective action* as a functional of $g_{\mu\nu}(x)$. The effective action is obtained in the Feynman path integral in which we functionally integrate over the embedding functions $\eta^a(x)$ of V_4, so that what remains is a functional dependence on $g_{\mu\nu}(x)$. Such an effective action contains the Ricci curvature scalar R and its higher orders. This theory was discussed in more detail in the previous chapter.

In this chapter we are going to generalize the above approach. The main problem with any reparametrization invariant theory is the presence of constraints relating the dynamical variables. Therefore there exist equivalence classes of functions $\eta^a(x)$ —related by reparametrizations of the coordinates x^μ— such that each member of an equivalence class represents the same spacetime sheet V_4. This must be taken into account in the quantized theory, e.g., when performing, for instance, a functional integration over $\eta^a(x)$. Though elegant solutions to such problems were found in string theories [105], the technical difficulties accumulate in the case of a p-dimensional membrane (p-brane) with p greater than 2 [80].

In Chapters 4–6 we discussed the possibility of removing constraints from a membrane (p-brane) theory. Such a generalized theory possesses additional degrees of freedom and contains the usual p-branes of the Dirac–Nambu–Goto type [106] as a special case. It is an extension, from a point particle to a p-dimensional membrane, of a theory which treats a relativistic particle without constraint, so that all coordinates x^μ and the conjugate momenta p_μ are independent dynamical variables which evolve along the invariant evolution parameter τ [1]–[16]. A membrane is then considered as a continuum of such point particles and has no constraints. It was shown [53] that the extra degrees of freedom are related to variable stress and fluid velocity on the membrane, which is therefore, in general, a *"wiggly membrane"*. We shall now apply the concept of a relativistic membrane without constraints to the embedding model of induced gravity in which the whole spacetime is considered as a membrane in a flat embedding space.

In Sec. 10.1 we apply the theory of unconstrained membranes to the concept of an $(n-1)$-dimensional membrane $\mathcal{V}_{n-1}$ moving in an N-dimensional embedding space V_N and thus sweeping a spacetime sheet V_n.

In Sec. 10.2 we consider the theory in which the whole space-time is an n-dimensional[1] unconstrained membrane $\mathcal{V}_n$. The theory allows for *motion* of $\mathcal{V}_n$ in the embedding space V_N. When considering the quantized theory it turns out that a particular wave packet functional exists such that:

(i) it approximately represents evolution of a simultaneity surface $\mathcal{V}_{n-1}$ (also denoted $\mathcal{V}_\Sigma$), and

(ii) all possible space-time membranes $\mathcal{V}_n$ composing the wave packet are localized near an average space-time membrane $\mathcal{V}_n^{(c)}$ which corresponds to a classical space-time unconstrained membrane.

This approach gives both: the evolution of a state (to which classically there corresponds the progression of time slice) and a fixed spacetime as the expectation value. The notorious problem of time, as it occurs in a reparametrization invariant theory (for instance in general relativity), does not exist in our approach.

---
[1] Usually $n=4$, but if we wish to consider a Kaluza–Klein like theory then $n>4$

10.1. SPACE AS A MOVING 3-DIMENSIONAL MEMBRANE IN V_N

The ideas which we have developed in Part II may be used to describe elementary particles as extended objects —unconstrained p-dimensional membranes $\mathcal{V}_p$— living in spacetime. In this Part we are following yet another application of p-branes: to represent spacetime itself! Spacetime is considered as a surface —also called a *spacetime sheet*— V_n embedded in a higher-dimensional space V_N. For details about some proposed models of such a kind see refs. [99]–[101], [88, 90, 54]. In this section we consider a particular model in which an $(n-1)$-dimensional membrane, called a *simultaneity surface* $\mathcal{V}_\Sigma$, moves in the embedding space according to the unconstrained theory of Part II and sweeps an n-dimensional spacetime sheet V_n. In particular, the moving membrane is 3-dimensional and it sweeps a 4-dimensional surface.

Since we are now talking about spacetime which is conventionally parametrized by coordinates x^μ, the notation of Part II is not appropriate. For this particular application of the membrane theory we use different notation. Coordinates denoting position of a spacetime sheet V_n (alias worldsheet) in the embedding space V_N are

$$\eta^a, \qquad a = 0, 1, 2, ..., N-1, \tag{10.1}$$

whilst parameters denoting positions of points on V_n are

$$x^\mu, \qquad \mu = 0, 1, 2, ..., n-1. \tag{10.2}$$

The parametric equation of a spacetime sheet is[2]

$$\eta^a = \eta^a(x) \tag{10.3}$$

Parameters on a simultaneity surface $\mathcal{V}_\Sigma$ are

$$\sigma^i, \qquad i = 1, 2, ..., n-1, \tag{10.4}$$

and its parametric equation is $\eta^a = \eta^a(\sigma)$. A moving $\mathcal{V}_\Sigma$ is described by the variables $\eta^a(\tau, \sigma)$.

The formal theory goes along the similar lines as in Sec. 7.1. The action is given by

$$I[\eta^a(\tau,\sigma)] = \tfrac{1}{2} \int \omega \, d\tau \, d^{n-1}\sigma \sqrt{f} \left(\frac{\dot\eta^a \dot\eta_a}{\Lambda} + \Lambda \right), \tag{10.5}$$

---

[2] To simplify notation we use the same symbol η^a to denote coordinates of an arbitrary point in V_N and also to to denote the embedding variables (which are functions of x^μ).

where

$$\bar{f} \equiv \det \bar{f}_{ij}, \qquad \bar{f}_{ij} \equiv \frac{\partial \eta^a}{\partial \sigma^i} \frac{\partial \eta_a}{\partial \sigma^j}, \qquad (10.6)$$

is the determinant of the induced metric on $\mathcal{V}_\Sigma$, and $\Lambda = \Lambda(\tau, \sigma)$ a fixed function. The tension κ is now replaced by the symbol ω. The latter may be a constant. However, in the proposed embedding model of spacetime we admit ω to be a function of position in V_N:

$$\omega = \omega(\eta). \qquad (10.7)$$

In the case in which ω is a constant we have a spacetime without "matter" sources. When ω is a function of η^a, we have in general a spacetime with sources (see Chapter 8)

A solution to the equations of motion derived from (10.5) represents a motion of a simultaneity surface $\mathcal{V}_\Sigma$. This is analogous to the motion of an unconstrained membrane discussed in Part II. Here again we see a big advantage of such an *unconstrained* theory: it predicts actual *motion* of $\mathcal{V}_\Sigma$ and *evolution* of a corresponding quantum state with τ being the evolution parameter or historical time [7]. The latter is a distinct concept from the coordinate time $t \equiv x^0$. The existence (and progression) of a time slice is automatically incorporated in our unconstrained theory. It not need be separately postulated, as it is in the usual, constrained relativistic theory[3]. Later, when discussing the quantized theory, we shall show how to take into account that setting up data on an entire (infinite) spacelike hypersurface is an idealistic situation, since it would require an infinite time span which, in practice, is not available to an observer.

The theory based on the action (10.5) is satisfactory in several respects. However, it still cannot be considered as a complete theory, because it is not manifestly invariant with respect to general coordinate transformations of spacetime coordinates (which include Lorentz transformations). In the next section we shall "improve" the theory and explore some of its consequences. We shall see that the theory of the motion of a time slice $\mathcal{V}_\Sigma$, based on the action (10.5), comes out as a particular case (solution) of the generalized theory which is fully relativistic, i.e., invariant with respect to reparametrizations of x^μ. Yet it incorporates the concept of state evolution.

---

[3] More or less explicit assumption of the existence of a time slice (associated with the perception of "now") is manifest in conventional relativistic theories from the very fact that the talk is about "point particles" or "strings" which are objects in three dimensions.

10.2. SPACETIME AS A MOVING 4-DIMENSIONAL MEMBRANE IN V_N

GENERAL CONSIDERATION

The experimental basis[4] on which rests the special relativity and its generalization to curved spacetime clearly indicates that spacetime is a continuum in which events are existing. On the contrary, our subjective experience clearly tells us that not the whole 4-dimensional[5] spacetime, but only a 3-dimensional section of it is accessible to our immediate experience. How to reconcile those seemingly contradictory observations?

It turns out that this is naturally achieved by joining the formal theory of membrane motion (7.1) with the concept of spacetime embedded in a higher-dimensional space V_N (Chapter 8). Let us assume that the spacetime is an unconstrained 4-dimensional membrane $\mathcal{V}_4$ which evolves (or moves) in the embedding space V_N. Positions of points on $\mathcal{V}_4$ at a given instant of the evolution time τ are described by embedding variables $\eta^a(\tau, x^\mu)$. The latter now depend not only on the spacetime sheet parameters (coordinates) x^μ, but also on τ. Let us, for the moment, just accept such a possibility that $\mathcal{V}_4$ evolves, and we shall later see how the quantized theory brings a physical sense to such an evolution.

The action, which is analogous to that of eq. (10.5), is

$$I[\eta^a(\tau, x)] = \frac{1}{2} \int \omega d\tau d^4 x \sqrt{|f|} \left(\frac{\dot{\eta}^a \dot{\eta}_a}{\Lambda} + \Lambda \right), \quad (10.8)$$

$$f \equiv \det f_{\mu\nu}, \quad f_{\mu\nu} \equiv \partial_\mu \eta^a \partial_\nu \eta_a, \quad (10.9)$$

where $\Lambda = \Lambda(\tau, x)$ is a fixed function of τ and x^μ (like a "background field") and $\omega = \omega(\eta)$.

The action (10.8) is invariant with respect to arbitrary transformations of the spacetime coordinates x^μ. But it is not invariant under reparametrizations of the evolution parameter τ. Again we use analogous reasoning as in Chapter 4. Namely, the freedom of choice of parametrization on a given initial $\mathcal{V}_4$ is trivial and it does not impose any constraints on the dynamical variables η^a which depend also on τ. In other words, we consider spacetime $\mathcal{V}_4$ as a physical continuum, the points of which can be identified and their τ-evolution in the embedding space V_N followed. For a chosen parametrization x^μ of the points on $\mathcal{V}_4$ different functions $\eta^a(x), \eta'^a(x)$ (at arbitrary τ)

---

[4] Crucial is the fact that simultaneity of events is relative to an observer. Different observers (in relative motion) determine as simultaneous different sets of events. Hence all those events must exist in a 4-dimensional spacetime in which time is just one of the coordinates.

[5] When convenient, in order to specify the discussion, let us specify the dimension of spacetime and take it to be 4.

represent different *physically deformed* spacetime continua $\mathcal{V}_4$, $\mathcal{V}_4'$. Different functions $\eta^a(x)$, $\eta'^a(x)$, even if denoting positions on the same mathematical surface V_4 will be interpreted as describing physically distinct spacetime continua, $\mathcal{V}_4$, $\mathcal{V}_4'$, locally deformed in different ways. An evolving physical spacetime continuum $\mathcal{V}_4$ is not a concept identical to a mathematical surface V_4. [6]

If, on the other hand, we focus attention to the mathematical manifold V_4 belonging to $\mathcal{V}_4$, then different functions $\eta^a(x)$, $\eta'^a(x)$ which represent the same manifold V_4 are interpreted as belonging to an equivalence class due to reparametrizations of x^μ. There are thus two interpretations (see also Sec. 4.1) of the transformations relating such functions $\eta^a(x)$, $\eta'^a(x)$: (i) the passive transformations, due to reparametrizations of x^μ, and (ii) the active transformations, due to deformations of the physical continuum $\mathcal{V}_4$ into $\mathcal{V}_4'$. In the first case, the induced metric on V_4, which is isometrically embedded in V_N, is given by $g_{\mu\nu} = \partial_\mu \eta^a \partial_\nu \eta_a$. The same expression also holds for the metric on the physical continuum $\mathcal{V}_4$. If the latter is deformed into $\mathcal{V}_4'$, then the metric tensor of $\mathcal{V}_4'$ is $g'_{\mu\nu} = \partial_\mu \eta'^a \partial_\nu \eta'_a$, which is still the expression for the induced metric on a mathematical manifold isometrically embedded in V_N.

Let us now start developing some basic formalism. The canonically conjugate variables belonging to the action (10.8) are

$$\eta^a(x), \qquad p_a(x) = \frac{\partial L}{\partial \dot\eta^a} = \omega\sqrt{|f|}\,\frac{\dot\eta_a}{\Lambda}. \qquad (10.10)$$

The Hamiltonian is

$$H = \frac{1}{2}\int d^4x \sqrt{|f|}\,\frac{\Lambda}{\omega}\left(\frac{p^a p_a}{|f|} - \omega^2\right). \qquad (10.11)$$

The theory can be straightforwardly quantized by considering $\eta^a(x)$, $p_a(x)$ as operators satisfying the equal τ commutation relations

$$[\eta^a(x),\, p_b(x')] = \delta^a{}_b \delta(x - x'). \qquad (10.12)$$

In the representation in which $\eta^a(x)$ are diagonal the momentum operator is given by the functional derivative

$$p_a = -i\,\frac{\delta}{\delta \eta^a(x)}. \qquad (10.13)$$

---

[6] A strict notation would then require a new symbol, for instance $\tilde\eta^a(x)$ for the variables of the physical continuum $\mathcal{V}_4$, to be distinguished from the embedding functions $\eta^a(x)$ of a mathematical surface V_4. We shall not use this distinction in notation, since the meaning will be clear from the context.

A quantum state is represented by a wave functional $\psi[\tau, \eta^a(x)]$ which depends on the evolution parameter τ and the coordinates $\eta^a(x)$ of a physical spacetime sheet $\mathcal{V}_4$, and satisfies the functional Schrödinger equation[7]

$$i \frac{\partial \psi}{\partial \tau} = H \psi. \qquad (10.14)$$

The Hamiltonian operator is given by eq. (10.11) in which p_a are now operators (10.13). According to the analysis given in Sec. 7.1 a possible solution to eq. (10.14) is a linear superposition of states with definite momentum $p_a(x)$ which are taken as constant functionals of $\eta^a(x)$, so that $\delta p_a / \delta \eta^a = 0$:

$$\psi[\tau, \eta(x)] = \int \mathcal{D}p \, c(p) \exp\left[i \int d^4x \, p_a(x)(\eta^a(x) - \eta_0^a(x))\right]$$

$$\times \exp\left[-\frac{i}{2} \int d^4x \sqrt{|f|} \frac{\Lambda}{\omega} \left(\frac{p^a p_a}{|f|} - \omega^2\right)\tau\right], \qquad (10.15)$$

where $p_a(x)$ are now eigenvalues of the momentum operator, and $\mathcal{D}p$ is the invariant measure in momentum space.

A PHYSICALLY INTERESTING SOLUTION

Let us now pay attention to eq. (10.15). It defines a wave functional packet spread over a continuum of functions $\eta^a(x)$. As discussed in more detail in Section 7.1, the expectation value of $\eta^a(x)$ is

$$\langle \eta^a(x) \rangle = \eta_C^a(\tau, x) \qquad (10.16)$$

where $\eta_C^a(\tau, x)$ represents motion of the *centroid* spacetime sheet $\mathcal{V}_4^{(c)}$ which is the "centre" of the wave functional packet. This is illustrated in Fig. 10.1.

In general, the theory admits an arbitrary motion $\eta_C^a(\tau, x)$ which is a solution of the classical equations of motion derived from the action (10.8). But, in particular, a wave packet (10.15) which is a solution of the Schrödinger equation (10.14) can be such that its centroid spacetime sheet is either

(i) at rest in the embedding space V_N, i.e. $\dot\eta_C^a = 0$;

or, more generally:

(ii) it moves "within itself" so that its shape does not change with τ. More precisely, at every τ and x^μ there exists a displacement ΔX^μ along a

---

[7] This is the extension, from point particle to membrane, of the equation proposed by Stueckelberg [2]. Such an equation has been discussed several times throughout this book.

curve $X^\mu(\tau)$ such that $\eta^a_C(\tau+\Delta\tau, x^\mu) = \eta^a_C(\tau, x^\mu + \Delta X^\mu)$, which implies $\dot\eta^a_C = \partial_\mu \eta^a_C X^\mu$. Therefore $\dot\eta^a_C$ is always tangent to a fixed mathematical surface V_4 which does not depend on τ.

Existence of a membrane $\eta^a(\tau, x^\mu)$, satisfying the requirement (ii) and the classical equations of motion, can be demonstrated as follows. The relation $\dot\eta^a = \dot X^\mu \partial_\mu \eta^a$ can be written (after expressing $\dot X^\mu = \dot\eta_a \partial^\mu \eta^a$) also in the form $\dot\eta^c b_c{}^a = 0$, where $b_c{}^a \equiv \delta_c{}^a - \partial^\mu \eta_c \, \partial_\mu \eta^a$. At an initial time $\tau = 0$, the membrane $\eta^a(0, x^\mu)$ and its velocity $\dot\eta^a(0, x^\mu)$ can be specified arbitrarily. Let us choose the initial data such that

$$\dot\eta^c(0) b_c{}^a(0) = 0, \qquad (10.17)$$

$$\ddot\eta^c(0) b_c{}^a(0) + \dot\eta^c(0) \dot b_c{}^a(0) = 0, \qquad (10.18)$$

where $\ddot\eta^a(0)$ can be expressed in terms of $\eta^a(0)$ and $\dot\eta^a(0)$ by using the classical equations of motion [54, 55]. At an infinitesimally later instant $\tau = \Delta\tau$ we have $\eta^a(\Delta\tau) = \eta^a(0) + \dot\eta^a(0)\Delta\tau$, $\dot\eta^a(\Delta\tau) = \dot\eta^a(0) + \ddot\eta^a(0)\Delta\tau$, and

$$\dot\eta^c(\Delta\tau) b_c{}^a(\Delta\tau) = \dot\eta^c(0) b_c{}^a(0) + \left[\ddot\eta^c(0) b_c{}^a(0) + \dot\eta^c(0)\dot b_c{}^a(0)\right]\Delta\tau. \qquad (10.19)$$

Using (10.17), (10.18) we find that eq. (10.19) becomes

$$\dot\eta^c(\Delta\tau) b_c{}^a(\Delta\tau) = 0. \qquad (10.20)$$

From the classical equations of motion [54, 55] it follows that also

$$\ddot\eta^c(\Delta\tau) b_c{}^a(\Delta\tau) + \dot\eta^c(\Delta\tau) \dot b_c{}^a(\Delta\tau) = 0. \qquad (10.21)$$

By continuing such a procedure step by step we find that the relation $\dot\eta^c b_c{}^a = 0$ holds at every τ. This proves that a membrane satisfying the initial conditions (10.17),(10.18) and the equations of motion indeed moves "within itself".

Now let us consider a special form of the wave packet as illustrated in Fig. 10.2. Within the effective boundary B a given function $\eta^a(x)$ is admissible with high probability, outside B with low probability. Such is, for instance, a Gaussian wave packet which, at the initial $\tau = 0$, is given by

$$\psi[0, \eta^a(x)] = A \exp\left[-\int d^4x \sqrt{|f|}\, \frac{\omega}{\Lambda}\, (\eta^a(x) - \eta^a_0(x))^2 \, \frac{1}{2\sigma(x)}\right]$$

$$\times \exp\left[i\int d^4x\, p_{0a}(x)\, (\eta^a(x) - \eta^a_0(x))\right], \qquad (10.22)$$

On the resolution of time problem in quantum gravity

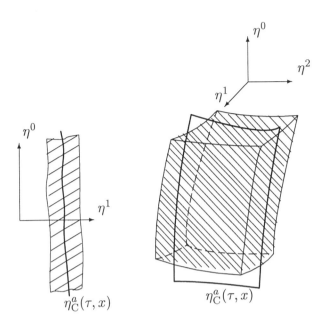

Figure 10.1. Quantum mechanically a state of our space–time membrane $\mathcal{V}_4$ is given by a wave packet which is a functional of $\eta^a(x)$. Its "centre" $\eta_C^a(\tau,x)$ is the expectation value $\langle \eta^a(x)\rangle$ and moves according to the classical equations of motion (as derived from the action (10.8)).

where $p_{0c}(x)$ is momentum and $\eta_0^a(x)$ position of the "centre" of the wave packet at $\tau = 0$. The function $\sigma(x)$ vary with x^μ so that the wave packet corresponds to Fig. 10.2.

Of special interest in Fig. 10.2 is the region P around a spacelike hypersurface Σ on $\mathcal{V}_4^{(c)}$. In that region the wave functional is much more sharply localized than in other regions (that is, at other values of x^μ). This means that in the neighborhood of Σ a spacetime sheet $\mathcal{V}_4$ is relatively well defined. On the contrary, in the regions that we call *past* or *future*, space-time is not so well defined, because the wave packet is spread over a relatively large range of functions $\eta^a(x)$ (each representing a possible spacetime sheet $\mathcal{V}_4$).

The above situation holds at a certain, let us say initial, value of the evolution parameter τ. Our wave packet satisfies the Schrödinger equation and is therefore subjected to evolution. The region of sharp localization depends on τ, and so it moves as τ increases. In particular, it can move within the mathematical spacetime surface $\mathcal{V}_4$ which corresponds to such a "centroid" (physical) spacetime sheet $\mathcal{V}_4^{(c)} = \langle \mathcal{V}_4 \rangle$ which "moves within itself" (case (ii)

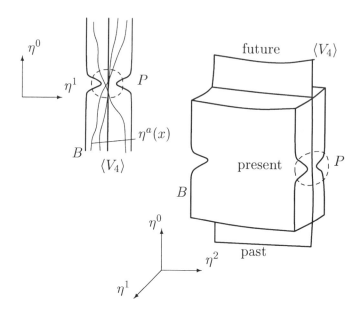

Figure 10.2. The wave packet representing a quantum state of a space–time membrane is localized within an effective boundary B. The form of the latter may be such that the localization is significantly sharper around a space-like surface Σ.

above). Such a solution of the Schrödinger equation provides, on the one hand, the existence of a fixed spacetime V_4, defined within the resolution of the wave packet (see Fig. 10.2), and, on the other hand, the existence of a moving region P in which the wave packet is more sharply localized. The region P represents the "present" of an observer. We assume that an observer measures, in principle, the embedding positions $\eta^a(x)$ of the entire spacetime sheet. Every $\eta^a(x)$ is possible, in principle. However, in the practical situations available to us a possible measurement procedure is expected to be such that only the embedding positions $\eta^a(x_\Sigma^\mu)$ of a simultaneity hypersurface Σ are measured with high precision[8], whereas the embedding positions $\eta^a(x)$ of all other regions of spacetime sheet are measured with low precision. As a consequence of such a measurement a wave packet like one of Fig. 10.2 and eq. (10.22) is formed and it is then subjected to the unitary τ-evolution given by the covariant functional Schrödinger equation (10.14).

---

[8] Another possibility is to measure the induced metric $g_{\mu\nu}$ on V_4, and measure $\eta^a(x)$ merely with a precision at a cosmological scale or not measure it at all.

Using our theory, which is fully covariant with respect to reparametrizations of spacetime coordinates x^μ, we have thus arrived in a natural way at the existence of a time slice Σ which corresponds to the "present" experience and which progresses forward in spacetime. The theory of Sec.10.1 is just a particular case of this more general theory. This can be seen by taking in the wave packet (10.22) the limit

$$\frac{1}{\sigma(x)} = \frac{\delta(x^0 - x_\Sigma^0)}{\sigma(x^i)(\partial_0 \eta^a \partial_0 \eta_a)^{1/2}}, \qquad i = 1, 2, 3, \qquad (10.23)$$

and choosing

$$p_{0a}(x^\mu) = p_{0a}^{(3)}(x^i)\delta(x^0 - x_\Sigma^0). \qquad (10.24)$$

Then the integration over the δ-function gives in the exponent the expression

$$\int d^3x \sqrt{|\bar{f}|} \frac{\omega}{\Lambda} \left(\eta^a(x^i) - \eta_0^a(x^i) \right)^2 / 2\sigma(x^i) + i \int d^3x\, p_{0a}(x^i) \left(\eta^a(x^i) - \eta_0^a(x^i) \right)$$

so that eq. (10.22) becomes a wave functional of a 3-dimensional membrane $\eta^a(x^i)$

So far we have taken the region of sharp localization P of a wave functional packet situated around a spacelike surface Σ, and so we have obtained a time slice. But there is a difficulty with the concept of "time slice" related to the fact that an observer in practice never has access to the experimental data on an entire spacelike hypersurface. Since the signals travel with the final velocity of light there is a delay in receiving information. Therefore, the greater is a portion of a space-like hypersurface, the longer is the delay. This imposes limits on the extent of a space-like region within which the wave functional packet (10.22) can be sharply localized. The situation in Fig. 10.2 is just an idealization. A more realistic wave packet, illustrated in Fig. 10.3, is sharply localized around an finite region of spacetime. It can still be represented by the expression (10.22) with a suitable width function $\sigma(x)$.

A possible interpretation is that such a wave packet of Fig. 10.3 represents a state which is relative (in the Everett sense[9]) to a state of a registering device with memory of past records. The region of sharp localization of such a relative state is centered within the registering device.

In summary, our model predicts:

---

[9]The Everett interpretation [107] of quantum mechanics which introduces the concept of a wave function relative to a registering device is not so unpopular among cosmologists. Namely, in quantum gravity the whole universe is considered to be described (of course, up to a sensible approximation) by a single wave function, and there is no outside observer who could measure it.

294   THE LANDSCAPE OF THEORETICAL PHYSICS: A GLOBAL VIEW

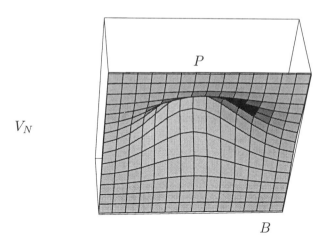

Figure 10.3. Illustration of a wave packet with a region of sharp localization P.

(i) existence of a spacetime continuum V_4 without evolution in τ (such is a spacetime of the conventional special and general relativity);

(ii) a region P of spacetime which changes its position on V_4 while the evolution time τ increases.

Feature (i) comes from the expectation value of our wave packet, and feature (ii) is due to its peculiar shape (Fig. 10.2 or Fig. 10.3) which evolves in τ.

INCLUSION OF SOURCES

In the previous chapter we have included point particle sources in the embedding model of gravity (which was based on the usual constrained p-brane theory). This was achieved by including in the action for a spacetime sheet a function $\omega(\eta)$ which consists of a constant part and a δ-function part. In an analogous way we can introduce sources into our unconstrained embedding model which has explicit τ-evolution.

For ω we can choose the following function of the embedding space coordinates η^a:

$$\omega(\eta) = \omega_0 + \sum_i \int m_i\, \delta^N(\eta - \hat{\eta}_i) \sqrt{|\hat{f}|}\, \mathrm{d}^m \hat{x}, \qquad (10.25)$$

where $\eta^a = \hat{\eta}_i^a(\hat{x})$ is the parametric equation of an m-dimensional surface $\widehat{V}_m^{(i)}$, called *matter sheet*, also embedded in V_N; $\hat{x}^{\hat{\mu}}$ are parameters (coordinates) on $\hat{V}_m^{(i)}$ and $\hat{f}$ is the determinant of the induced metric tensor on $\hat{V}_m^{(i)}$. If we take $m = N - 4 + 1$, then the intersection of V_4 and $\hat{V}_m^{(i)}$ can be a (one-dimensional) line, i.e., a worldline C_i on V_4. If V_4 moves in V_N, then the intersection C_i also moves. A moving spacetime sheet was denoted by V_4 and described by τ-dependent coordinate functions $\eta^a(\tau, x^\mu)$. Let a moving worldline be denoted C_i. It can be described either by the coordinate functions $\eta^a(\tau, u)$ in the embedding space V_N or by the coordinate functions $X^\mu(\tau, u)$ in the moving spacetime sheet V_4. Besides the evolution parameter τ we have also a 1-dimensional worldline parameter u which has an analogous role as the spacetime sheet parameters x^μ in $\eta^a(\tau, x^\mu)$. At a fixed τ, $X^\mu(\tau, u)$ gives a 1-dimensional worldline $X^\mu(u)$. If τ increases monotonically, then the worldlines continuously change or *move*. In the expression (10.25) $m - 1$ coordinates $\hat{x}^{\hat{\mu}}$ can be integrated out and we obtain

$$\omega = \omega_0 + \sum_i \int m_i \frac{\delta^4(x - X_i)}{\sqrt{|f|}} \left(\frac{\mathrm{d}X_i^\mu}{\mathrm{d}u} \frac{\mathrm{d}X_i^\nu}{\mathrm{d}u} f_{\mu\nu} \right)^{1/2} \mathrm{d}u, \quad (10.26)$$

where $x^\mu = X_i^\mu(\tau, u)$ is the parametric equation of a (τ-dependent) worldline C_i, u an arbitrary parameter on C_i, $f_{\mu\nu} \equiv \partial_\mu \eta^a \partial_\nu \eta_a$ the induced metric on V_4 and $f \equiv \det f_{\mu\nu}$. By inserting eq. (10.26) into the membrane's action (10.8) we obtain the following action

$$I[X^\mu(\tau, u)] = I_0 + I_m \quad (10.27)$$

$$= \frac{\omega_0}{2} \int \mathrm{d}\tau \, \mathrm{d}^4x \, \sqrt{|f|} \left(\frac{\dot{\eta}^a \dot{\eta}_a}{\Lambda} + \Lambda \right)$$

$$+ \frac{1}{2} \int \mathrm{d}\tau \, \mathrm{d}^4x \sqrt{|f|} \sum_i \left(\frac{\dot{X}_i^\mu \dot{X}_i^\nu f_{\mu\nu}}{\Lambda} + \Lambda \right)$$

$$\times m_i \frac{\delta^4(x - X_i)}{\sqrt{|f|}} \left(\frac{\mathrm{d}X_i^\mu}{\mathrm{d}u} \frac{\mathrm{d}X_i^\nu}{\mathrm{d}u} f_{\mu\nu} \right)^{1/2} \mathrm{d}\lambda.$$

In a special case where the membrane V_4 is static with respect to the evolution in τ, i.e., all τ derivatives are zero, then we obtain, taking $\Lambda = 2$, the usual Dirac–Nambu–Goto 4-dimensional membrane coupled to point particle sources

$$I[X^\mu(u)] = \omega_0 \int \mathrm{d}^4x \, \Lambda \sqrt{|f|} + \int \mathrm{d}u \sum_i m_i \left(\frac{\mathrm{d}X_i^\mu}{\mathrm{d}u} \frac{\mathrm{d}X_i^\nu}{\mathrm{d}u} f_{\mu\nu} \right)^{1/2}. \quad (10.28)$$

However, the action (10.27) is more general than (10.28) and it allows for solutions which evolve in τ. The first part I_0 describes a 4-dimensional membrane which evolves in τ, whilst the second part I_m describes a system of (1-dimensional) worldlines which evolve in τ. After performing the integration over x^μ, the 'matter' term I_m becomes —in the case of one particle— analogous to the membrane's term I_0:

$$I_\mathrm{m} = \frac{1}{2} \int \mathrm{d}\tau \, \mathrm{d}u \, m \left(\frac{\mathrm{d}X^\mu}{\mathrm{d}u} \frac{\mathrm{d}X^\nu}{\mathrm{d}u} f_{\mu\nu} \right)^{1/2} \left(\frac{\dot{X}^\mu \dot{X}^\nu f_{\mu\nu}}{\Lambda} + \Lambda \right). \tag{10.29}$$

Instead of 4 parameters (coordinates) x^μ we have in (10.29) a single parameter u, instead of the variables $\eta^a(\tau, x^\mu)$ we have $X^\mu(\tau, u)$, and instead of the determinant of the 4-dimensional induced metric $f_{\mu\nu} \equiv \partial_\mu \eta^a \partial_\nu \eta_a$ we have $(\mathrm{d}X^\mu/\mathrm{d}u)(\mathrm{d}X_\mu/\mathrm{d}u)$. All we have said about the theory of an unconstrained n-dimensional membrane evolving in τ can be straightforwardly applied to a worldline (which is a special membrane with $n = 1$).

After inserting the matter function $w(\eta)$ of eq. (10.25) into the Hamiltonian (10.11) we obtain

$$H = \frac{1}{2} \int \mathrm{d}^4 x \sqrt{|f|} \frac{\Lambda}{\omega_0} \left(\frac{p^{(0)a}(x) p_a^{(0)}(x)}{|f|} - \omega_0^2 \right)\bigg|_{x \notin \mathcal{C}_i}$$

$$+ \frac{1}{2} \sum_i \int \mathrm{d}u \left(\frac{\mathrm{d}X_i^\mu}{\mathrm{d}u} \frac{\mathrm{d}X_{i\mu}}{\mathrm{d}u} \right)^{1/2} \frac{\Lambda}{m_i} \left(P_\mu^{(i)} P^{(i)\mu} - m_i^2 \right), \tag{10.30}$$

where $p_a^{(0)} = \omega_0 \Lambda^{-1} \sqrt{|f|}\, \dot\eta_a$ is the membrane's momentum everywhere except on the intersections $V_4 \cap \widehat{V}_m^{(i)}$, and $P_\mu^{(i)} = m_i \dot X_{i\mu}/\Lambda$ is the membrane momentum on the intersections $V_4 \cap \widehat{V}_m^{(i)}$. In other words, $P_\mu^{(i)}$ is the momentum of a worldline $\mathcal{C}_i$. The contribution of the wordlines is thus explicitly separated out in the Hamiltonian (10.30).

In the quantized theory a membrane's state is represented by a wave functional which satisfies the Schrödinger equation (10.14). A wave packet (e.g., one of eq. (10.22)) contains, in the case of $w(\eta)$ given by eq. (10.25), a separate contribution of the membrane's portion outside and on the intersection $V_4 \cap \widehat{V}_m^{(i)}$:

$$\psi[0, \eta(x)] = \psi_0[0, \eta(x)] \psi_m[0, X(u)] \tag{10.31}$$

$$\psi_0[0, \eta^a(x)] = A_0 \exp\left[-\int \mathrm{d}^4 x \sqrt{|f|} \frac{\omega}{\Lambda} \left(\eta^a(x) - \eta_0^a(x) \right)^2 \frac{1}{2\sigma(x)} \right]$$

$$\times \exp\left[i \int \mathrm{d}^4 x \, p_{0a}(x) \left(\eta^a(x) - \eta_0^a(x) \right) \right]\bigg|_{x \notin \mathcal{C}_i} \tag{10.32}$$

$$\psi_m[0, X(u)] = A_m \exp\left[-\sum_i \int du \frac{m_i}{\Lambda} \left(\frac{dX_i^\mu}{du}\frac{dX_i^\nu}{du} f_{\mu\nu}\right)^{1/2}\right.$$

$$\times (X_i^\mu(u) - X_{0i}^\mu(u))^2 \frac{1}{2\sigma_i(u)}\right]$$

$$\times \exp\left[i \int du \left(\frac{dX_i^\mu}{du}\frac{dX_i^\nu}{du} f_{\mu\nu}\right)^{1/2}\right.$$

$$\left.\times P_{0\mu}^{(i)}(u)\, (X_i^\mu(u) - X_{0i}^\mu(u))\right]. \qquad (10.33)$$

In the second factor of eq. (10.31) the wave packets of worldlines are expressed explicitly (10.33). For a particular $\sigma(x)$, such that a wave packet has the form as sketched in Fig. 10.2 or Fig. 10.3, there exists a region P of parameters x^μ at which the membrane $\mathcal{V}_4$ is much more sharply localized than outside P. The same is true for the intersections (which are worldlines): any such worldline $\mathcal{C}_i$ is much more sharply localized in a certain interval of the worldline parameter u. With the passage of the evolution time τ the region of sharp localization on a worldline moves in space-time $\mathcal{V}_4$. In the limit (10.23), (10.24) expression (10.33) becomes a wave packet of a point particle (event) localized in space-time. The latter particle is just an unconstrained point particle, a particular case (for $p = 0$) of a generic unconstrained p-dimensional membrane described in Sec. 7.1.

The unconstrained theory of point particles has a long history. It was considered by Fock, Stueckelberg, Schwinger, Feynman, Horwitz, Fanchi, Enatsu, and many others [1]–[16]. Quantization of the theory had appeared under various names, for instance the Schwinger proper time method or the parametrized relativistic quantum theory. The name unconstrained theory is used in Ref. [15, 16] both for the classical and the quantized theory. In the last few years similar ideas in relation to canonical quantization of gravity and Wheeler DeWitt equation were proposed by Greensite and Carlini [39]. In a very lucid paper Gaioli and Garcia-Alvarez [40] have convincingly explained why the invariant evolution parameter is necessary in quantum gravity (see also ref. [41]).

CONCLUSION

We have formulated a reparametrization invariant and Lorentz invariant theory of p-dimensional membranes without constraints on the dynamical variables. This is possible if we assume a generalized form of the Dirac–Nambu–Goto action, such that the dependence of the dynamical variables on an extra parameter, the evolution time τ, is admitted. In Sec. 4.2

we have seen that the unconstrained theory naturally arises within the larger framework of a constrained theory formulated by means of geometric calculus based on Clifford algebra which incorporates polyvectors (Clifford aggregates) as representing physical quantities.

Such a membrane theory manifests its full power in the embedding model of gravity, in which space-time is treated as a 4-dimensional unconstrained membrane, evolving in an N-dimensional embedding space. The embedding model was previously discussed within the conventional theory of constrained membranes [88]. Release of the constraints and introduction of the τ evolution brings new insight into the quantization of the model. Particularly interesting is a state represented by a functional of 4-dimensional membranes $\mathcal{V}_4$, localized around an average space-time membrane $\mathcal{V}_4^{(c)}$, and even more sharply localized around a finite segment of a space-like surface Σ on $\mathcal{V}_4^{(c)}$. Such a state incorporates the existence of a classical space-time continuum and the evolution in it. The notorious problem of time [108] is thus resolved in our approach to quantum gravity. The space-time coordinate $x^0 = t$ is not time[10] at all! Time must be separately introduced, and this has been achieved in our theory in which the action depends on the evolution time τ. The importance of the evolution time has been considered, in the case of a point particle, by many authors [1]–[16].

Our embedding model incorporates sources in a natural way. Worldlines occur in our embedding model as intersections of space–time membranes $\mathcal{V}_4$ with $(N-4+1)$-dimensional "matter" sheets or as the intersections of $\mathcal{V}_4$ with itself. In the quantized theory the state of a worldline can be represented by a wave functional $\psi_m[\tau, X^\mu(u)]$, which may be localized around an average worldline (in the quantum mechanical sense of the expectation value). Moreover, at a certain value $u = u_P$ of the worldline parameter the wave functional may be much more sharply localized than at other values of u, thus approximately imitating the wave function of a point particle (or event) localized in space-time. And since $\psi_m[\tau, X^\mu(u)]$ evolves with τ, the point u_P also changes with τ.

The embedding model, based on the theory of unconstrained membranes satisfying the action (10.8), appears to be a promising candidate for the

---

[10]In this sentence 'time' stands for the parameter of evolution. Such is the meaning of the word 'time' adopted by the authors who discuss the problem of time in general relativity. What they say is essentially just that there is a big problem because the coordinate x^0 does not appear at all in the Wheeler DeWitt equation and hence cannot have the role of an evolution parameter (or 'time,' in short). In our work, following Horwitz [7], we make explicit distinction between the coordinate x^0 and the parameter of evolution τ (which indeed takes place in our classical and quantum equations of motions). These two distinct concepts are usually mixed and given the same name 'time'. In order to distinguish them, we use the names '*coordinate time*' and '*evolution time*'.

theoretical formulation of quantum gravity including bosonic sources. Incorporation of fermions is expected to be achieved by taking into account the Grassmann coordinates or by employing the polyvector generalization of the underlying formalism.

IV

BEYOND THE HORIZON

Chapter 11

THE LANDSCAPE OF THEORETICAL PHYSICS: A GLOBAL VIEW

In the last Part, entitled "Beyond the Horizon", I am going to discuss conceptual issues and the foundations of theoretical physics. I shall try to outline a broader[1] view of the theoretical physics landscape as I see it, and, as seems to me, is becoming a view of an increasing number of researchers. The introductory chapter of Part IV, bearing the same title as the whole book, is an overview aimed at being understandable to the widest possible circle of readers. Therefore use of technical terminology and jargon will be avoided. Instead, the concepts and ideas will be explained by analogies and illustrative examples. The cost, of course, is a reduced scientific rigor and precision of expression. The interested reader who seeks a more precise scientific explanation will find it (but without much maths and formulas) in the next chapters, where many concepts will be discussed at a more elaborate level.

Throughout history people have been always inventing various cosmological models, and they all have always turned out to be wrong, or at least incomplete. Can we now be certain that a similar fate does not await the current widely accepted model, according to which the universe was born in a "big bang"? In 1929 an american astronomer Edwin Hubble discovered that light coming from galaxies is shifted towards the red part of the spectrum, and the shift increases with galactic distance. If we ascribe the red shift to galactic velocity, then Hubble's discovery means that the universe is expanding, since the more distant a galaxy is from us the greater is its velocity; and this is just a property of expansion. Immediately after that discovery Einstein recognized that his equation for gravity admitted

---

[1] The outline of the view will in many respects be indeed "broader" and will go beyond the horizon.

precisely such a solution which represented the expansion of a universe uniformly filled with matter. In fact, he had already come to just such a result in 1917, but had rejected it because he had considered it a nonsense, since an expanding universe was in disagreement with the static model of the universe widely accepted at that time. In 1917 he had preferred to modify his equation by adding an extra term containing the so called "cosmological constant". He had thus missed the opportunity of predicting Hubble's discovery, and later he proclaimed his episode with the cosmological constant as the biggest blunder in his life.

General relativity is one of the most successful physical theories. It is distinguished by an extraordinary conceptual elegance, simplicity of the basic postulates, and an accomplished mathematical apparatus, whilst numerous predictions of the theory have been tested in a variety of important and well known experiments. No experiment of whatever kind has been performed so far that might cast doubt on the validity of general relativity. The essence of the theory is based on the assumption (already well tested in special relativity) that space and time form a four-dimensional continuum named *spacetime*. In distinction with special relativity, which treats spacetime as a *flat* continuum, in general relativity spacetime can be curved, and curvature is responsible for gravitational phenomena. How spacetime is curved is prescribed by Einstein's equation. Strictly speaking, Einstein's equations determine only in which many different possible ways spacetime can be curved; how it is actually curved we have to find out at "the very place". But how do we find this? By observing particles in their motion. If we are interested in spacetime curvature around the Sun, then such particles are just planets, and if we are interested in the curvature of the Universe as the whole, then such particles are galaxies or clusters of galaxies. In flat spacetime, in the absence of external forces, all particles move uniformly along straight lines, whilst in a curved spacetime particles move non-uniformly and in general along curved lines. By measuring the relative acceleration and velocity of one particle with respect to another, nearby, particle we can then calculate the curvature of spacetime in a given point (occupied by the particle). Repeating such a procedure we can determine the curvature in all sample points in a given region of spacetime. The fact that a planet does not move along a straight line, but along an elliptic trajectory, is a consequence of the curvature of spacetime around the Sun. The gravitational "force" acting on a planet is a consequence of the curvature. This can be illustrated by an example of a curved membrane onto which we throw a tiny ball. The ball moves along a curved trajectory, hence a force is acting on the ball. And the latter force results from the membrane's curvature.

Another very successful theory is quantum mechanics. Without quantum mechanics we would not be able to explain scattering of electrons by crystals, nor the ordered stable crystal structure itself, nor the properties of electromagnetic waves and their interactions with matter. The widely known inventions of today, such as the laser, semiconductors, and transistors, have developed as a result of understanding the implications of quantum mechanics. Without going into too much detail, the essence of quantum mechanics, or at least one of its essential points, can be summarized in the following simplified explanation. There exists a fundamental uncertainty about what the universe will be like at a future moment. This uncertainty is the bigger, as more time passes after a given moment. For instance, it is impossible to predict precisely at which location an electron will be found, after leaving it to move undisturbed for some time. When we finally measure its position it will be, in principle, anywhere in space; however, the probability of finding the electron will be greater at some places than at others. To everyone of those possible results of measurements there corresponds a slightly different universe. In classical, Newtonian, physics the uncertainty about the future evolution of the universe is a consequence of the uncertainty about the present state of the universe. If the present state could be known precisely, then also the future evolution of the universe could be precisely calculated. The degree of precision about the prediction of the future is restricted by the degree of precision with which the initial conditions are determined. (I am intentionally speaking about the whole universe, since I wish to point out that the size of the observed system and its complexity here does not, in principle, play any role.) In quantum mechanics, on the contrary, such uncertainty is of quite a different kind from that in classical mechanics. No matter how precisely the present state of an observed system is known, the uncertainty about what position of the particles we shall measure in the future remains. A generic state of a system can be considered as a superposition of a certain set of basis states. It can be described by the *wave function* which enables calculation of *the probability* to observe a definite quantum state upon measurement. The latter state is just one amongst the states belonging to the set of basis states, and the latter *set* itself is determined by the measurement situation. Such a probability or statistical interpretation of quantum mechanics was unacceptable for Einstein, who said that "God does not play dice". And yet everything points to him having been wrong. So far no experiment, no matter how sophisticated, has disproved the probability interpretation, whilst many experiments have eliminated various rival interpretations which assume the existence of some "hidden variables" supposedly responsible for the unpredictable behavior of quantum systems.

* * *

We thus have two very successful theories, *general relativity* on the one hand, and *quantum mechanics* on the other, which so far have not been falsified by any experiment. What is then more natural than to unify those two theories into a single theory? And yet such a unification has not yet been successfully achieved. The difficulties are conceptual as well as mathematical and technical. As it appears now, final success will not be possible without a change of paradigm. Some of the basic principles the two theories rest on will have to be changed or suitably generalized. Certain significant moves in this direction have already been made. In the following I will briefly, and in a simplified way, discuss some of those, in my opinion, very important approaches. Then I will indicate how those seemingly unconnected directions of research lead towards a possible solution of the problem of quantum gravity, and hence towards an even more profound understanding of the universe and the role of an intelligent observer in it.

Before continuing, let me point out that some epochs in history are more ready for changes, other less. The solution of a certain basic scientific problem or a significantly improved insight into the nature of Nature is nearly always a big shock for those who have been used to thinking in the old terms, and therefore do their best to resist the changes, while regretfully they do not always use the methods of scientific argument and logic only. Copernicus did not publish his discoveries until coming close to his death, and he had reason for having done so. The idea that the whole Earth, together with the oceans, mountains, cities, rivers, is moving around the Sun, was too much indeed! Just as were Wegener's theory about the relative motions of the continents, Darwin's theory about the origin and evolution of the species, and many other revolutionary theories. I think that we could already have learned something from the history of science and be now slightly more prudent while judging new ideas and proposals. At least the "arguments" that a certain idea is much too fantastic or in disagreement with common sense should perhaps not be used so readily. The history of science has taught us so many times that many successful ideas were just such, namely at first sight crazy, therefore in the future we should avoid such a "criterion" of judging the novelties and rather rely less on emotional, and more on scientific criteria. The essence of the latter is a cold, strictly rational investigation of the consequences of the proposed hypotheses and verification of the consequences by experiments. However, it is necessary to have in mind that a final elaboration of a successful theory takes time. Many researchers may participate in the development and every contribution is merely a piece of the whole. Today it is often stressed that a good theory has to able to incorporate all the known phenomena and predict new ones, not yet discovered. This is, of course, true, but it holds for a finished theory, and not for the single contributions of scientists who enabled the development of the theory.

In 1957 the American physicist Hugh Everett [107] successfully defended his PhD thesis and published a paper in which he proposed that all the possibilities, implicit in the wave function, actually exist. In other words, all the possible universes incorporated in the wave function actually exist, together with all the possible observers which are part of those universes. In addition to that, Everett developed the concept of *relative state*. Namely, if a given physical system consists of two mutually interacting subsystems, then each of them can be described by a wave function which is relative to the possible states of the other subsystem. As one subsystem we can take, for example, an intelligent observer, and as the other subsystem the rest of the universe. The wave function of the remaining universe is relative to the possible states of the observer. The quantum mechanical correlation, also known under the name "entanglement", is established amongst the possible quantum states of the observer and the possible quantum states of the remaining universe. As an example let us consider an observer who measures the radioactive gamma decay of a low activity source with short life time. A Geiger counter which detects the particles (in our example these are photons, namely gamma rays) coming from the source will then make only single sounds, e.g., one per hour. Imagine now that we have isolated a single atom containing the nucleus of our radioactive source. At a given moment the wave function is a superposition of two quantum states: the state with photon emission and the state without the photon emission. The essence of Everett's thesis (for many still unacceptable today) lies in assuming that the states of the Geiger counter, namely the state with the sound and the state without the sound, also enter the superposition. Moreover, even the states of the observer, i.e., the state in which the observer has heard the sound and the state in which the observer has not heard the sound, enter the superposition. In this example the quantum correlation manifests itself in the following. To the state in which the observer became aware[2] that he has heard the sound there corresponds the state in which the detector has detected a photon, and to the latter state, in turn, there correspond the state in which the excited nucleus has emitted the photon. And similarly, to the state in which the observer has not heard the sound, there corresponds the state in which the detector has not detected and the

---

[2]In this example we are using a *male* observer and the source of gamma rays. In some other example we could use a *female* observer and laser beams instead. In fact, throughout the book I am using female or male observers interchangely for doing experiments for my illustrations. So I avoid using rather cumbersome (especially if frequently repeated) "he or she", but use "he" or "she" instead. When necessary, "he" may stand for a generic observer. Similarly for "she".

source has not emitted a photon. Each of those two chains of events belongs to a different universe: in one universe the decay has happened and the observer has perceived it, whilst in the other universe at the given moment there was no decay and the observer has not perceived the decay. The total wave function of the universe is a superposition of those two chains of events. In any of the chains, from the point of view of the observer, there is no superposition.

The Everett interpretation of quantum mechanics was strongly supported by John Archibald Wheeler [109]. Somewhat later he was joined by many others, among them also Bryce DeWitt who gave the name "many worlds interpretation", that is, the interpretation with many worlds or universes. Today the majority of physicists is still opposed to the Everett interpretation, but it is becoming increasingly popular amongst cosmologists.

Later on, Wheeler distanced himself from the Everett interpretation and developed his own theory, in which he put the quantum principle as the basis on which rests the creation and the functioning of the universe [111]. The *observer* is promoted to the *participator*, who not only perceives, but is actively involved in, the development of the universe. He illustrated his idea as follows. We all know the game "twenty questions". Person A thinks of an object or a concept—and person B poses questions to which the answer is *yes* or *no*. Wheeler slightly changed the rules of the game, so that A may decide what the object is *after* B asks the first question. After the second question A may change the idea and choose another object, but such that it is in agreement with his first answer. This continues from question to question. The object is never completely determined, but is only determined within the set of possible objects which are in agreement with the questions posed (and the answers obtained) so far. However, with every new question the set of possible objects is narrowed, and at the end it may happen that only one object remains. The player who asked questions, with the very choice of her questions, has herself determined the set of possible answers and thus the set of possible objects. In some way reality is also determined by the question we ask it. The observer observes the universe by performing various measurements or experiments. With the very choice of experiment she determines what the set of possible results of measurement is, and hence what the set of possible universes at a given moment is. The observer is thus involved in the very creation of the universe she belongs to. In my opinion Wheeler's approach is not in disagreement with Everett's, but completes it, just as it also completes the commonly accepted interpretation of quantum mechanics.

Nowadays a strong and influential supporter of the Everett interpretation is an Oxford professor David Deutsch. In his book *The Fabric of Reality* [112] he developed the concept of *multiverse*, which includes all possible universes that are admitted by a wave function. In a 1991 *Physical Review* article [113] he proved that the paradoxes of so called *time machines* can be resolved by means of the Everett interpretation of quantum mechanics. Many theoretical physicists study in detail some special kinds of solutions to the Einstein equation, amongst them the best known are *wormwholes* [114]. These are special, topologically non-trivial, configurations of space-time which under certain conditions allow for *causal loops*. Therefore such solutions are called *time machines*. A particle which enters a time machine will go back in time and meet itself in the past. Such a situation is normally considered paradoxical and the problem is how to avoid it. On the one hand, if we believe the Einstein equations such time machines are indeed possible. On the other hand, they are in conflict with the principle of causality, according to which it is impossible to influence the past. Some researchers, therefore, have developed a hypothesis of a self-consistent arrangement of events which prevents a particle from meeting itself in the past; the time machine may exist and a particle may enter it and travel back into the past, but there is no means by which it can arrive at a point in spacetime at which it had already been. Others, with Stephen Hawking as the leader, on the contrary, are proving that quantum mechanics forbids the formation of time machines, since the quantum fluctuations in the region of the supposed formation of a time machine are so strong that they prevent the formation of the time machine. However, Deutsch has shown that, exactly because of quantum mechanics and the Everett interpretation, causal loops are not paradoxical at all! Namely, a particle never travels a well defined trajectory, but its quantum mechanical motion is spread around an average trajectory. According to the Everett interpretation this means that there exist many copies of the particle, and hence many universes which distinguish between themselves by the slightly different positions the particle occupies in each and every of those universes. If a particle travels in a time machine and meets its copy in the past, the result of such a collision will be quantum mechanically undetermined within the range of spreading of the wave function. To every possible pair of directions to which the two particles can recoil after the collision there corresponds a different universe. We have a causal paradox only if we assume the existence of a single universe. Then the collision of a particle with its own copy in the past necessarily changes *the initial history*, which is the essence of the causal paradox. But if we assume that a set of universe exists, then there also exists a set of histories, and hence a journey of a particle into the past does not imply any paradox at all. A similar resolution [115] of the causal paradox has also been proposed for *tachyons*. Tachyons are so far unobserved particles moving with a speed faster than light. The equations of relativity in principle admit not only the existence of *bradyons* (moving slower than light) and *photons* (moving with the speed of light), but also of tachyons. But tachyons appear problematic in several respects[3], mainly because they allow for the formation of causal loops. This is one of the main arguments employed against the possibility that tachyons could be found in nature. However, the latter argument no longer holds after assuming the validity of the Everett interpretation of quantum mechanics, since then causal loops are not paradoxical, and in fact are not "loops" at all.

We have arrived at the following conclusion. If we take seriously the equations of general relativity, then we have also to take seriously their solutions. Amongst the solutions there are also such configurations of spacetime which allow for the formation of causal loops. We have mentioned wormholes. Besides, there also exists the well known Gödel solution for spacetime around a rotating mass. If such solutions are in fact realized in nature, then we have to deal with time machines and such experimental situations, which enables us to *test* the Everett interpretation of quantum mechanics. In this respect the Everett interpretation distinguishes itself from the other interpretations, including the conventional *Copenhagen interpretation*. In the other experimental situations known so far the Everett interpretation gives the same predictions about the behavior of physical systems as the rival interpretations (including the Copenhagen interpretation).

* * *

Life, as we know it, requires the fulfilment of certain strict conditions. It can develop only within a restricted temperature interval, and this can be realized only on a planet which is at just the right distance from a star with just the right activity and sufficiently long life time. If the fundamental constants determining the strength of the gravitational, electromagnetic, weak and strong forces were slightly different those conditions would not have been met, the universe would be different to the extent that a life of our kind would not be possible in it. In physics so far no a reliable principle or law has been discovered according to which the values of the fundamental constant could be determined. Just the contrary, all values of those constants are possible in principle. The fact that they are "chosen" just as they are, has been attempted to be explained by the so called *anthropic principle* [116]. According to that principle there exists a fundamental relationship between the values of the fundamental constants and our existence; our existence in the universe conditions the values of those constants. Namely, the world must be such that we the observers can exist

---

[3]Some more discussion about tachyons is provided in Sec. 13.1.

in it and observe it. However, by this we have not explained much, since the question remains, why is the universe just such that it enables our life. Here we can again help ourselves by employing the Everett interpretation which says that everything which physically can happen actually does happen— in some universe. The physical reality consists of a collection of universes. There exist all sorts of universes, with various values of the fundamental constants. In a vast majority of the universes life is not possible, but in few of them life is, nevertheless, possible, and in some universes life actually develops. In one such universe we live. We could say as well that "in one of *those* universes we live". The small probability of the occurrence of life is not a problem at all. It is sufficient that the emergence of life is possible, and in some universes life would have actually developed. Hence in the Everett interpretation the anthropic principle is automatically contained.

* * *

It is typical for general relativity that it deals merely with the intrinsic properties of spacetime, such as its metric and the intrinsic curvature. It disregards how spacetime looks "from the outside". The practitioners of general relativity are not interested in an eventual existence of an embedding space in which our spacetime is immersed. At the same time, paradoxically, whenever they wish to illustrate various solutions of the Einstein equations they actually draw spacetime as being embedded in a higher-dimensional space. Actually they draw spacetime as a 2-dimensional surface in 3-dimensional space. If they had known how to do it they would have drawn it as a 4-dimensional surface in a higher-dimensional space, but since this is not possible[4] they help themselves by suppressing two dimensions of spacetime.

How can we talk at all about a fourth, fifth, or even higher dimension, if we are unable to perceive them. For a description of a point in a three dimensional space we need three numbers, i.e., coordinates. In order to describe its motion, that is the trajectories, we need three equations. There is an isomorphism between the algebraic equations and geometric objects, for instance curves in space. This we can generalize, and instead of three equations take four or more equations; we then talk about four- or higher-dimensional spaces.

Instead of considering the embedding of spacetime in a higher-dimensional space merely as a usefull tool for the illustration of Einstein's equations, some physicists take the embedding space seriously as an "arena" in which

---

[4]By using suitable projection techniques this might be in fact possible, but such drawings would not be understandable to an untrained person.

312   *THE LANDSCAPE OF THEORETICAL PHYSICS: A GLOBAL VIEW*

lives the 4-dimensional surface representing spacetime. Distribution of matter on this surface is determined by the distribution of matter in the embedding space[5]. The motion of the latter surface (actually the motion of a 3-brane which sweeps a 4-dimensional surface, called a *worldsheet*) can be considered as being a classical motion, which means that the surface and its position in the embedding space are well determined at every moment. However, such a classical description does not correspond to the reality. The motion of the 3-brane has to obey the laws of quantum mechanics, hence a generic state of the brane is represented by a wave function. The latter function in general does not represent a certain well determined brane's worldsheet, but is "spread" over various worldsheets. More precisely, a wave function is, in general, a superposition of the particular wave functions, every one of them representing some well defined worldsheet. Such a view automatically implies that our spacetime worldsheet is not the only possible one, but that there exist other possible worldsheets which represent other possible universes, with different configurations of geometry and matter, and thus with different possible observers. But they all stay in a quantum mechanical superposition! How can we then reconcile this with the fact that at the macroscopic level we observe a well determined spacetime, with a well determined matter configuration? We again employ the Everett interpretation. According to Everett all those spacetime worldsheets together with the corresponding observers, which enter the superposition, are not merely *possible*, but they actually exist in the multiverse. Relative to every one of those observers the wave function represents a state with a well determined universe, of course up to the accuracy with which the observer monitors the rest of his universe. This is the "objective" point of view. From a "subjective" point of view the situation looks as follows. If I "measure" the position of a single atom in my surroundings, then the positions of all the other atoms, say in a crystal, will be irrevocably determined forever and I shall never observe a superposition of that crystal. Moreover, since the crystal is in the interaction with its surroundings and indirectly also with the entire universe, I shall never be able to observe a superposition of the universe, at least not at the macroscopic level[6]. Of

---

[5] In Chapter 8 we have developed a model in which our spacetime surface is a worldsheet of a brane. Assuming that there are many other similar branes of various dimensionality which can intersect our world brane we obtain, as a result of the intersection, the matter on our world brane in the form of point particles, strings, 2-branes and 3-branes (i.e., space filling branes). All those other branes together with our world brane form the matter in the embedding space. Moreover, we have shown that the embedding space is actually identified with all those branes. Without the branes there is no embedding space.

[6] In fact, I measure the position of an atom in a crystal by the very act of looking at it. So my universe actually is no longer in such a macroscopic superposition after the moment I looked at it (or even touched it) for the first time. *Relative to me* the universe certainly was in a superposition

course, a superposition of the universe at the microscopic level remains, but is reduced every time we perform a corresponding measurement.

According to the conventional Copenhagen interpretation of quantum mechanics it is uncertain which of the *possible* universes will be realized after a measurement of a variable. According to the Everett interpretation, however, all those universes actually exist. This is an 'objective' point of view. By introducing the concept of *relative wave function* Everett explains that from a "subjective" point of view it is uncertain in which of those universes the observer will happen to "find himself". In this respect the Everett interpretation coincides with the Copenhagen interpretation. The questions "what universe?" and "which universe?" are intertwined in the Everett interpretation, depending on whether we look at it from an "objective" or a "subjective" point of view.

* * *

The quantum theory of the spacetime worldsheet in an embedding space, outlined in rough contours in this chapter, is in my opinion one of the most promising candidates for the quantum description of gravity. In its future development it will be necessary to include the other interactions, such as the electromagnetic, weak and strong interactions. This could be achieved by following the Kaluza–Klein idea and extend the dimensionality of the spacetime sheet from four to more dimensions. Also fermions could be included by performing a supersymmetric generalization of the theory, that is by extending the description to the anticommuting Grassmann coordinates, or perhaps by taking a polyvector generalization of the theory.

(and consequently I was not aware of anything) before my embryo started to evolve, and will be again in a superposition after my death. The latter metaphor attempts to illustrate that a conscious observer and the corresponding definite macroscopic universe are in a tight relationship.

Chapter 12

NOBODY REALLY UNDERSTANDS QUANTUM MECHANICS

Quantum mechanics is a theory about the relative information that subsystems have about each other, and this is a complete description about the world
—Carlo Rovelli

The motto from a famous sentence by Feynman [117] will guide us through this chapter. There are many interpretations of quantum mechanics (QM) described in some excellent books and articles. No consensus about which one is "valid", if any, has been established so far. My feeling is that each interpretation has its own merits and elucidates certain aspects of QM. Let me briefly discuss the essential points (as I see them) of the three main interpretations[1].

Conventional (Copenhagen) interpretation. The wave function ψ evolves according to a certain evolution law (the Schrödinger equation). ψ carries the information about *possible* outcomes of a measurement process. Whenever a measurement is performed the wave function collapses into one of its eigenstates. The absolute square of the scalar product of ψ with its eigenfunctions are the probabilities (or probability densities) of the occurrence of these particular eigenvalues in the measurement process [120, 121].

Collapse or the reduction of the wave function occurs in an observer's mind. In order to explain how the collapse, which is extraneous

---
[1]Among modern variants of the interpretations let me mention the *relational quantum mechanics* of Rovelli [118], and the *many mind interpretation* of Butterfield [119]

to the Schrödinger evolution of ψ, happens at all, one needs something more. If one postulates that the collapse occurs in a (say, macroscopic) measuring apparatus the problem is not solved at all, since also the interaction of our original system (described by ψ) with the measuring apparatus is governed by the Schrödinger evolution for the combined system–apparatus wave function. Therefore also the measuring apparatus is in a state which is a superposition of different eigenstates corresponding to different results of measurement[2]. This is true even if the result of measurement is registered by a magnetic tape, or punched tape, etc.. A conscious observer has to look at the result of measurement; only at that moment is it decided which of various possibilities actually occurs [122]. Meanwhile, the tape has been in a state which is a superposition of states corresponding to the eigenvalues in question.

Everett, Wheeler, Graham many worlds interpretation. Various quantum possibilities actually occur, but in different branches of the world [107, 109, 110]. Every time a measurement is performed the observed world splits into several (often many) worlds corresponding to different eigenvalues of the measured quantities. All those worlds coexist in a higher universe, the *multiverse*. In the multiverse there exists a (sufficiently complicated) subsystem (e.g., an automaton) with memory sequences. To a particular branching path there corresponds a particular memory sequence in the automaton, and vice versa, to a particular memory sequence there belongs a particular branching path. No collapse of the wave function is needed. All one needs is to decide which of the possible memory sequences is the one to follow. (My interpretation is that there is no collapse in the multiverse, whilst a particular memory sequence or stream of consciousness experiences the collapse at each branching point.) A particular memory sequence in the automaton actually defines a possible life history of an observer (e.g., a human being). Various well known paradoxes like that of Einstein–Podolsky–Rosen, which are concerned with correlated, non-interacting systems, or that of Schrödinger's cat, etc., are easily investigated and clarified in this scheme [107].

Even if apparently non-related the previous three interpretations in fact illuminate QM each from its own point of view. In order to introduce the reader to my way of looking at the situation I am now going to describe some of my earlier ideas. Although not being the final word I have to say about QM, these rough ideas might provide a conceptual background which will facilitate understanding the more advanced discussion (which will also

---

[2] For a more detailed description of such a superposition and its duration see the section on decoherence.

take into account the modern *decoherence* approach) provided later in this chapter. A common denominator to the three views of QM discussed above we find in the assumption that a 3-dimensional simultaneity hypersurface Σ moves in a higher-dimensional space of *real* events[3]. Those events which are intersected by a certain Σ-motion are observed by a corresponding observer. Hence we no longer have a conflict between realism and idealism. There exists a certain physical reality, i.e., the world of events in a higher-dimensional space. In this higher universe there exist many 4-dimensional worlds corresponding to different quantum possibilities (see also Wheeler [123]). A particular observer, or, better, his mind chooses by an act of free will one particular Σ-surface, in the next moment another Σ-surface, etc.. A sequence of Σ-surfaces describes a 4-dimensional world[4]. A consequence of the act of free will which happens in a particular mind is the wave function reduction (or collapse). Before the observation the mind has certain information about various *possible* outcomes of measurement; this information is incorporated in a certain wave function. Once the measurement is performed (a measurement procedure terminates in one's mind), one of the *possible* outcomes has become the *actual* outcome; the term actual is relative to a particular stream of consciousness (or memory sequence in Everett's sense). Other possible outcomes are actual relative to the other possible streams of consciousness.

So, which of the possible quantum outcomes will happen is—as I assume—indeed decided by mind (as Wigner had already advocated). But this fact does not require from us to accept an idealistic or even solipsistic interpretation of the world, namely that the external worlds is merely an illusion of a mind. The duty of mind is merely *a choice of a path* in a higher-dimensional space, i.e., a choice of a sequence of Σ-hypersurfaces (the three dimensional "nows"). But various possible sequences exist independently of a mind; they are real and embedded in a timeless higher-dimensional world.

However, a strict realism alone, independent of mind or consciousness is also no more acceptable. There does not exist a *motion* of a real external object. The external "physical" world is a static, higher-dimensional structure of events. One gets a dynamical (external) 4-dimensional world by postulating the existence of a new entity, a *mind*, with the property of *moving* the simultaneity surface Σ into any permissible direction in the higher space. This act of Σ-*motion* must be separately postulated; a consequence

---

[3] We shall be more specific about what the "higher-dimensional space" is later. It can be either the usual higher-dimensional configuration space, or, if we adopt the brane world model then there also exists an infinite-dimensional membrane space $\mathcal{M}$. The points of $\mathcal{M}$-space correspond to the "coordinate" basis vectors of a Hilbert space which span an arbitrary brane state.
[4] This is elaborated in Sec. 10.1.

of this motion is the subjective experience that the (3-dimensional) external world is continuously changing. *The change* of the (3-dimensional) external world is in fact an illusion; what really changes with time is an observer's mind, while the external world —which is more than (3+1)-dimensional— is real and static (or timeless).

Let us stress: only *the change* of an external (3-dimensional) world is an illusion, not *the existence* of an external world as such. Here one must be careful to distinguish between the concept of time as a coordinate (which enters the equations of special and general relativity) and the concept of time as a subjective experience of change or becoming. Unfortunately we often use the same word 'time' when speaking about the two different concepts[5].

One might object that we are introducing a kind of metaphysical or non physical object —mind or consciousness— into the theory, and that a physical theory should be based on *observable* quantities only. I reply: how can one dismiss mind and consciousness as something non-observable or irrelevant to nature, when, on the contrary, our own consciousness is the most obvious and directly observable of all things in nature; it is through our consciousness that we have contacts with the external world (see also Wigner [122]).

12.1. THE 'I' INTUITIVELY UNDERSTANDS QUANTUM MECHANICS

If we think in a really relaxed way and unbiased with preconcepts, we realize the obvious, that the wave function is consciousness. In the following I will elaborate this a little. But before continuing let me say something about the role of extensive verbal explanations and discussions, especially in our attempts to clarify the meaning of quantum mechanics. My point is that we actually need as much such discussion as possible, in order to develop our inner, intuitive, perception of what quantum mechanics is about. In the case of Newtonian (classical) mechanics we already have such an intuitive perception. We have been developing our perception since we are born. Every child intuitively understands how objects move and what the consequences are of his actions, for instance what happens if he throws a ball. Imagine our embarrassment, if, since our birth, we had no direct contact with the physical environment, but we had nevertheless been indirectly taught about the existence of such an environment. The precise situation

---

[5]One of the goals of the present book is to formalize such a distinction; see the previous three parts of the book.

is not important for the argument, just imagine that we are born in a space ship on a journey to a nearby galaxy, and remain fixed in our beds with eyes closed all the time and learning only by listening. Even if not seeing and touching the objects around us, we would eventually nevertheless learn indirectly about the functioning of the physical world, and perhaps even master Newtonian mechanics. We might have become very good at solving all sorts of mechanical problem, and thus be real experts in using rigorous techniques. We might even be able to perform experiments by telling the computer to "throw" a stone and then to tell us about what has happened. And yet such an expertise would not help us much in *understanding* what is behind all the theory and "experiments" we master so well. Of course, what is needed is a direct contact with the environment we model so well. In the absence of such a direct contact, however, it will be indispensable for us to discuss as much as possible the functioning of the physical environment and the meaning of the theory we master so well. Only then would we have developed to a certain extent an intuition, although indirect, about the physical environment.

An analogous situation, of course, should be true for quantum mechanics. The role of extensive verbalization when we try to understand quantum mechanics can now be more appreciated. We have to read, discuss, and think about quantum mechanics as much as we are interested. When many people are doing so the process will eventually crystalize into a very clear and obvious picture. At the moment we see only some parts of the picture. I am now going to say something about how I see my part of the picture.

Everything we know about the world we know through consciousness. We are describing the world by a wave function. Certain simple phenomena can be described by a simple wave function which we can treat mathematically. In general, however, phenomena are so involved that a mathematical treatment is not possible, and yet conceptually we can still talk about the wave function. The latter is our information about the world. Information does not exist *per se*, information is relative to consciousness [124]. Consciousness has information about something. This could be pushed to its extreme and it be asserted that information is consciousness, especially when information refers to itself (self-referring information). On the other hand, a wave function is information (which is at least a certain very important aspect of wave function). Hence we may conclude that a wave function has a very close relation with consciousness. In the strongest version we cannot help but conclude that a wave function should in fact be identified with consciousness. Namely, *if, on the one hand, the wave function is everything I can know about the world, and, on the other, the content of my consciousness is everything I can know about the world, then consciousness is a wave function.* In certain particular cases the content of my consciousness can be very clear: after having prepared an experiment I *know* that an electron is localized in a given box. This situation can be described precisely by means of a mathematical object, namely, the wave function. If I open the box then I know that the electron is no longer localized within the box, but can be anywhere around the box. Precisely how the probability of finding it in some place evolves with time I can calculate by means of quantum mechanics. Instead of an electron in a box we can consider electrons around an atomic nucleus. We can consider not one, but many atoms. Very soon we can no longer do maths and quantum mechanical calculation, but the fact remains that our knowledge about the world is encoded in the wave function. We do not know any longer a precise mathematical expression for the wave function, but we still have a perception of the wave function. The very fact that we see definite macroscopic objects around us is a signal of its existence: so we know that the atoms of the objects are localized at the locations of the object. Concerning single atoms, we know that electrons are localized in a well defined way around the nuclei, etc.. Everything I know about the external world is encoded in the wave function. However, consciousness is more than that. It also knows about its internal states, about the memories of past events, about its thoughts, etc.. It is, indeed, a very involved self-referring information system. I cannot touch upon such aspects of consciousness here, but the interesting reader will profit from reading some good works [125, 126].

The wave function of an isolated system evolves freely according to the Schrödinger evolution. After the system interacts with its sorroundings, the system and its surroundings then become entangled and they are in a quantum mechanical superposition. However, there is, in principle, a causal connection with my brain. For a distant system it takes some time until the information about the interaction reaches me. The collapse of the wave function happens at the moment when the information arrives in my brain. Contrary to what we often read, the collapse of the wave function does not spread with infinite speed from the place of interaction to the observer. There is no collapse until the signal reaches my brain. Information about the interaction need not be explicit, as it usually is when we perform a controlled experiment, e.g., with laser beams. Information can be implicit, hidden in the many degrees of freedom of my environment, and yet the collapse happens, since my brain is coupled to the environment. But why do I experience the collapse of the wave function? Why does the wave function not remain in a superposition? The collapse occurs because the information about the content of my consciousness about the measured system cannot be in superposition. Information about an external degree of freedom can be in superposition. *Information about the degrees of freedom which are the carriers of the very same information cannot remain in a superposi-*

tion. This would be a logical paradox, or the Gödel knot [125, 126]: it is resolved by the collapse of the wave function. My consciousness "jumps" into one of the possible universes, each one containing a different state of the measured system and my different knowledge about the measurement result. However, from the viewpoint of an external observer no collapse has happened until the information has arrived in his brain. Relative to him the measured system and my brain have both remained in a superposition.

In order to illustrate the situation it is now a good point to provide a specific example.

A single electron plane wave hits the screen. Suppose an electron described by a wide wave packet hits a screen. Before hitting the screen the electron's position was undetermined within the wave packet's localization. What happens after the collision with the screen? If we perform strictly quantum mechanical calculations by taking into account the interaction of the electron with the material in the screen we find that the location of the traces the interaction has left within the screen is also undetermined. This means that the screen is in a superposition of the states having a "spot" at different places of the screen. Suppose now that an observer $\mathcal{O}$ looks at the screen. Photons reflected from the screen bear the information about the position of the spot. They are, according to quantum mechanical calculations, in a superposition. The same is true for an observer who looks at the screen. His eyes' retinas are in a superposition of the states corresponding to different positions of the spot, and the signal in the nerves from the retina is in a superposition as well. Finally, the signal reaches the visual center in the observer's brain, which is also in the superposition. Before the observer has looked at the screen the latter has been in a superposition state. After having looked, the screen state is still in a superposition, but at the same time there is also a superposition of the brain states representing different states of consciousness of the observer $\mathcal{O}$.

Read carefully again: different brain (quantum mechanical) states *represent* different *consciousness* states. And what is the content of those consciousness state? Precisely the information about the location of the spot on the screen. But the latter information is, in fact, *the wave function* of the screen, more precisely the collapsed wave function. So we have a direct piece of evidence about the relation between the wave function about an external state and a conscious state. The external state is *relative* to the brain state, and the latter state in turn represents a state of consciousness. At this point it is economical to identify the relative "external" state with the corresponding consciousness state.

Relative to the observer $\mathcal{O}$'s consciousness states there is no superposition of the screen states. "Subjectively", a collapse of the wave function has occurred relative to the observer's consciousness state, but "objectively" there is no collapse.

The term *objective* implies that there should exist an "objective" wave function of the universe which never collapses. We now ask "is such a concept of an objective, universal, wave function indeed necessary?" Or, put it differently, what is "the universal wave function"? Everett himself introduced the concept of *the relative wave function*, i.e., the wave function which is relative to another wave function. In my opinion the relative wave function suffices, and there is no such a thing as an objective or universal wave function. This will become more clear after continuing with our discussion.

Now let us investigate how I experience the situation described above. Before I measure the position of the electron, it was in a superposition state. Before I had any contact with the screen, the observer $\mathcal{O}$, or their environment, they were altogether in a superposition state. After looking at the screen, or after communicating with the observer $\mathcal{O}$, there was no longer superposition relative to my consciousness. However, relative to another observer $\mathcal{O}'$ the combined state of the screen S, $\mathcal{O}$, and my brain can remain in superposition until $\mathcal{O}'$ himself gets in contact with me, $\mathcal{O}$, S, or the environment of S, $\mathcal{O}$, and me. A little more thought in such a direction should convince everybody that a wave function is always *relative* to something, or, better, to somebody. There can be no "objective" wave function.

If I contemplate the electron wave packet hitting the screen I know that the wave packet implies the existence of the multiverse, but I also know, after looking at the screen, that I have found myself in one of those many universes. I also know that according to some other observer my brain state can be a superposition. But I do not know how my brain state could *objectively* be a superposition. Who, then is this objective observer? Just think hard enough about this and you will start to realize that there can be no objective wave function, and if so, then a wave function, being always relative to someone's consciousness, can in fact be identified with someone's consciousness. The phrase "wave function is relative to someone's consciousness" could be replaced by "wave function is (someone's) consciousness". All the problems with quantum mechanics, also the difficulties concerning the Everett interpretation, then disappear at once.

I shall, of course, elaborate this a little bit more in due course. At the moment let me say again that the difficulties concerning the understanding of QM can be avoided if we consider a wave function as a measure of the information an observer has about the world. A wave function, in a sense, *is* consciousness. We do not yet control all the variables which are relevant to consciousness. But we already understand some of those variables, and

12.2. DECOHERENCE

Since the seminal work by Zurek [127] and Zeh [128] it has becomes very clear why a macroscopic system cannot be in a superposition state. A system S which we study is normally coupled to its environment E. As a consequence S no longer behaves as a quantum system. More precisely, the partial wave function of S relative to E is no longer a superposition of S's eigenstates. The combined system SE, however, still behaves as a quantum system, and is in a superposition state. Zurek and Zeh have demonstrated this by employing the description with *density matrices*.

The density matrix. A quantum state is a vector $|\psi\rangle$ in *Hilbert space*. The projection of a generic state onto the position eigenstates $|x\rangle$ is *the wave function*

$$\psi(x) \equiv \langle x|\psi\rangle. \tag{12.1}$$

Instead of $|\psi\rangle$ we can take the product

$$|\psi\rangle\langle\psi| = \hat{\rho}, \tag{12.2}$$

which is called *the density operator*. The description of a quantum system by means of $|\psi\rangle$ is equivalent to description by means of $\hat{\rho}$.

Taking the case of a single particle we can form the sandwich

$$\langle x|\hat{\rho}|x'\rangle \equiv \rho(x, x') = \langle x|\psi\rangle\langle\psi|x'\rangle = \psi(x)\psi^*(x'). \tag{12.3}$$

This is *the density matrix* in the coordinate representation. Its diagonal elements

$$\langle x|\hat{\rho}|x\rangle = \rho(x,x) \equiv \rho(x) = |\psi(x)|^2 \tag{12.4}$$

form *the probability density* of finding the particle at the position x. However, the off-diagonal elements are also different from zero, and they are responsible for interference phenomena. If somehow the off-diagonal terms vanish, then the interference also vanishes.

Consider, now, a state $|\psi\rangle$ describing a spin $\frac{1}{2}$ particle coupled to a detector:

$$|\psi\rangle = \sum_i \alpha_i |i\rangle\langle d_i|, \tag{12.5}$$

where
$$|i\rangle = |\tfrac{1}{2}\rangle,\ |-\tfrac{1}{2}\rangle \tag{12.6}$$
are spin states, and
$$|d_i\rangle = |d_{1/2}\rangle,\ |d_{-1/2}\rangle \tag{12.7}$$
are the detector states.

The density operator is
$$|\psi\rangle\langle\psi| = \sum_{ij} \alpha_i \alpha_j^* |i\rangle|d_i\rangle\langle j|\langle d_j|. \tag{12.8}$$

It can be represented in some set of basis states $|m\rangle$ which are rotated relative to $|i\rangle$:
$$|m\rangle = \sum_k |k\rangle\langle k|m\rangle,\quad |d_m\rangle = \sum_{d_k} |d_k\rangle\langle d_k|d_m\rangle \tag{12.9}$$

We then obtain the density matrix
$$\langle d_m, m|\psi\rangle\langle\psi|n, d_n\rangle = \sum_{ij} \alpha_i \alpha_j^* \langle d_m, m|i, d_i\rangle\langle j, d_j|n, d_n\rangle. \tag{12.10}$$

which has non-zero off diagonal elements. Therefore the combined system *particle–detector* behaves quantum mechanically.

Let us now introduce yet another system, namely, the environment. After interacting with the environment the evolution brings the system to the state
$$|\psi\rangle = \sum_i \alpha_i |i\rangle|d_i\rangle|E_i\rangle, \tag{12.11}$$
where
$$|E_i\rangle = |E_{1/2}\rangle,\ |E_{-1/2}\rangle \tag{12.12}$$
are the environment states after the interaction with the *particle–detector* system.

The density operator is
$$|\psi\rangle\langle\psi| = \sum_{ij} \alpha_i \alpha_j^* |i\rangle|d_i\rangle|E_i\rangle\langle j|\langle d_j|\langle E_j| \tag{12.13}$$

The combined system *particle–detector–environment* is also in a *superposition state*. The density matrix has non zero off-diagonal elements.

Whilst the degrees of freedom of the particle and the detector are under the control of an observer, those of the environment are not. The observer cannot distinguish $|E_{1/2}\rangle$ from $|E_{-1/2}\rangle$, therefore he cannot know the total density matrix. We can define *the reduced density operator* which takes into account the observer's ignorance of $|E_i\rangle$. This is achieved by summing over the environmental degrees of freedom:

$$\sum_k \langle E_k|\psi\rangle\langle\psi|E_k\rangle = \sum_i |\alpha_i|^2 |i\rangle|d_i\rangle\langle i|\langle d_i|. \tag{12.14}$$

We see that the reduced density operator, when represented as a matrix in the states $|i\rangle$, has only the diagonal terms different from zero. This property is preserved under rotations of the states $|i\rangle$.

We can paraphrase this as follows. With respect to the environment the density matrix is diagonal. Not only with respect to the environment, but with respect to any system, the density matrix is diagonal. This has already been studied by Everett [107], who introduced the concept of *relative state*. The reduced density matrix indeed describes the relative state. In the above specific case the state of the system *particle–detector* is relative to the environment. Since the observer is also a part of the environment the state of the system *particle–detector* is relative to the observer. The observer *cannot* see a superposition (12.5), since very soon the system evolves into the state (12.11), where $|E_i\rangle$ includes the observer as well. After the interaction with environment the system *particle–detector* loses the interference properties and behaves as a classical system. However, the total system *particle–detector–environment* remains in a superposition, but nobody who is coupled to the environment can observe such a superposition after the interaction reaches him. This happens very soon on the Earth, but it may take some time for an observer in space.

The famous *Schrödinger's cat* experiment [129] can now be easily clarified. In order to demonstrate that the probability interpretation of quantum mechanics leads to paradoxes Schrödinger envisaged a box in which a macroscopic object —a cat— is linked to a quantum system, such as a low activity radioactive source. At every moment the source is in a superposition of the state in which a photon has been emitted and the state in which no photon has been emitted. The photons are detected by a Geiger counter connected to a device which triggers the release of a poisonous gas. Schrödinger considered the situation as paradoxical, as the cat should remain in a superposition state, until somebody looks into the box. According to our preceding discussion, however, the cat could have remained in a superposition only if completely isolated from the environment. This is normally not the case, therefore the cat remains in a superposition for a very short time, thereafter the combined system *cat–environment* is in a superposition state. The environment includes me as well. But I cannot be in a superposition, therefore my consciousness jumps into one of the two branches of the superposition (i.e., the cat alive and the cat dead). This happens even *before* I look into the box. Even before I look into the box it is already decided into which of the two branches my consciousness resides. This is so because I am coupled to the environment, to which also the cat is coupled. Hence, I am already experiencing one of the branches. My consciousness, or, better subconsciousness, has already decided to choose one of the branches, even before I became aware of the cat's state by obtaining the relevant information (e.g., by looking into the box). What counts here is that the necessary information is available in principle: it is implicit in the environmental degrees of freedom. The latter are different if the cat is alive or dead.

12.3. ON THE PROBLEM OF BASIS IN THE EVERETT INTERPRETATION

One often encounters an objection against the Everett interpretation of quantum mechanics that is known under a name such as "the problem of basis". In a discussion group on internet (Sci.Phys., 5 Nov.,1994) I have found a very lucid discussion by Ron Maimon (Harvard University, Cambridge, MA) which I quote below.

> It's been about half a year since I read Bell's analysis, and I don't have it handy. I will write down what I remember as being the main point of his analysis and demonstrate why it is incorrect.
>
> Bell claims that Everett is introducing a new and arbitrary assumption into quantum mechanics in order to establish collapse, namely the "pointer basis". His claim is that it is highly arbitrary in what way you split up the universe into a macroscopic superposition and the way to do it is in no way determined by quantum mechanics. For example, if I have an electron in a spin eigenstate, say $|+\rangle$ then I measure it with a device which has a pointer, the pointer should (if it is a good device) be put into an eigenstate of its position operator.
>
> This means that if we have a pointer which swings left when the electron has spin up, it should be put into the state "pointer on the left" if the electron was in the state $|+\rangle$. If it similarly swings right when the electron is in the state $|-\rangle$ then if the electron is in the state $|-\rangle$ the pointer should end up in the state "pointer on the right".
>
> Now, says Bell, if we have the state $(1/\sqrt{2})(|+\rangle + |-\rangle)$ then the pointer should end up in the state $(1/\sqrt{2})(|\text{right}\rangle + |\text{left}\rangle)$. According to Bell, Everett says that this is to be interpreted as two universes, distinct and noninteracting, one in which the pointer is in the state "right" and one in which the pointer is in the state "left".
>
> But aha! says Bell, this is where that snaky devil Everett gets in an extra hypothesis! We don't have to consider the state $1/\sqrt{2}(|\text{right}\rangle + |\text{left}\rangle)$ as a superposition—I mean it is a state in its own right. Why not say that there has been no split at all, or that the split is into two universes, one in which the pointer is in the state
>
> $$a_1|\text{right}\rangle + a_2|\text{left}\rangle \qquad (12.15)$$

and one where it is in the state

$$b_1|\text{right}\rangle + b_2|\text{left}\rangle \tag{12.16}$$

So long as $a_1 + b_1 = a_2 + b_2 = 1/\sqrt{2}$ this is allowed. Then if we split the universe along these lines we again get those eerie macroscopic superpositions.

In other words, Everett's unnatural assumption is that the splitting of the universes occurs along the eigenstates of the pointer position operator. Different eigenstates of the pointer correspond to different universes, and this is arbitrary, unnatural, and just plain ugly.

Hence Everett is just as bad as anyone else.

Well this is WRONG.

The reason is that (as many people have mentioned) there is no split of the universe in the Everett interpretation. The state

$$\frac{1}{\sqrt{2}}(|\text{right}\rangle + |\text{left}\rangle) \tag{12.17}$$

is no more of a pair of universe than the state $1/\sqrt{2}(|+\rangle + |-\rangle)$ of spin for the electron.

Then how comes we never see eerie superpositions of position eigenstates?

Why is it that the "pointer basis" just happens to coincide with him or her self. This is the "state of mind" basis. The different states of this basis are different brain configurations that correspond to different states of mind, or configurations of thoughts.

Any human being, when thrown into a superposition of state of mind will split into several people, each of which has a different thought. Where before there was only one path of mind, after there are several paths. These paths all have the same memories up until the time of the experiment, and these all believe different events have occurred. This is the basis along which the universe *subjectively seems to split*.

There is a problem with this however—what guarantees that eigenstates of my state of mind are the same as eigenstates of the pointer position. If this wasn't the case, then a definite state of mind would correspond to an eerie neither here nor there configuration of the pointer.

The answer is, NOTHING. It is perfectly possible to construct a computer with sensors that respond to certain configurations by changing the internal state, and these configurations are not necessarily eigenstates of position of a needle. They might be closer to eigenstates of momentum of the needle. Such a computer wouldn't see weird neither-here-nor-there needles, it would just "sense" momenta, and won't be able to say to a very high accuracy where the needle is.

So why are the eigenstates of our thoughts the same as the position eigenstates of the needle?

They aren't!

They are only very approximately position eigenstates of the needle.

This can be seen by the fact that when we look at a needle it doesn't start to jump around erratically, it sort of moves on a smooth trajectory. This means that when we look at a needle, we don't "collapse" it into a position eigenstate, we only "collapse it into an approximate position eigenstate. In Everett's language, we are becoming correlated with a state that is neither an eigenstate of the pointer's position, nor its momentum, but approximately an eigenstate of

328   THE LANDSCAPE OF THEORETICAL PHYSICS: A GLOBAL VIEW

both, constrained by the uncertainty principle. This means that we don't have such absurdly accurate eyes that can see the location of a pointer with superhigh accuracy.

If we were determining the *exact* position of the needle, we would have gamma ray sensor for eyes and these gamma rays would have enough energy to visibly jolt the needle whenever we looked at it.

In order to determine exactly what state we are correlated with, or if you like, the world (subjectively) collapses to, you have to understand the mechanism of our vision.

A light photon bouncing off a needle in a superposition

$$\frac{1}{\sqrt{2}}(|\text{right}\rangle + |\text{left}\rangle) \tag{12.18}$$

will bounce into a superposition of the states $|1\rangle$ or $|2\rangle$ corresponding to the direction it will get from either state. The same photon may then interact with our eyes. The way it does this is to impinge upon a certain place in our retina, and this place is highly sensitive to the direction of the photon's propagation. The response of the pigments in our eyes is both higly localized in position (within the radius of a cell) and in momentum (the width of the aperture of our pupil determines the maximal resolution of our eyes). So it is not surprising that our pigment excitation states become correlated with approximate position and approximate momentum eigenstates of the needle. Hence we see what we see.

If we had good enough mathematical understanding of our eye we could say in the Everett interpretation *exactly* what state we seem to collapse the needle into. Even lacking such information it is easy to see that we will put it in a state resembling such states where Newton's laws are seen to hold, and macroscopic reality emerges.

A similar reasoning holds for other information channels that connect the outside world with our brane (e.g., ears, touch, smell, taste). The problem of choice of basis in the Everett interpretation is thus nicely clarified by the above quotation from Ron Maimon.

12.4. BRANE WORLD AND BRAIN WORLD

Let us now consider the model in which our world is a 3-brane moving in a higher-dimensional space. How does it move? According to the laws of quantum mechanics. A brane is described by a wave packet and the latter is a solution of the Schrödinger equation. This was more precisely discussed in Part III. Now I will outline the main ideas and concepts. An example of a wave packet is sketched in Fig. 12.1.

If the brane self-intersects we obtain *matter* on the brane (see Sec. 8.3). When the brane moves it sweeps a surface of one dimension more. A 3-brane sweeps a 4-dimensional surface, called a *world sheet* or a *spacetime sheet*.

We have seen in Sec. 10.2 that instead of considering a 3-brane we can consider a 4-brane. The latter brane is assumed to be a possible spacetime sheet (and thus has three space-like and one time-like intrinsic dimensions). Moreover, it is assumed that the 4-brane is subjected to dynamics along an invariant evolution parameter τ. It is one of the main messages of this book to point out that such a dynamics naturally arises within the description of geometry and physics based on Clifford algebra. Then a scalar and a pseudoscalar parameter appear naturally, and evolution proceeds with respect to such a parameter.

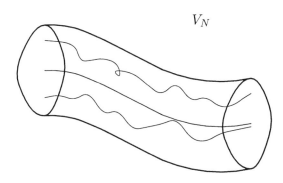

Figure 12.1. An illustration of a wave packet describing a 3-brane. Within the effective region of localization any brane configuration is possible. The wavy lines indicate such possible configurations.

A 4-brane state is represented by a wave packet localized around an average 4-surface (Fig. 12.2)

It can be even more sharply localized within a region P, as shown in Fig. 10.2 or Fig. 12.3. (For convenience we repeat Fig. 10.3.)

All these were mathematical possibilities. We have a Hilbert space of 4-brane kinematic states. We also have the Schrödinger equation which a dynamically possible state has to satisfy. As a dynamically possible state we obtain a wave packet. A wave packet can be localized in a number of possible ways, and one is that of Fig. 10.3, i.e., localization within a region P. How do we interpret such a localization of a wave packet? What does it mean physically that a wave packet is localized within a 4-dimensional region (i.e., it is localized in 3-space and at "time" $t \equiv x^0$)? This means that the 4-brane configuration is better known within P than elsewhere. Since the 4-brane represents spacetime and matter (remember that the 4-brane's self-intersections yield matter on the 4-brane), such a localized wave packet tells us that spacetime and matter configuration are better known within P than elsewhere. Now recall *when*, according to quantum mechanics, a matter configuration (for instance a particle's position) is better known than otherwise. It is better known after a suitable measurement. But we have also seen that a measurement procedure terminates in one's brain, where it is decided —relative to the brain state— about the outcome of the measurement. Hence the 4-brane wave packet is localized within P, because an observer has measured the 4-brane's configuration. Therefore the wave packet (the wave function) is *relative* to that observer.

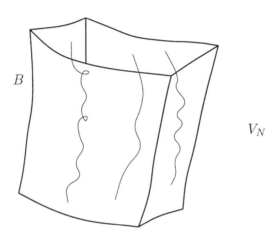

Figure 12.2. A 4-brane wave packet localized within an effective boundary B. A wavy line represents a possible 4-brane.

The 4-brane configuration after the measurement is not well known at every position on the 4-brane, but only at the positions within P, i.e., within a certain 3-space region and within a certain (narrow) interval of the coordinate x^0. Such a 4-brane configuration (encompassing a matter configuration as well) can be very involved. It can be involved to the extent that it forms the structure of an observer's brain contemplating the "external" world by means of sense organs (eyes, ears, etc.).

We have arrived at a very important observation. *A wave packet localized within P can represent the brain structure of an observer $\mathcal{O}$ and his sense organs, and also the surrounding world!* Both the observer and the surrounding world are represented by a single (very complicated) wave packet. Such a wave packet represents the observer's knowledge about his brain's state and the corresponding surrounding world—all together. It represents the observer's consciousness! This is the most obvious conclusion; without explicitly adopting it, the whole picture about the meaning of QM remains foggy.

One can now ask, "does not the 4-brane wave packet represent the brain structure of another observer $\mathcal{O}'$ too?" Of course it does, but not as completely as the structure of $\mathcal{O}$. By "brain" structure I mean here also the content of the brain's thought processes. The thought processes of $\mathcal{O}'$ are not known very well to $\mathcal{O}$. In contrast, his own thought processes are very well known to $\mathcal{O}$, at the first person level of perception. Therefore, the 4-brane wave packet is well localized within $\mathcal{O}$'s head and around it.

Such a wave packet is relative to $\mathcal{O}$. There exists, of course, another possible wave packet which is relative to the observer $\mathcal{O}'$, and is localized around $\mathcal{O}'$'s head.

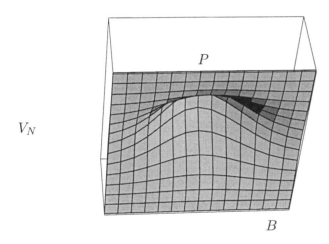

Figure 12.3. Illustration of a wave packet with a region of sharp localization P.

Different initial conditions for a wave function mean different initial conditions for consciousness. A wave function can be localized in another person's head: my body can be in a superposition state with respect to that person (at least for a certain time allowed by decoherence). If I say (following the Everett interpretation) that there are many Matejs writing this page, I have in mind a wave function relative to another observer. Relative to me the wave function is such that I am writing these words right now. In fact, I am identical with the latter wave function. Therefore at the basic level of perception I intuitively understand quantum mechanics.

An 'I' intuitively understands quantum mechanics. After clarifying this I think that I have acquired a deeper understanding of quantum mechanics. The same, I hope, holds for the careful reader. I hope, indeed, that after reading these pages the reader will understand quantum mechanics, not only at the lowest, intuitive, level, but also at a higher cognitive level of perception. An ultimate understanding, however, of what is really behind quantum mechanics and consciousness will probably never be reached by us, and according to the Gödel incompletness theorem [125, 126] is even not possible.

Box 12.1: Human language and multiverse[a]

In the proposed brane world model spacetime, together with matter, is represented by a 4-dimensional self-intersecting surface V_4. An observer associated with a V_4 distinguishes present, past, and future events. Because of the quantum principle an observer is, in fact, associated not with a definite V_4, but with a corresponding wave function. The latter takes into account all *possible* V_4s entering the superposition.

We see that within the conceptual scheme of the proposed brane world model all the principal tenses of human language —present, past, future tenses, and *conditional*— are taken into account. In our human conversations we naturally talk not only about the actual events (present, past, future), but also about possible events, i.e., those which could have occurred (conditional). According to Piaget [135] a child acquires the ability of formal logical thinking, which includes use of alternatives and conditional, only at an advanced stage in his mental development. Reasoning in terms of possible events is a sign that an individual has achieved the highest stage on the Piaget ladder of conceptual development.

Now, since the emergence of quantum mechanics, even in physics, we are used to talking about possible events which are incorporated in the wave function. According to the Everett interpretation of quantum mechanics as elaborated by Deutsch, those possible events (or better states) constitute the *multiverse*.

---

[a]This idea was earlier discussed in ref. [88]. Later it was also mentioned by Deutsch [112].

12.5. FINAL DISCUSSION ON QUANTUM MECHANICS, AND CONCLUSION

In classical mechanics different initial conditions give different *possible* trajectories of a dynamical system. Differential equations of motion tell us only what is a possible set of solutions, and say nothing about which one is actually realized. Selection of a particular trajectory (by specifying initial conditions) is an *ad hoc* procedure.

The property of classical mechanics admitting many possible trajectories is further developed by Hamilton–Jacobi theory. The latter theory naturally suggests its generalization—quantum mechanics. In quantum mechanics different possible trajectories, or better, a particle's positions, are described by means of a wave function satisfying the Schrödinger equation of motion.

In quantum mechanics different initial conditions give different *possible* wave functions. In order to make discussion more concrete it turns out to be convenient to employ a brane world model in which spacetime together with matter in it is described by a self-intersecting 4-dimensional sheet, a worldsheet V_4. According to QM such a sheet is not definite, but is described by a wave function[6]. It is spread around an average spacetime sheet, and is more sharply localized around a 3-dimensional hypersurface Σ on V_4. Not all the points on Σ are equally well localized. Some points are more sharply localized within a region P (Fig. 10.3), which can be a region around an observer on V_4. Such a wave function then evolves in an invariant evolution parameter τ, so that the region of sharp localization P moves on V_4.

Different *possible* wave functions are localized around different observers. QM is a mechanics of consciousness. Differently localized wave functions give different possible consciousnesses and corresponding universes (worlds).

My brain and body can be a part of somebody's else consciousness. The wave function relative to an observer $\mathcal{O}'$ can encompass my body and my brain states. Relative to $\mathcal{O}'$ my brain states can be in a superposition (at least until decoherence becomes effective). Relative to $\mathcal{O}'$ there are many Matejs, all in a superposition state. Relative to me, there is always one Matej only. All the others are already out of my reach because the wave function has collapsed.

According to Everett a wave function never does collapse. Collapse is subjective for an observer. My point is that subjectivity is the essence of wave function. A wave function is always relative to some observer, and hence is subjective. So there is indeed collapse, call it subjective, if you

---

[6]For simplicity we call it a 'wave function', but in fact it is a wave functional—a functional of the worldsheet embedding functions $\eta^a(x^\mu)$.

wish. Relative to me a wave function is collapsing all the time: whenever the information (direct or indirect—through the environmental degrees of freedom) about the outcome of measurement reaches me.

There is no collapse[7] if I contemplate other observers performing their experiments.

Let us now consider, assuming the brane world description, a wave packet of the form given in Fig. 12.2. There is no region of sharp localization for such a wave packet. It contains a superposition of all the observers and worlds within an effective boundary B. Is this then the *universal wave function*? If so, why is it not spread a little bit more, or shaped slightly differently? The answer can only make sense if we assume that such a wave function is relative to a super-observer $\mathcal{O}_S$ who resides in the embedding space V_N. The universe of the observer $\mathcal{O}_S$ is V_N, and the wave packet of Fig. 12.2 is a part of the wave function, relative to $\mathcal{O}_S$, describing $\mathcal{O}_S$'s consciousness and the corresponding universe.

To be frank, we have to admit that the wave packet itself, as illustrated in Fig. 10.3, is relative to a super-observer $\mathcal{O}_S$. In order to be specific in describing our universe and a conscious observer $\mathcal{O}$ we have mentally placed ourselves in the position of an observer $\mathcal{O}_S$ outside our universe, and envisaged how $\mathcal{O}_S$ would have described the evolution of the consciousness states of $\mathcal{O}$ and the universe belonging to $\mathcal{O}$. The wave packet, relative to $\mathcal{O}_S$, representing $\mathcal{O}$ and his world could be so detailed that the super-observer $\mathcal{O}_S$ would have identified himself with the observer $\mathcal{O}$ and his world, similarly as we identify ourselves with a hero of a novel or a movie.

At a given value of the evolution parameter τ the wave packet represents in detail the state of the observer $\mathcal{O}$'s brain and the belonging world. With evolution the wave packet spreads. At a later value of τ the wave packet might spread to the extent that it no longer represents a well defined state of $\mathcal{O}$'s brain. Hence, after a while, such a wave packet could no longer represent $\mathcal{O}$'s consciousness state, but a superposition of $\mathcal{O}$'s consciousness states. This makes sense relative to some other observer $\mathcal{O}'$, but not relative to $\mathcal{O}$. From the viewpoint of $\mathcal{O}$ the wave packet which describes $\mathcal{O}$'s brain state cannot be in a superposition. Otherwise $\mathcal{O}$ would not be conscious. Therefore when the evolving wave packet spreads too much, it collapses relative to $\mathcal{O}$ into one of the well defined brain states representing well defined states of $\mathcal{O}$'s consciousness . Relative to another observer $\mathcal{O}'$, however, no collapse need happen until decoherence becomes effective.

---

[7]There is no collapse until decoherence becomes effective. If I am very far from an observer $\mathcal{O}'$, e.g., on Mars, then $\mathcal{O}'$ and the states of his measurement apparatus are in a superposition relative to me for a rather long time.

If the spreading wave packet would not collapse from time to time, the observer could not be conscious. The quantum states that represent $\mathcal{O}$'s consciousness are given in terms of certain basis states. The same wave packet can be also expanded in terms of some other set of basis states, but those states need not represent (or support) consciousness states. This explains why collapse happens with respect to a certain basis, and not with respect to some other basis.

We have the following model. An observer's consciousness and the world to which he belongs are defined as being represented by an evolving wave packet. At every moment τ the wave packet says which universes ($\equiv$ consciousness state + world belonged to) are at disposal. A fundamental postulate is that from the first person viewpoint the observer (his consciousness) necessarily finds himself in one of the available universes implicit in the spreading wave packet. During the observer's life his body and brain retain a well preserved structure, which poses strict constraints on the set of possible universes: a universe has to encompass one of the available consciousness states of $\mathcal{O}$ and the "external" worlds coupled to those brain states. This continues until $\mathcal{O}$'s death. At the moment of $\mathcal{O}$'s death $\mathcal{O}$'s brain no longer supports consciousness states. $\mathcal{O}$'s body and brain no longer impose constraints on possible universes. The set of available universes increases dramatically: every possible world and observer are in principle available! If we retain the fundamental postulate, and I see no logical reason why not to retain it, then the consciousness has to find itself in one of the many available universes. Consciousness jumps into one of the available universes and continues to evolve. When I am dead I find myself born again! In fact, every time my wave packets spreads too much, I am dead; such a spread wave packet cannot represent my consciousness. But I am immediately "reborn", since I find myself in one of the "branches" of the wave packet, representing my definite consciousness state and a definite "external" world.

A sceptical reader might think that I have gone too far with my discussion. To answer this I wish to recall how improbable otherwise is the fact that I exist. (From the viewpoint of the reader 'I' refers to himself, of course.) Had things gone slightly differently, for instance if my parents had not met each other, I would not have been born, and my consciousness would not not have existed. Thinking along such lines, the fact that I exist is an incredible accident!. Everything before my birth had to happen just in the way it did, in order to enable the emergence of my existence. Not only my parents, but also my grandparents had to meet each other, and so on back in time until the first organisms evolved on the Earth! And the fact that my parents had become acquainted was not sufficient, since any slightly different course of their life together would have led to the birth not of me, but of my brother or sister (who do not exist in this world). Any sufficiently deep reasoning in such a direction leads to an unavoidable conclusion that (i) the multiverse in the Everett–Wheeler–DeWitt–Deutsch sense indeed exist, and (ii) consciousness is associated (or identified) with the wave function which is relative to a sufficiently complicated information processing system (e.g., an observer's brain), and *evolves* according to (a) the Schrödinger evolution and experiences *collapse* at every measurement situation. In an extreme situation (death) available quantum states (worlds) can include those far away from the states (the worlds) I have experienced so far. My wave function (consciousness) then collapses into one of those states (worlds), and I start experiencing the evolution of my wave functions representing my life in such a "new world".

All this could, of course, be put on a more rigorous footing, by providing precise definitions of the terms used. However, I think that before attempting to start a discussion on more solid ground a certain amount of heuristic discussion, expounding ideas and concepts, is necessary.

A reader might still be puzzled at this point, since, according to the conventional viewpoint, in Everett's many worlds interpretation of quantum mechanics there is no collapse of the wave function. To understand why I am talking both about the many worlds interpretation (the multiverse) and collapse one has to recall that according to Everett and his followers collapse is a subjective event. Precisely that! Collapse of the wave function is a subjective event for an observer, but such also is the wave function itself. The wave function is always relative and thus subjective. Even the Everett "universal" wave function has to be relative to some (super-) observer.

In order to strengthen the argument that (my) consciousness is not necessarily restricted to being localized just in my brain, imagine the following example which might indeed be realized in a not so remote future. Suppose that my brain is connected to another person's brain in such a way that I can directly experience her perceptions. So I can experience what she sees, hears, touches, etc.. Suppose that the information channel is so perfect that I can also experience her thoughts and even her memories. After experiencing her life in such a way for a long enough time my personality would become split between my brain and her brain. The wave function representing my consciousness would be localized not only in my brain but also in her brain. After long time my consciousness would become completely identified with her life experience; at that moment my body could die, but my consciousness would have continued to experience the life of her body.

The above example is a variant of the following thought experiment which is often discussed. Namely, one could gradually install into my brain small electronic or bioelectronic devices which would resume the functioning of my brain components. If the process of installation is slow enough my biological brain can thus be replaced by an electronic brain, and I would not have noticed much difference concerning my consciousness and my experience of 'I'.

Such examples (and many others which can be easily envisaged by the reader) of the transfer of consciousness from one physical system to another clearly illustrate the idea that (my) consciousness, although currently associated (localized) in my brain, could in fact be localized in some other brain too. Accepting this, there is no longer a psychological barrier to accepting the idea that the wave function (of the universe) is actually closely related, or even identified, with the consciousness of an observer who is part of that universe. After becoming habituated with such, at first sight perhaps strange, wild, or even crazy ideas, one necessarily starts to realize that quantum mechanics is not so mysterious after all. It is a mechanics of consciousness.

With quantum mechanics the evolution of science has again united two pieces, matter and mind, which have been put apart by the famous Cartesian cut. By separating mind from matter[8] —so that the natural sciences have disregarded the question of mind and consciousness— Descartes set the ground for the unprecedented development of physics and other natural sciences. The development has finally led in the 20th century to the discovery of quantum mechanics, which cannot be fully understood without bringing mind and consciousness into the same arena.

---

[8]There is an amusing play of words[130]:

What is matter? — Never mind!
What is mind? — No matter!

Chapter 13

FINAL DISCUSSION

We are now at the concluding chapter of this book. I have discussed many different topics related to fundamental theoretical physics. My emphasis has been on the exposition of ideas and concepts rather than on their further development and applications. However, even the formulation of the basic principles has often required quite involved formalism and mathematics. I expect that that has been stimulating to mathematically oriented readers, whilst others could have skipped the difficult passages, since the ideas and concepts can be grasped to certain extent also by reading many non-technical descriptions, especially in Part IV. In the following few sections I will discuss some remaining open questions, without trying to provide precisely formulated answers.

13.1. WHAT IS WRONG WITH TACHYONS?

In spite of many stimulating works [131] tachyons nowadays have predominantly the status of impossible particles. Such an impossibility has its roots in the current mainstream theoretical constructions, especially quantum field theory and special relativity. Not only that tachyons are generally considered as violating causality (a sort of the "grandfather paradox") [18], there are also some other well known problems. It is often taken for granted that the existence of tachyons, which may have negative energies, destabilizes the vacuum and thus renders the theory unreasonable. Moreover, when one tries to solve a relativistic equation for tachyons, e.g., the Klein–Gordon equation, one finds that (i) localized tachyon disturbances are subluminal, and (ii) superluminal disturbances are non-local, and supposedly cannot be used for information transmission.

There is now a number of works [132], experimental and theoretical, which point out that the superluminal disturbances (case (ii)) are X-shaped and hence, in a sense, localized after all. Therefore they could indeed be used to send information faster than light.

Moreover, by using the Stueckelberg generalization of the Klein–Gordon equation, discussed at length in this book, the question of tachyon localization and its speed acquires a new perspective. The tachyon's speed is no longer defined with respect to the coordinate time $x^0 \equiv t$, but with respect to the invariant evolution parameter τ. Also the problem of negative energy and vacuum destabilization has to be reformulated and re-examined within the new theory. According to my experience with such a Stueckelberg-like quantum field theory, the vacuum is not unstable in the case of tachyons, on the contrary, tachyons are necessary for the consistency of the theory.

Finally, concerning causality, by adopting the Everett interpretation of quantum mechanics, transmission of information into the past or future, is not paradoxical at all! A signal that reaches an observer in the past, merely "splits" the universe into two branches: in one branch no tachyon signal is detected and the course of history is "normal", whilst in the other branch the tachyonic signal is observed, and hence the course of history is altered. This was discussed at length by Deutsch [113] in an example where not tachyon signals, but ordinary matter was supposed sent into the past by means of *time machines*, such as wormholes[1].

My conclusion is that tachyons are indeed theoretically possible; actually they are predicted by the Stueckelberg theory [133]. Their discovery and usage for information transmission will dramatically extend our perception of the world by providing us with a window into the multiverse.

13.2. IS THE ELECTRON INDEED AN EVENT MOVING IN SPACETIME?

First let me make it clear that by "point-like object moving in spacetime" I mean precisely the particle described by the Stueckelberg action (Chapter 1). In that description a particle is considered as being an "event" in spacetime, i.e., an object localized not only in 3-space, but also in the time coordinate $x^0 \equiv t$. We have developed a classical and quantum theory of the *relativistic* dynamics of such objects. That theory was just a first step, in fact an idealization—a studying example. Later we generalized the Stueckelberg theory to extended objects (strings and membranes of any

---

[1] Although Deutsch nicely explains why traveling into the past is not paradoxical from the multiverse point of view, he nevertheless maintains the view that tachyons are impossible.

Final discussion

dimension) and assumed that in reality physical objects are extended. We assumed that a generic membrane need not be space-like. It can be time-like as well. For instance, a string may extend into a time-like direction, or into a space-like direction.

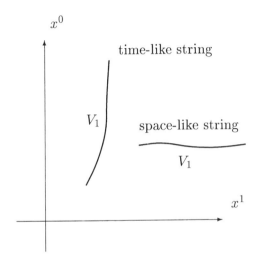

Figure 13.1. Illustration of a time-like and a space-like string. They both move in τ, and the picture is taken at a fixed value of τ (which is an invariant evolution parameter).

If we calculate —at fixed τ— the electromagnetic interaction between two time-like strings we find that it is just that of the familiar Maxwell theory. A string is an extended object, is described by embedding functions $X^\mu(u)$, and can be considered as a collection of point-like objects (the Stueckelberg particles), each one being specified by a value of the parameter u. The electromagnetic interaction between the time-like strings is of the Maxwell type, and can be obtained by summing the contributions of all the Stueckelberg particles constituting the string. A pre-Maxwell field is a function of the parameter u (telling us which Stueckelberg particle); the integration over u gives the corresponding Maxwell field[2]. The latter field does not depend on the string parameter u, but it still depends on the evolution parameter τ.

---

[2] In the literature [7] Maxwell fields are obtained by considering not strings, but point-like Stueckelberg particles, and by integrating the corresponding pre-Maxwell fields over the evolution parameter τ. Such a procedure is called concatenation, and is in fact a sort of averaging over τ.

Between the time-like strings we thus obtain the Maxwell interaction, more precisely, a generalized Maxwell interaction since it depends on τ.

The first and the second quantized theory of the Stueckelberg particle (Chapter 1) is in fact an example for study. In order to apply it to the electron the theory has to be generalized to strings, more precisely, to time-like strings. Namely, the electromagnetic properties of electrons clearly indicate that electrons cannot be point-like Stueckelberg particles, but rather time-like strings, obeying the generalized Stueckelberg theory (the unconstrained theory of strings, a particular case of the unconstrained theory of membranes of any dimension discussed in Chapter 4). The wave function $\psi(\tau, x^\mu)$ is generalized to the string wave functional $\psi[\tau, X^\mu(u)]$. A state of an electron can be represented by ψ which is a functional of the time-like string $X^\mu(u)$, and evolves in τ.

The conventional field theory of the electron is expected to be a specific case when $\psi[\tau, X^\mu(u)]$ is *stationary* in τ. More generally, for a membrane of any dimension we have obtained *the theory of conventional p-branes* as a particular case, when the membrane wave functional is *stationary* in τ (see Sec. 7.1).

13.3. IS OUR WORLD INDEED A SINGLE HUGE 4-DIMENSIONAL MEMBRANE?

At first sight it may seem strange that the whole universe could be a 4-dimensional membrane. The quantum principle resolves the problem: at every scale it is undetermined what the membrane is. Besides that, in quantum field theory there is not only one membrane, but a system of (many) membranes. The intersections or self-intersections amongst those membranes behave as bosonic sources (matter). In the case of supermembranes the intersections presumably behave as fermionic sources (matter). A single membrane can be very small, and also closed. Many such small membranes of various dimensionalities can constitute a large 4-dimensional membrane of the size of the universe.

The universe is thus a system of intersecting and self-intersecting membranes of various dimensions. Compactification of the embedding space is not necessary. The wave functional can be localized around some average (centroid) 4-dimensional membrane. The latter membrane is our classical spacetime. The Einstein gravity described by the action $I_{\text{eff}} = \int \sqrt{|g|} \mathrm{d}^4 R$ is an approximation valid at the scale of the solar system. At larger scales the effect of embedding becomes significant and the effective action deviates from the Einstein–Hilbert action.

Final discussion

Figure 13.2. A system of many small membranes of various dimensionalities constitutes a large membrane at the macroscopic scale.

In short, the picture that our universe is a big 4-dimensional membrane is merely an idealization: the universe on average (the "expectation value"). Actually the universe is a system of many membranes, described by a many membranes wave functionals $f[\eta_1, \eta_2, ...]$ forming a generic state of the universe

$$|A\rangle = \left[\int f[\eta]\Phi^\dagger[\eta]\mathcal{D}\eta + \int f[\eta_1, \eta_2]\Phi^\dagger[\eta_1]\Phi^\dagger[\eta_2]\mathcal{D}\eta_1\,\mathcal{D}\eta_2 \right. \quad (13.1)$$

$$\left. + \int f[\eta_1, \eta_2, \eta_3]\Phi^\dagger[\eta_1]\Phi^\dagger[\eta_2]\Phi^\dagger[\eta_3]\mathcal{D}\eta_1\,\mathcal{D}\eta_2\mathcal{D}\eta_3 + ... \right]|0\rangle,$$

where $\phi^\dagger[\eta]$ is the creation operator for a membrane $\eta \equiv \eta^a(x^\mu)$.

13.4. HOW MANY DIMENSIONS ARE THERE?

In an important paper entitled *Space–time dimension from a variational principle* [134] D. Hochberg and J.T. Wheeler generalize the concept of coordinates and dimension. Instead of coordinates x^μ with a discrete set of indices $\mu = 1, 2, ..., n$ they take a set of coordinates

$$x^\mu \to x(z), \quad (13.2)$$

where z is a continuous real index. They also introduce a weight function $\rho(z)$ and define the dimension p as

$$p = \int \rho(z)\,\mathrm{d}z. \quad (13.3)$$

In general $\rho(z)$ is an arbitrary function of z. In particular, it may favor a certain subset of z-values. For instance, when

$$\rho(z) = \sum_{\mu=1}^{n} \delta(z - z_\mu) \qquad (13.4)$$

the weight favors a set of discrete values z_μ. We then have

$$p = \int \sum_{\mu=1}^{n} \delta(z - z_\mu) \, dz = n. \qquad (13.5)$$

A space of a discrete set of dimensions $\mu = 1, 2, ..., n$ is thus a particular case of a more general space which has a continuous set of dimensions z.

A problem then arises of what is $\rho(z)$. It should not be considered as a fixed function, given once for all, but has to arise dynamically as a consequence of a chosen solution to certain dynamical equations. Formulation of a theory in which dimensions are not *a priori* discrete and their number fixed remains one of the open problems for the future. A first step has been made in [134], but a generalization to membranes is still lacking.

13.5. WILL IT EVER BE POSSIBLE TO FIND SOLUTIONS TO THE CLASSICAL AND QUANTUM BRANE EQUATIONS OF MOTION AND MAKE PREDICTIONS?

My answer is YES. Only it will be indispensable to employ the full power of computers. Consider the equation of a minimal surface

$$D_\mu D^\mu \eta^a = 0. \qquad (13.6)$$

Choosing a gauge $\eta^\mu = x^\mu$ the above equation becomes

$$D_\mu D^\mu \eta^{\bar{a}} = 0, \qquad (13.7)$$

where $\eta^{\bar{a}}(x)$, $\bar{a} = n+1, n+2, ..., N$, are the transverse embedding functions.

What are solutions to the equation (13.7)? We shall write a general solution as

$$\eta^{\bar{a}}(x) = \text{Surf}^{\bar{a}}(x^\mu, \alpha), \qquad (13.8)$$

where α is a set of parameters. But what is $\text{Surf}^{\bar{a}}$? It is *defined* as a general solution to the set of second order non-linear differential equations (13.7). Yes, but what are its numerical values, what are the graphs for various

Final discussion

choices of parameters α? Well, ask a computer! A computer program is required which, at a press of button will give you the required answer on the display. Fine, but this is not an exact solution.

Consider again what you mean by an exact solution. Suppose we have studied a dynamical system and found that its equation of motion is

$$\ddot{x} + \omega^2 x = 0. \tag{13.9}$$

What is the general solution to the second order differential equation (13.9)? That is easy, it is

$$x(t) = A \sin \omega t + B \cos \omega t \equiv \text{Osc}(t, A, B, \omega), \tag{13.10}$$

where A, B are arbitrary parameters. Yes, but what are the numerical values of $\text{Osc}(t, A, B, \omega)$, what do the graphs for various choices of A, B, ω look like? This is something we all know from school. We used to look up the tables, but nowadays we obtain the answer easily by computer or calculator. There is a number of algorithms which enable us to compute $\sin \omega t$ and $\cos \omega t$ to arbitrary precision. Similarly for solutions to other second order differential equations, like $\ddot{x} - \omega^2 x = 0$, etc.. If so, why do we not develop algorithms for computing $\text{Surf}^{\bar{a}}$? Once the algorithm is installed in the computer we could treat $\text{Surf}^{\bar{a}}$ in an analogous way as $\text{Osc}(t, A, B, \omega)$ (i.e., as $\sin \omega t$ and $\cos \omega t$). Similarly we could develop algorithms for computing numerical values and plotting the graphs corresponding to the solutions of other differential equations of motion describing various classical and quantum dynamical systems of branes and their generalizations. Once we can simply visualize a solution as a graph, easily obtained by means of a suitable computer program, it is not so difficult to derive predictions of various competing theories and to propose strategies for their experimental verification.

13.6. HAVE WE FOUND A UNIFYING PRINCIPLE?

We have discussed numerous ideas, theories, techniques and approaches related to fundamental theoretical physics. Can we claim that we have found a unifying principle, according to which we could formulate a final theory, incorporating all the fundamental interactions, including gravity? No, we cannot. If anything, we have found a sort of meta-principle, according to which, in principle, there is no limit to the process of understanding the Nature. A final observer can never completely understand and describe reality. Whenever we may temporarily have an impression of understanding the relationship between gravity and quantum theory, some new experiments and theoretical ideas will soon surpass it. In this book I have aimed, amongst other things, to point out how rich and vast is the arsenal of conceptual, theoretical, and technical possibilities. And there seems to be no limit to the numerous ways in which our theoretical constructions could be generalized and extended in a non-trivial and elegant way. No doubt there is also no limit to surprises that cleverly and ingeniously designed experiments can and will bring us. Theory and experiment will feed each other and bring us towards greater and greater progress. And this is a reason why we insist on our road of scientific investigation, in spite of being aware that we shall never come to the end, and never grasp the whole landscape.

And yet, at the current stage of our progress a nice picture is starting to emerge before our eyes. We have found that there is much more to geometry than it is usually believed. By fully employing the powerful language of Clifford algebra we have arrived at a very rich geometrical structure which offers an elegant formulation of current fundamental theories in a unified way, and also provides a natural generalization. I sincerely hope that I have succeeded in showing the careful reader, if nothing more, at least a glimpse of that magnificent picture without which, in my opinion, there can be no full insight into contemporary fundamental theoretical physics.

Appendix A
The dilatationally invariant system of units

> *That an electron here has the same mass as an electron there is also a triviality or a miracle. It is a triviality in quantum electrodynamics because it is assumed rather than derived. However, it is a miracle on any view that regards the universe as being from time to time "reprocessed".*
> —Charles W. Misner, Kip S. Thorne and John Archibald Wheeler[1]

We shall show how all the equations of physics can be cast in the system of units in which $\hbar = c = G = 4\pi\epsilon_0 = 1$. In spite of its usefulness for all sorts of calculations such a sytem of units is completely unknown.

Many authors of modern theoretical works use the system of units in which either $\hbar = c = 1$ or $c = G = 1$, etc.. This significantly simplifies equations and calculations, since various inessential $\hbar^3$, c^2, etc., are no longer present in formal expressions. But I have never seen the use of the next step, namely the units in which "all" fundamental constant are 1, that is $\hbar = c = G = 4\pi\epsilon_0 = 1$. Let us call such a system *the dilatationally invariant system of units*, briefly, the system D. It is introduced with the aid of the fine structure constant α, the Planck mass M_P, the Planck time T_P and the Planck length L_P by setting $\hbar = c = G = 4\pi\epsilon_0 = 1$ in the usual MKSA expression for these quantities (Table A.1). That is, in the system D all quantities are expressed relative to the Planck units, which are dimensionless; the unit is 1. For practical reasons sometimes we will formally add the symbol D: so there holds $1 = 1D$. With the aid of the formulas in Table A.1 we obtain the relation between the units MKSA and the units D (Table A.2).

---

[1] See ref. [136]

Table A.1. Physical constants in two systems of units

Description	Symbol	MKSA	D
Planck's constant/2π	$\hbar$	$1.0545887 \times 10^{-34}$ Js	1
Speed of light	c	2.99792458×10^{8} ms^{-1}	1
Gravitational constant	G	6.6720×10^{-11} kg^{-1}m^{3}s^{-2}	1
Dielectric constant of vacuum	ϵ_0	$8.8541876 \times 10^{-12}$ kg^{-1}A^{2}s^{4}m^{-3}	$\frac{1}{4\pi}$
Induction constant of vacuum	μ_0	1.2566371×10^{-6} kg m s^{-2}A^{-2}	4π
Electron's charge	e	$1.6021892 \times$ A s	$\alpha^{1/2}$
Electron's mass	m_e	9.109534×10^{-31} kg	$\kappa_0 \alpha^{1/2}$
Boltzman constant	k_B	1.380622×10^{-23} J/°K	1
	Fine structure constant	$\alpha = 1/137.03604$	
	'Fundamental scale'	$\kappa_0 = 0.489800 \times 10^{-21}$	
$e = \alpha^{1/2}(4\pi\epsilon_0 \hbar c)^{1/2}$	$M_P = (\hbar c/G)^{1/2}$	$T_P = (\hbar G/c^5)^{1/2}$	$L_P = (\hbar G/c^3)^{1/2}$

Let us now investigate more closely the system D. The Planck length is just the Compton wavelength of a particle with the Planck mass M_P

$$L_P = \frac{\hbar}{M_P c} \text{ (system MKSA)}, \qquad L_P = \frac{1}{M_P} = M_P = 1 \text{ (system D)} \quad (A.1)$$

In addition to M_P and L_P we can introduce

$$M = \alpha^{1/2} M_P, \qquad L = \alpha^{1/2} L_P. \quad (A.2)$$

In the system D it is

$$e = M = L = \alpha^{1/2}. \quad (A.3)$$

The length L is the classical radius that a particle with mass M and charge e would have:

$$L = \frac{e^2}{4\pi\epsilon_0 M c^2} \text{ (system MKSA)}, \qquad L = \frac{e^2}{M} \text{ (system D)}. \quad (A.4)$$

The classical radius of electron (a particle with the mass m_e and the charge e) is

$$r_c = \frac{e^2}{4\pi\epsilon_0 m_e c^2} \text{ (system MKSA)}, \qquad r_c = \frac{e^2}{m_e} \text{ (system D)}. \quad (A.5)$$

From (A.4) and (A.5) we have in consequence of (A.1) and Table A.1

$$\frac{r_c}{L} = \frac{M}{m_e} = \frac{e}{m_e}(4\pi\epsilon_0 G)^{-1/2} \equiv \kappa_0^{-1} \quad (A.6)$$

Therefore

$$m_e = \kappa_0 M = \kappa_0 \alpha^{1/2} = \kappa_0 e \text{ (system D)}. \quad (A.7)$$

APPENDIX A: The dilatationally invariant system of units

The ratio L/r_c represents the scale of the electron's classical radius relative to the length L. As a consequence of (A.4)–(A.7) we have $r_c = \kappa_0^{-1} e$.

Table A.2. Translation between units D and units MKSA

$$1D = (\hbar c/G)^{1/2} = 2.1768269 \times 10^{-8} \text{ kg}$$
$$1D = (\hbar G/c^5)^{1/2} = 5.3903605 \times 10^{-44} \text{ s}$$
$$1D = (\hbar G/c^3)^{1/2} = 1.6159894 \times 10^{-35} \text{ m}$$
$$1D = (4\pi\epsilon_0 \hbar c)^{1/2} = \alpha^{-1/2} e = 1.8755619 \times 10^{-18} \text{ As}$$
$$1D = c^3(4\pi\epsilon_0/G)^{1/2} = 3.47947723^{25} \text{ A}$$
$$1D = c^2(4\pi\epsilon_0 G)^{-1/2} = 1.0431195 \times 10^{27} \text{ V}$$
$$1D = c^2(\hbar c/G)^{1/2} = 1.9564344^9 \text{ J}$$
$$1D = 1.41702 \times 10^{32} \, {}^\circ\text{K}$$

At this point let us observe that the fundamental constants $\hbar$, c, G and ϵ as well as the quantities L_P, M_P, T_P, L, M, e, are by definition invariant under dilatations. The effect of a dilatation on various physical quantities, such as the spacetime coordinates x^μ, mass m, 4-momentum p_μ, 4-force f^μ, and 4-acceleration a^μ is [137]–[139]:

$$
\begin{aligned}
x^\mu &\to x'^\mu = \rho x^\mu, \\
p_\mu &\to p'_\mu = \rho^{-1} p_\mu, \\
m &\to m' = \rho^{-1} m, \\
f^\mu &\to f'^\mu = \rho^{-2} f^\mu, \\
a^\mu &\to a'^\mu = \rho^{-1} a^\mu.
\end{aligned}
\quad (A.8)
$$

Instead of the *inhomogeneous coordinates* x^μ one can introduce [137]–[139] the *homogeneous coordinates* $\tilde{x}^\mu = \kappa x^\mu$ which are invariant under dilatations provided that the quantity κ transforms as

$$\kappa \to \kappa' = \rho^{-1}\kappa. \quad (A.9)$$

For instance, if initially $x^0 = 1$ sec, then after applying a dilatation, say by the factor $\rho = 3$, we have $x'^0 = 3$ sec, $\kappa = \frac{1}{3}$, $\tilde{x}'^0 = \tilde{x}^0 = 1$ sec. The quantity κ is the *scale* of the quantity x^μ relative to the corresponding invariant quantity $\tilde{x}^\mu$.

If we write a given equation we can check its consistency by comparing the dimension of its left hand and right hand side. In the MKSA system the dimensional control is in checking the powers of meters, kilograms, seconds and ampères on both sides of the equation. In the system D one has to verify that both sides transform under dilatations as the same power of

Table A.3. Some basic equations in the two systems of units

Description	Symbol	MKSA	D
Electric force between two electrons	F_e	$\dfrac{e^2}{4\pi\epsilon_0 r^2}$	$\dfrac{e^2}{r^2}$
Electric force between two electrons at the distance r_c	F_e	$\dfrac{e^2}{4\pi\epsilon_0 r_c^2}$	κ_0^2
Gravitational force between two electrons at the distance r_c	F_G	$\dfrac{\kappa^{-2} G m_e^2}{r_c^2}$	$\kappa^{-2}\kappa_0^4$
Ratio between electric and gravitational force	$\dfrac{F_e}{F_G}$	$\dfrac{e^2}{4\pi\epsilon_0 G m_e^2}$	$\kappa^2\kappa_0^{-2}$
Bohr radius	a_0	$\dfrac{4\pi\epsilon_0 \hbar^2}{m_e e^2}$	$\kappa_0^{-1} e^{-3}$
Potential energy of electron at the distance a_0 from the centre	E_c	$\dfrac{e^2}{4\pi\epsilon_0 a_0}$	$\kappa_0 e^5$
Rydberg constant	Ry	$\dfrac{m_e e^4}{2(4\pi\epsilon_0 \hbar)^2}$	$\dfrac{1}{2}\kappa_0 e^5$

ρ. For instance, eq. (A.7) is consistent, since $[m_e] = \rho^{-1}$, $[\kappa_0] = \rho^{-1}$ and $[e] = 1$, where $[A]$ denotes the dimension of a generic quantity A.

In Table A.3 some well known equations are written in both systems of units. They are all covariant under dilatations. Taking G invariant, the equation $F = Gm^2/r^2$ is not dilatationally covariant, as one can directly check from (A.8). The same is true for the Einstein equations $G_{\mu\nu} = -8\pi G T^{\mu\nu}$ with $T^{\mu\nu} = (\rho + p)u^\mu u^\nu - p g^{\mu\nu}$, from which the Newtonian gravitation equation is derivable. Usually this non-covariance is interpreted as the fact that the gravitational coupling constant G is not dimensionless. One can avoid this difficulty by using the homogeneous coordinates $\tilde{x}^\mu$ and express the Einstein tensor $G^{\mu\nu}$, the rest mass density ρ and all other relevant quantities in terms of these homogeneous coordinates [139, 140]. Then the Einstein equations become $\tilde{G}^{\mu\nu} = -8\pi G \tilde{T}^{\mu\nu}$ with $\tilde{T}^{\mu\nu} = (\tilde{\rho} + \tilde{p})\tilde{u}^\mu \tilde{u}^\nu - \tilde{p}\tilde{g}^{\mu\nu}$, where the quantities with tildes are invariant under dilatations. If the homogeneous Einstein equations are written back

APPENDIX A: The dilatationally invariant system of units

in terms of the inhomogeneous quantities, we have $G^{\mu\nu} = -8\pi G \kappa^{-2} T^{\mu\nu}$, which is covariant with respect to dilatations. If we choose $\kappa = 1$ then the equations for this particular choice correspond to the usual Einstein equations, and we may use either the MKSA system or the D system. Further discussion of this interesting and important subject would go beyond the scope of this book. More about the dilatationally and conformally covariant theories the reader will find in refs. [137]–[141].

Using the relations of Table A.2 all equations in the D system can be transformed back into the MKSA system. Suppose we have an equation in the D system:

$$a_0 = \frac{1}{m_e e^2} = \kappa_0^{-1} e^{-3} = \kappa_0^{-1} \alpha^{-3/2} = \kappa_0^{-1} \alpha^{-3/2} \, \text{D}. \qquad (A.10)$$

We wish to know what form the latter equation assumes in the MKSA system. Using the expression for the fine structure constant $\alpha = e^2 (4\pi \epsilon_0 \hbar c)^{-1}$ and eq. (A.6) we have

$$a_0 = \frac{e}{m_e (4\pi \epsilon_0 G)^{1/2}} \left(\frac{e^2}{4\pi \epsilon_0 \hbar c} \right)^{-3/2} \text{D}. \qquad (A.11)$$

If we put 1D = 1 then the right hand side of the latter equation is a dimensionless quantity. If we wish to obtain a quantity of the dimension of length we have to insert $1\text{D} = (\hbar G/c^3)^{1/2}$, which represents translation from meters to the D units. So we obtain

$$a_0 = \frac{4\pi \epsilon_0 \hbar^2}{m_e e^2}, \qquad (A.12)$$

which is the expression for Bohr's radius.

Instead of rewriting equations from the D system in the MKSA system, we can retain equations in the D system and perform all the algebraic and numerical calculations in the D units. If we wish to know the numerical results in terms of the MKSA units, we can use the numbers of Table A.2. For example,

$$a_0 = \kappa_0^{-1} e^{-3} = 2.04136 \times 10^{21} \times 137.03604^{3/2} \, \text{D}. \qquad (A.13)$$

How much is this in meters? From Table A.2 we read $1\,\text{D} = 1.615989^{-35}$ m. Inserting this into (A.13) we have $a_0 = 0.529177 \times 10^{-10}$ m which is indeed the value of Bohr's radius.

Equations in the D system are very simple in comparison with those in the MKSA system. Algebraic calculations are much easier, since there are no inessential factors like $\hbar^2$, c^3, etc., which obsure legibility and clarity of equations. The transformation into the familiar MKSA units is quick with the aid of Table A.2 (and modern pocket calculators, unknown in the older times from which we inherit the major part of present day physics). However, I do not propose to replace the international MKSA system with the D system. I only wish to recall that most modern theoretical works do not use the MKSA system and that it is often very tedious to obtain the results in meters, seconds, kilograms and ampères. What I wish to point out here is that even when the authors are using the units in which, for example, $\hbar = c = 1$, or similar, we can easily transform their equations into the units in which $\hbar = c = G = 4\pi\epsilon_0 = 1$ and use Table A.2 to obtain the numerical results in the MKSA system.

To sum up, besides the Planck length, Planck time and Planck mass, which are composed of the fundamental constants $\hbar$, c and G, we have also introduced (see Table A.2) the corresponding electromagnetic quantity, namely the charge $E_P = (4\pi\epsilon_0 \hbar c)^{1/2}$ (or, equivalently, the current and the potential difference), by bringing into play the fundamental constant ϵ_0. We have then extended *the Planck system* of units [136] in which $c = \hbar = G = 1$ to the system of units in which $c = \hbar = G = 4\pi\epsilon_0 = 1$, in order to incorporate all known sorts of physical quantities.

Finally, let me quote the beautiful paper by Levy-Leblond [142] in which it is clearly stated that our progress in understanding the unity of nature follows the direction of eliminating from theories various (inessential) numerical constants with the improper name of "fundamental" constants. In fact, those constants are merely the constants which result from our unnatural choice of units, the choice due to our incomplete understanding of the unified theory behind.

References

[1] V. Fock, *Phys. Z. Sowj.* **12**, 404 (1937).

[2] E.C.G. Stueckelberg, *Helv. Phys. Acta*, **14**, 322 (1941); **14**, 588 (1941); **15**, 23 (1942)

[3] R. P. Feynman *Phys. Rev*, **84**, 108 (1951)

[4] J. Schwinger, *Phys. Rev*, **82**, 664 (1951)

[5] W.C. Davidon, *Physical Review* **97**, 1131 (1955); **97**, 1139 (1955)

[6] L.P. Horwitz and C. Piron, *Helv. Phys. Acta*, **46**, 316 (1973); L.P. Horwitz and F. Rohrlich, *Physical Review D* **24**, 1528 (1981); **26**, 3452 (1982)

[7] L.P. Horwitz, R.I. Arshansky and A.C. Elitzur *Found. Phys* **18**, 1159 (1988); R.I. Arshansky, L.P. Horwitz and Y. Lavie, *Foundations of Physics* **13**, 1167 (1983)

[8] H.Enatsu, *Progr. Theor. Phys* **30**, 236 (1963); *Nuovo Cimento A* **95**, 269 (1986);

[9] E. R. Collins and J.R. Fanchi, *Nuovo Cimento A*, **48**, 314 (1978);

[10] J.R.Fanchi, *Phys. Rev. D* **20**, 3108 (1979); see also the review J.R.Fanchi, *Found. Phys.* **23**, 287 (1993), and many references therein; J.R. Fanchi *Parametrized Relativistic Quantum Theory* (Kluwer, Dordrecht, 1993)

[11] A. Kyprianidis *Physics Reports* **155**, 1 (1987)

[12] F. Reuse, *Foundations of Physics* **9**, 865 (1979)

[13] R. Kubo, *Nuovo Cimento A* , 293 (1985);

[14] N. Shnerb and L.P. Horwitz, *Phys. Rev. A* **48**, 4068 (1993).

[15] M. Pavšič, *Found. Phys.* **21**, 1005 (1991)

[16] M. Pavšič, *Nuovo Cim.* **A104**, 1337 (1991); *Doga, Turkish Journ. Phys.* **17**, 768 (1993)

[17] M.B. Mensky and H. von Borzeszkowski, *Physics Letters A* **208**, 269 (1995); J.P. Aparicio, F.H. Gaioli and E.T. Garcia-Alvarez, *Physical Review A* **51**, 96 (1995); *Physics Letters A* **200**, 233 (1995); L. Hannibal, *International Journal of Theoretical Physics* **30**, 1445 (1991)

[18] W.B. Rollnick, *Physical Review* **183**, 1105 (1969); F.A.E. Pirani, *Physical Review D* **1**, 3224 (1970); R.G. Root, J.S. Trefil, *Letter al Nuovo Cimento* **3**, 412 (1970); L.S. Schulman, *American Journal of Physics* **39**, 481 (1971); G.A. Benford, D.L. Book, W.A. Newcomb, *Physical Review D* **2**, 263 (1970)

[19] L.P. Horwitz, in *Old and New Questions in Physics, Cosmology, Phylosophy and Theoretical Biology* (Editor Alwyn van der Merwe, Plenum, New York, 1983); L.P. Horwitz and Y. Lavie, *Physical Review D* **26**, 819 (1982); see also ref. [6]

[20] L. Burakovsky, L.P. Horwitz and W.C. Schieve, *Physical Review D* **54**, 4029 (1996); L.P. Horwitz and W.C. Schieve, *Annals of Physics* **137**, 306 (1981)

[21] B.S. DeWitt, *Phys. Rep.* **19**, 295 (1975); S.M. Christensen, *Phys. Rev. D* **14**, 2490 (1976); L.S. Brown, *Phys. Rev. D* **15**, 1469 (1977); T.S. Bunch and L. Parker, *Phys. Rev. D* **20**, 2499 (1979); H. Boschi-Filho and C.P. Natividade, *Phys. Rev. D* **46**, 5458 (1992); A. Follacci, *Phys. Rev. D* **46**, 2553 (1992); A. Sugamoto, *Nuclear Physics B* **215**, 381 (1983)

[22] D. Hestenes, *Space–Time Algebra* (Gordon and Breach, New York, 1966); D. Hestenes *Clifford Algebra to Geometric Calculus* (D. Reidel, Dordrecht, 1984)

[23] W.M. Pezzaglia Jr, *Classification of Multivector Theories and Modification of the Postulates of Physics*, e-Print Archive: gr-qc/9306006;
W.M. Pezzaglia Jr, *Polydimensional Relativity, a Classical Generalization of the Automorphism Invariance Principle*, e-Print Archive: gr-qc/9608052;
W.M. Pezzaglia Jr, *Physical Applications of a Generalized Clifford Calculus: Papapetrou Equations and Metamorphic Curvature*, e-Print Archive: gr-qc/9710027;
W.M. Pezzaglia Jr nad J. J. Adams, *Should Metric Signature Matter in Clifford Algebra Formulation of Physical Theories?*, e-Print Archive: gr-qc/9704048;
W.M. Pezzaglia Jr and A.W. Differ, *A Clifford Dyadic Superfield from Bilateral Interactions of Geometric Multispin Dirac Theory*, e-Print Archive: gr-qc/9311015;
W.M. Pezzaglia Jr, *Dimensionally Democratic Calculus and Principles of Polydimensional Physics*, e-Print Archive: gr-qc/9912025

[24] W.M. Pezzaglia Jr, *Generalized Fierz Identities and the Superselection Rule for Geometric Multispinors*, e-Print Archive: gr-qc/9211018;
W.M. Pezzaglia Jr, *Polydymensional Supersymmetric Principles*, e-Print Archive: gr-qc/9909071

[25] B.S. DeWitt, in *Gravitation: An Introduction to Current Research* (Editor L. Witten, Wiley, New York, 1962)

[26] C. Rovelli, *Classical and Quantum Gravity* **8**, 297 (1991); **8** 317 (1991)

[27] M. Riesz, in *Comptes Rèndus du Dixiéme Congrés des Mathematiques des Pays Scandinaves* (Copenhagen, 1946), p. 123; M. Riesz, in *Comptes Réndus du Dixième Congrès des Mathematiques des Pays Scandinaves* (Lund, 1953), p. 241; M. Riesz,

in *Lecture Series No 38* (University of Maryland, 1958), Chaps. I-IV; see also refs. [37]; P.Rashevskiy, *Uspekhi Matematicheskikh Nauk* **10**, 3, (1955)

[28] D. Cangemi, R. Jackiw and B. Zwiebach, *Annals of Physics* **245**, 408 (1996); E. Benedict, R. Jackiw and H.-J. Lee, *Physical Review D* **54**, 6213 (1996)

[29] P.A.M. Dirac, *Lecture Notes on Quantum Field Theory* (Yeshiva University, New York, 1965)

[30] H. Rund, *The Hamilton–Jacobi Theory in the Calculus of Variations* (Van Nostrand, London, 1966)

[31] P.S. Howe and R.W. Tucker, *J. Phys. A: Math. Gen.* **10**, L155 (1977); A. Sugamoto, *Nuclear Physics B* **215**, 381 (1983); E. Bergshoeff, E. Sezgin and P.K. Townsend, *Physics Letter B* **189**, 75 (1987); A. Achucarro, J.M. Evans, P.K. Townsend and D.L. Wiltshire, *Physics Letters B* **198**, 441 (1987); M. Pavšič, *Class. Quant. Grav.* **5**, 247 (1988); see also M. Pavšič, *Physics Letters B* **197**, 327 (1987)

[32] See, e.g., M. Kaku, *Introduction to Superstrings* (Springer-Verlag, New York, N.Y., 1988).

[33] B.S. DeWitt, *Reviews of Modern Physics* **29**, 377 (1957); see also B.S. DeWitt *Phys. Rev.* **85**, 653 (1952)

[34] A.O. Barut, *Electrodynamics and Classical Theory of Fields and Particles* (Macmillan, New York, 1964)

[35] I. Białynycki-Birula and Z. Białynicki-Birula, *Quantum Electrodynamics* (Pergamon Press, Oxford, 1975)

[36] K. Greider, *Physical Review Letters* **44**, 1718 (1980); *Foundations of Physics* **4**, 467 (1984); D. Fryberger, *Foundations of Physics* **19**, 125 (1989); J. Keller, *International Journal of Theoretical Physics* **30**, 137 (1991); V. de Sabbata and L. Ronchetti, *Foundations of Physics* **29**, 1000 (1999); C. Doran, A. Lasenby and S. Gull, *Foundations of Physics* **23**, 1175 (1993); **23**, 1239 (1993); **23**, 1295 (1993); W.A. Rodrigues, Jr., J. Vaz and E. Recami *Foundations of Physics* **23**, 469 (1993); J.R. Zeni and W.A. Rodrigues, Jr, *International Jornal of Modern Physics A* **7**, 1793 (1992); W.A. Rodrigues, Jr., J. Vaz and M. Pavšič, *The Clifford bundle and the dynamics of superpaticle*, in *Generalization of Complex Analysis and their Applications in Physics* (Editor J. Lawrynowicz, Banach Center Publication **37**, The Institute of Mathematics, Polish Academy of Sciences, Warszawa, 1996); P. Lounesto, *Clifford Algebras and Spinors* (Cambridge University Press, Cambridge, 1997); A. Garett Lisi, *It's all in GR: spinors, time, and gauge symmetry*, e-Print Archive: gr-qc/9804033; C. Doran, A. Lasenby, A. Challinors and S. Gull, *Journal of Mathematical Physics* **39**, 3303 (1998)

[37] S. Teitler, *Supplemento al Nuovo Cimento* **III**, 1 (1965); *Supplemento al Nuovo Cimento* **III**, 15 (1965); *Journal of Mathematical Physics* **7**, 1730 (1966); *Journal of Mathematical Physics* **7**, 1739 (1966)

[38] B. Hatfield, *Quantum Field Theory of Point Particles and Strings* (Adison-Wesley, Redwood City, 1992)

[39] J. Greenstie, *Classical and Quantum Gravity* **13**, 1339 (1996); *Physical Review D* **49**, 930 (1994); A. Carlini and J. Greensite, *Physical Review D* **52**, 936 (1995); **52**, 6947 (1955); **55**, 3514 (1997)

[40] F.H. Gaioli and E.T. Garcia-Alvarez, *General Relativity and Gravitation* **26**,1267 (1994)

[41] J. Brian and W.C. Schieve, *Foundations of Physics* **28**, 1417 (1998)

[42] See modern discussion in ref [26]

[43] S. Weinberg, *Review of Modern Physics* **61**, 1 (1989)

[44] Y.S. Kim and Marilyn E. Noz, *Phys. Rev. D* **8**, 3521 (1973) 12 (1975) 122.

[45] D. Cangemi, R. Jackiw and B. Zwiebach, *Annals of Physics* **245**, 408 (1996); E. Benedict, R. Jackiw and H.-J. Lee, *Phys. Rev. D* **54**, 6213 (1996)

[46] R P. Feynman, M. Kislinger and F. Ravndal, *Phys. Rev. D* **3**, 2706 (1971); H. Leutwyler and J. Stern, *Annals of Physics* **112**, 94 (1978)

[47] S.W. Hawking and G.F.R. Ellis, *The Large Scale Structure of Spacetime* (Cambridge University Press, Cambridge, 1973); F. J. Tipler, *Annals of Physics* **108**, 1 (1977)

[48] M.S. Morris, K.S. Thorne, and U. Yurtsever, *Phys. Rev. Lett.* **61**, 1446 (1988)

[49] M. Alcubierre, *Class. Quant. Grav.* **11**, L73 (1994)

[50] H.B.G. Casimir, *Proc. K. Ned. Akad. Wet.* **51**, 793 (1948) 793

[51] B. Green, *The Elegant Universe* (Norton, New York, 1999)

[52] E. Witten, *Physics Today*, April, 24 (1996)

[53] M. Pavšič, *Found. Phys.* **25**, 819 (1995);
M. Pavšič, *Nuovo Cimento A* **108**, 221 (1995)

[54] M. Pavšič, *Foundations of Physics* **26**, 159 (1996)

[55] M. Pavšič, *Nuovo Cimento A* **110**, 369 (1997)

[56] K. Bardakci, *Nucl. Phys.* **B271**, 561 (1986)

[57] B.S. DeWitt, *Supermanifolds* (Cambridge University Press, Cambridge 1984)

[58] K. Fujikawa, *Phys. Rev. Lett.* **42**, 1195 (1979); **44**, 1733 (1980); *Phys. Rev. D* **21**, 2448 (1980); **23**, 2262 (1981); *Nucl. Phys. B* **226**, 437 (1983); **245**, 436 (1984); for a review see M. Basler, *Fortschr. Phys.* **41**, 1 (1993).

[59] V. Moncrief and C. Teitelboim, *Physical Review D* **6**, 966 (1972)

[60] M. Pavšič, *Class. Quant. Grav.* **5**, 247 (1988)

[61] M. Pavšič, *Class. Quant. Grav.* **9**, L13 (1992)

REFERENCES

[62] M. Pavšič, *Nuovo Cimento A* **108**, 221 (1995)

[63] A. Schild, *Phys.Rev.* **D16**, 1722 (1977)

[64] A.O. Barut and M. Pavšič, *Modern Physics Letters A* **7**, 1381 (1992)

[65] A.O. Barut and M. Pavšič, *Physics Letters B* **306**, 49 (1993); **331**, 45 (1994)

[66] P.A.M. Dirac, *Proc. R. Soc. (London) A* **268**, 57 (1962)

[67] J.L. Synge, *Relativity: The General Theory* (North-Holland, Amsterdam, 1964)

[68] L.Smolin, *Physical Review D* **62**, 086001 (2000)

[69] M. Gaul and C. Rovelli, *Loop Quantum Gravity and the Meaning of Diffeomorphism Invariance*, e-Print Archive: gr-qc/9910079; C. Rovelli, *The Century of the Incomplete Revolution: Searching for General Relativistic Quantum Field Theory*, hep-th/9910131

[70] C. Rovelli and L. Smollin, *Nuclear Physics B* **331**, 80 (1990); *Nuclear Physics B* **442**, 593 (1995); Erratum: *Nuclear Physics B* **456**, 734 (1995);

[71] C. Rovelli and L. Smollin, *Physical Review D* **52**, 5743 (1995)

[72] J.C. Baez, *Classical and Quantum Gravity* **15**, 1827 (1998)

[73] See e.g. Vera C. Rubin, *Scientific America* **248**, 88 (1983); Vera C. Rubin, W. K. Ford, Jr. and N. Thonnard, *The Astrophysical Journa* **261**, 439 (1982); P. J. E. Peebles, *Nature* **321**,27 (1986)

[74] A. Ashtekar, *Physical Review Letters* **57**, 2244 (1986)

[75] C. Castro, *The String Uncertainty Relations follow from the New Relativity Principle*, e-Print Archive: hep-th/0001023;
C. Castro, *Is Quantum Spacetime Infinite Dimensional?*, e-Print Arhive: hep-th/0001134;
C. Castro, *Chaos, Solitons and Fractals* **11**, 1721 (2000);
C. Castro and A. Granik, *On M Theory, Quantum Paradoxes and the New Relativity*, e-Print Archive: physics/0002019

[76] C. Castro, *p-Branes as Composite Antisymmetric Tensor Field Theories*, e-ptint Archive: hep/9603117;
C. Castro, *p-Brane Quantum mechanical Wave Equations*, e-Print Archive: hep/th/9812189;
C. Castro, *The Search for the Origin of M-Theory: Loop Quantum Mechanics, Loops/Strings and Bulk/Boundary Dualities*, e-Print Archive: hep-th/9809102;
C. Castro, *Chaos, Solitons and Fractals* **11**, 1721 (2000)

[77] See e.g. M.B. Green, *Classical and Quantum Gravity* **16**, A77 (1999), and references therein

[78] A. Aurilia, A. Smailagic and E. Spallucci, *Physical Review D* **47**,2536 (1993); A. Aurilia and E. Spallucci, *Classical and Quantum Gravity* **10**, 1217 (1993); A. Aurilia, E. Spallucci and I. Vanzetta, *Physical Review D* **50**, 6490 (1994); S.

Ansoldi, A. Aurilia and E. Spallucci, *Physical Review D* **53**, 870 (1996); S. Ansoldi, A. Aurilia and E. Spallucci, *Physical Review D* **56**, 2352 (1997)

[79] M. Henneaux, *Phys. Lett.* B **120**, 179 (1983)
U. Marquard and M. Scholl, *Phys. Lett.* B **209**, 434 (1988)

[80] M. Kaku: *Introduction to Superstrings*, (Springer-Verlag, New York, 1988); J. Hughes, J. Liu and J. Polchinski, *Phys. Lett.* B **180**, 370 (1986); E. Bergshoeff, E. Sezgin and P. K. Townsend, *Phys. Lett.* B **189**, 75 (1987); **209** (1988); E. Bergshoeff, E. Sezgin and P. K. Townsend, *Annals of Physics* **185**, 330 (1988); M.P. Blencowe and M. J. Duff, *Nucl. Phys.* B **310**, 387 (1988); U. Marquard and M. Scholl, *Phys. Lett.* B **209**, 434 (1988); A. Karlhede and U. Lindström, *Phys. Lett.* B **209**, 441 (1988); M. Duff, *Class. Quant. Grav.* **6**, 1577 (1989); A.A. Bytsenko and S. Zerbini, *Mod. Phys. Lett.* A **8**, 1573 (1993); A. Bytsenko and S. Odintsov, *Fortsch. Physik* **41**, 233 (1993)

[81] J. Hong, J. Kim and P. Sikivie, *Phys.Rev.Lett.*, **69**, 2611 (1992)

[82] T. Eguchi, *Phys. Rev. Lett.* **44**, 126 (1980)

[83] S.N. Roshchupkin and A. A. Zheltukhin, *On a Possibility of Membrane Cosmology*, e-Print Archive hep-th/9607119

[84] J. Schwarz, *Lett. Math. Phys.* **34**, 309 (1995); J.A. Harvey and A. Strominger, *Nucl. Phys.* B **449**, 535 (1995); A. Sen, *Phys.Let.* B **329**, 217 (1994); *Nucl. Phys.* B **450**, 103 (1995)

[85] I.A. Batalin and G. A. Vilkovisky, *Nuclear Physics B* **234**, 106 (1984) G. A. Vilkovisky, *Nuclear Physics B* **234**, 125 (1984); Hing Tong Cho, *Physical Review D* **40**, 3302 (1989); W. F. Kao and Shih-Yuin Lin, *Modern Physics Letters A* **13**, 26 (1998)

[86] R. Jackiw, *The Unreasonable Effectiveness of Quantum Field Theory*, 4th Workshop in High Energy Phenomenology (WHEPP 4), Partha Ghose: S.N. Bose National Center for Basic Sciences: Salt Lake, Calcutta 700 064, India, 2–14 Jan 1996, Calcutta, India and Conference on Foundations of Quantum Field Theory, Boston, Mass., 1-3 Mar 1996. In *Boston 1996, Conceptual foundations of quantum field theory* 148-160, e-Print Archive: hep-th/9602122

[87] I.V. Kanatchikov, *On quantization of field theories in polymomentum variables*, in: *Particles, Fields and Gravitation* (Proc. Int. Conf. devoted to the memory of Prof. R. Rączka, Łodz, Poland, Apr. 1998) ed. J. Rembieliński, *AIP Conf. Proc.* vol. **453** (1998) 356–67, e-Print Archive hep-th/9811016; I.V. Kanatchikov, *De Donder-Weyl theory and a hypercomplex extension of quantum mechanics to field theory*, *Rep. Math. Phys.* **43** (1999) 157-70, e-Print Archive hep-th/9810165; I.V. Kanatchikov, *Canonical Structure of Classical Field Theory in the Polymomentum Phase Space*, *Rep. Math. Phys.* **41** (1998) 49–90, e-Print Archive hep-th/9709229

[88] M. Pavšič, *Foundations of Physics* **24**, 1495 (1994)

[89] M. Pavšič, *Gravitation & Cosmology* **2**, 1 (1996)

[90] M. Pavšič, *Physics Letters A* **116**, 1 (1986)

REFERENCES

[91] I.A. Bandos, *Modern Physics Letters A* **12**, 799 (1997); I. A. Bandos and W. Kummer, hep-th/9703099

[92] A.Sugamoto, *Nuclear Physics B* **215**, 381 (1983)

[93] R.Floreanini and R.Percacci, *Modern Physics Letters A* **5**, 2247 (1990)

[94] K. Akama, in *Proceedings of the Symposium on Gauge Theory and Gravitation*, Nara, Japan, (Editors K. Kikkawa, N. Nakanishi and H. Nariai, Springer-Verlag, 1983), e-Print Archive: hep-th/0001113; V.A. Rubakov and M. E. Shaposhnikov, *Physics Letters B* **125**, 136 (1983); **125**, 139 (1983); M. Visser, *Physics Letters B* **167**, 22 (1985); E.J. Squires, *Physics Letters B* **167**,286 (1985); G.W. Gibbons and D. L. Wiltshire, *Nuclear Physics B* **287**, 717 (1987)

[95] L. Randall and R. Sundrum, *Physical Review Letters* **83**, 3370 (1999); **83**, 4690 (1999)

[96] A. D. Sakharov, *Dok. Akad. Nauk. SSSR* **177**, 70 (1967) [*Sov. Phys. JETP* **12**, 1040 (1968)].

[97] S.L. Adler, *Rev. Mod. Phys.* **54**, 729 (1982), and references therein.

[98] K. Akama, *Progr. Theor. Phys.* **60**, 1900 (1978); **79**, 1299 (1988); K. Akama and H. Terazawa, *Prog. Theor. Phys.* **79**, 740 (1988); H. Terazawa, in *Proceedings of the First International A. D. Sakharov Conference on Physics*, L. V. Keldysh et al., eds. (Nova Science, New York, 1991); S. Naka and C. Itoi, *Progr. Theor. Phys.*, **70**, 1414 (1983).

[99] T. Regge and C. Teitelboim, in *Proceedings of the Marcel Grossman Meeting* (Trieste, 1975)

[100] M. Pavšič, *Class. Quant. Grav.* **2**, 869 (1985); *Phys. Lett.* **A107**, 66 (1985); V. Tapia, *Class. Quant. Grav.* **6**, L49 (1989); D.Maia, *Class. Quant. Grav.* **6**, 173 (1989)

[101] M. Pavšič, *Phys. Lett.* **A116**, 1 (1986); *Nuov. Cim.* **A95**, 297, (1986)

[102] K. Fujikawa, *Phys. Rev. Lett.* **42**, 1195 (1979); **44**, 1733 (1980); *Phys. Rev. D* **21**, 2448 (1980); **23**, 2262 (1981); *Nucl. Phys. B* **226**, 437 (1983); **245**, 436 (1984); for a review see M. Basler, *Fortschr. Phys.* **41**, 1 (1993).

[103] See e.g. B.S. De Witt, *Phys. Reports* **19**, 295 (1975); S. M. Christensen, *Phys. Rev.* **D14**, 2490 (1976); N. D. Birrell and P.C.W. Davies, *Quantum Fields in Curved Space* (Cambridge University Press, Cambridge, 1982); H. Boschi-Filho and C. P. Natividade, Phys. Rev. D, **46**, 5458 (1992)

[104] G. Dvali, G. Gabadadze and M. Porrati, *Physics Letters B* **485**, 208 (2000)

[105] See e.g. M.B. Green, J. H. Schwarz and E. Witten, *Superstrings* (Cambridge University Press, Cambridge, 1987); M. Kaku [80]

[106] P.A.M. Dirac, *Proc. R. Soc. (London) A* **268**, 57 (1962); Y. Nambu, *Lectures at the Copenhagen Symposium* 1970, unpublished; T. Goto, *Progr. Theor. Phys.* **46**, 1560 (1971)

[107] H. Everett III, *Rev. Mod. Phys.* **29**, 454 (1957); "The Theory of the Universal Wave Function", in *The Many-Worlds Interpretation of Quantum Mechanics*, B.S. DeWitt and N. Graham, eds. (Princeton University Press, Princeton, 1973).

[108] K.V. Kuchař, in *Proceedings of the 4th Canadian Conference on General Relativity and Relativistic Astrophysics*, Winnipeg, Manitoba, 1991, G. Constatter, D. Vincent, and J. Williams, eds. (World Scientific, Singapore, 1992); K. V. Kuchař, in *Conceptual problems of Quantum Mechanics*, A. Ashtekar and J. Stachel, eds, (Birkhäuser, Boston, 1991); M. Hennaux and C. Teitelboim, *Phys. Lett. B* **222**, 195 (1989); W. Unruh and R. Wald, *Phys. Rev D* **40**, 2598 (1989)

[109] J.A. Wheeler, *Review of Modern Physics* **29**, 463 (1957)

[110] B.S. DeWitt, "The Everett–Wheeler Interpretation of Quantum Mechanics",, in *Battelle Rencontres* **1**, 318 (1967); *Physical Review* **160**, 113 (1967); *Physics Today* **23** 30 (1970)

[111] J.A. Wheeler, "From Relativity to Mutability", in *The Physicist's Conception of Nature* (Editor J. Mehra, D.Reidel, Dordrecht, 1973); "Is Physics Legislated by Cosmology?", in *Quantum Gravity* (Editors C. J. Isham, R. Penrose and D. W. Sciama, Clarendon Press, Oxford, 1975)

[112] D. Deutsch, *The Fabric of Reality* (Penguin Press, London, 1997)

[113] D. Deutsch, *Physical Review D* **44**, 3197 (1991)

[114] V. P. Frolov and I. D. Novikov, *Physical ReviewD* **42**, 1057 (1990); J. Friedman, M. S. Morris, I. D. Novikov and U. Yurstever, *Physical Review D* **42**, 1915 (1990)

[115] M. Pavšič, *Lettere al Nuovo Cimento* **30**, 111 (1981)

[116] J D. Barrov and F. J. Tipler, *The Anthropic Cosmological Principle* (Oxford University Press, Oxford, 1986)

[117] R.P. Feynman, *The Character of Physical Law* (MIT Press, Cambridge, MA, 1965)

[118] C. Rovelli, *International Journal of Theoretical Physics* **35**, 1637 (1966)

[119] J. Butterfield, "Words, Minds and Quanta", in *Symposium on Quantum Theory* (Liverpool, 1995)

[120] J. Von Neumann, *Mathematical Foundations of Quantum Mechanics* (Princeton University Press, Princeton, 1955)

[121] C.F. Von Wezsacker, in *The Physicist's Conception of Nature* (Editor J. Mehra, Reidal Publ. Co., Boston, 1973)

[122] E.P. Wigner, in *Symmetries and Reflections* (Indiana University Press, Bloomington, 1967)

[123] J.A. Wheeler, in *The Physicist's Conception of Nature* (Editor J. Mehra, Reidal Publ. Co., Boston, 1973)

[124] M. Tribus and E.C. McIrvine, *Scientific American* **224**, 179 (1971)

REFERENCES

[125] D.R. Hofstadter, *Gödel, Escher, Bach: an Eternal Golden Braid* (Penguin, London, 1980); D. C. Denet and D. R. Hofstadter, eds., *The Mind's Eye* (Penguin, London, 1981)

[126] R. Rucker, *Infinity and the Mind* (Penguin, London, 1995)

[127] W.H. Zurek, *Physical Review D* **24**, 1516 (1981); **26**, 1862 (1982)

[128] H.D. Zeh, *Foundations of Physics* **1** 69 (1970); E. Joos and H. D. Zeh, *Zeitschrift für Physik B* **59**, 223 (1985)

[129] E. Schrödinger, *Naturwissenschaften* **23**, 807; 823; 844 (1935); see also J. Gribbin, *In Search of Schrödinger Cat* (Wildwood House, London, 1984)

[130] H. Atmanspacher, *Journal of Consciousness Studies* **1**, 168 (1994)

[131] J. P. Terletsky, *Doklady Akadm. Nauk SSSR*, **133**, 329 (1960); M. P. Bilaniuk, V. K. Deshpande and E.C. G. Sudarshan, *American Journal of Physics* **30**, 718 (1962); E. Recami and R. Mignani, *Rivista del Nuovo Cimento* **4**, 209 (1974); E. Recami, *Rivista del Nuovo Cimento* **9**, 1 (1986)

[132] E. Recami, *Physica A* **252**, 586 (1998), and references therein

[133] J.R. Fanchi, *Foundations of Physics* **20**, 189 (1990)

[134] D. Hochberg and J. T. Wheeler, *Physical Review D* **43**, 2617 (1991)

[135] J. Piaget, *The Origin of Inteligence in the Child* (Routledge & Kegan Paul, London, 1953)

[136] C.W. Misner, K. S. Thorne and J. A. Wheeler, *Gravitation*, Freeman, San Francisco, 1973), p. 1215

[137] H.A. Kastrup, *Annaled der Physik (Lpz.)* **7**, 388 (1962)

[138] A.O. Barut and R. B. Haugen, *Annals of Physics* **71**, 519 (1972)

[139] M. Pavšič, *Nuovo Cimento B* **41**, 397 (1977); M. Pavšič, *Journal of Physics A* **13**, 1367 (1980)

[140] P. Caldirola, M. Pavšič and E. Recami, *Nuovo Cimento B* **48**, 205 (1978); *Physics Letters A* **66**, 9 (1978)

[141] M. Pavšič, *International Journal of Theoretical Physics* **14**, 299 (1975)

[142] J. M. Levy-Leblond, *Riv. Nuovo Cimento* **7**, 188 (1977)

Index

Action
  first order form, 124
Action
  first order
    in $\mathcal{M}$-space, 130
  minimal surface, 126
Active transformations, 109
Affinity, 4, 168
  in $\mathcal{M}$-space, 114
Annihilation operators, 36
Anthropic principle, 310–311
Anticommutation relations
  for a fermionic field, 227
Antisymmetric field, 192
Ashtekar variables, 187
Background independent approach, 165
Background matter, 187
Background metric, 257
  in $\mathcal{M}$-space, 118
Background of many branes, 259
Bare vacuum, 234
Basis spinors, 78, 201
  matrix representation of, 80
Basis vector fields
  action, 236
Basis vectors, 59, 77
  in $\mathcal{M}$-space, 167, 181, 183, 196
  in phase space, 223–224
  uncountable set of, 226
Bivector, 56
  in phase space, 223
Bosonic field, 230
Bosonic matter field, 281
Bradyon, 310
Brane configuration, 261
Brane, 188, 249–250
  in curved background, 249
Bulk, 250, 261
Canonical momentum, 4, 120
  for point particle, 5

  point particle, 7, 12
Canonically conjugate variables, 288
Casimir effect, 102
Castro Carlos, 201–202
Causal loops, 309–310
Causality paradoxes, 51
Causality violation, 51
Charge current density, 41
Charge density, 41
Charge operator, 42
Charged particle, 61
Classical limit, 25, 207–208
Classical spin, 69
Clifford aggregates, 57, 90, 138
Clifford algebra valued line, 201
Clifford algebra, 55, 59
  in $\mathcal{M}_{(\Phi)}$-space, 199
Clifford manifold, 200
Clifford number, 57
Clifford numbers, 55, 137
Clifford product, 56
  in infinite-dimensional space, 178
Collaps of the wave function, 315, 317, 320–321, 326–328, 333–334, 336
Conformal factor, 267
Consciousness, 317–319, 322, 325, 331, 334, 336
Conservation law, 33, 41
Conservation
  of the constraint, 122–123
Constraint, 137, 143
  for point particle, 4, 6
Contact interaction, 260, 268
Continuity equation, 18–19, 24–25, 208, 210
Coordinate conditions
  in membrane space, 214
Coordinate frame, 27
Coordinate functions
  generalized, 199
  polyvector, 197

363

Coordinate representation, 16
Coordinate transformations, 214
  active, 108
  passive, 107
Coordinates
  of $\mathcal{M}$-space, 188
Copenhagen interpretation, 313, 315
Cosmological constant, 93, 278, 282
Covariant derivative, 30, 170
  in membrane space, 114
Creation operators, 36
  for bosons, 229
  for fermions, 228
Current density, 41
Curvature scalar
  in $\mathcal{M}$-space, 181
Curvature tensor, 169
  in $\mathcal{M}$-space, 181
Dark matter, 187
DeWitt
  Dynamical theory in curved spaces, 20
Decoherence, 323, 331
Density matrix, 323–324
Derivative
  of a vector, 169
  with resepct to a polyvector, 176
  with respect to a vector, 173
Diffeomorphism invariance, 107
Diffeomorphisms
  active, 108, 186
Dirac algebra, 59
Dirac equation, 77
Dirac field, 231
Dirac field
  parametrized
    quantization of, 236
Dirac matrices, 83
Dirac–Nambu–Goto action, 125–126
Dirac–Nambu–Goto membrane, 119, 143
Distance, 54, 111
  between membrane configurations, 161
Dynamics
  in spacetime, 138
  of spacetime, 138
Effective action, 272, 275, 277–279, 283
Effective metric, 257, 260
Effective spacetime, 186, 279
Einstein equations, 165, 185, 187, 191
  in $\mathcal{M}$-space, 183
Einstein tensor
  in $\mathcal{M}$-space, 159
Electric charge density, 150
Electric current density, 150
Electromagnetic current, 60–61
Embedding space, 261
Energy, 50
Energy–momentum operator, 38

Energy–momentum, 35
Equation of motion
  for a p-brane, 255
Equations of motion
  for basis vectors, 229
Euclidean signature, 87
Everett interpretation, 293
Evolution operator, 19, 206
Evolution parameter, 9, 31, 85, 119, 140, 143, 289, 291
Evolution time, 297–298
Exotic matter, 101
Expectation value, 22, 26, 28, 43, 220
Extrinsic coordinates, 163
Feynman path integral, 19
Feynman propagator, 19
Feynman, 275
Field equation
  in $\mathcal{M}$-space, 156
  in target space, 156
Field quantization
  nature of, 238
Field theory, 31
First quantization
  in curved spacetime, 20
  in flat spacetime, 16
Fluid
  of membranes, 165
Fock space, 242–243
Fractal, 214
Frame field, 170
Functional derivative, 35, 111, 179, 181
  covariant, 114
Functional differential equations, 160, 165
Functional transformations, 111
Gödel incompletness theorem, 332
Gödel knot, 321
Gauge condition, 214
  in membrane space, 216
Gauge field, 13
  in $\mathcal{M}$-space, 146, 155
Gauge fixing relation, 119
Gauge transformation, 13
Generator, 33, 37, 41–42
  of evolution, 33
  of spacetime translations, 33
Generators
  of the Clifford algebra, 87
Geodesic equation, 4, 171, 255
  in $\mathcal{M}$-space, 115, 160
Geodesic, 261
  as the intersection of two branes, 274
  in $\mathcal{M}$-space, 114, 145
Geometric calculus, 167
  in $\mathcal{M}$-space, 130, 137
Geometric principle
  behind string theory, 107, 145

INDEX

Geometric product
  in infinite-dimensional space, 178
Gradient operator, 60–61
Grassmann numbers, 90
Gravitational constant
  induced, 278
Gravitational field, 100
  repulsive, 101
Gravity around a brane, 264
Green's function, 20, 280
  inverse, 280
Gupta–Bleuler quantization, 6
'Hamiltonian' constraint, 123
Hamilton operator, 289
  for membrane, 207
Hamilton–Jacobi equation, 25
  for membrane, 206
Hamilton–Jacobi functional, 122
Hamiltonian, 33, 38, 52, 71, 84, 94, 100, 121, 242, 296
  expectation value of, 234
  for membrane, 204
  for point particle, 4, 6, 15
  in curved space, 23
Harmonic oscillator, 93
  pseudo-Euclidean, 94
Heisenberg equation, 38, 46
  for Dirac field, 238
Heisenberg equations, 85, 204, 242
Heisenberg picture, 85
Hestenes David, 57
Hidden mass, 187
Higher-dimensional membranes, 107
Historical time, 286
Hole argument, 186
Holographic projections, 200
Howe–Tucker action, 126–127, 130, 274, 276
  for point particle, 5
Ideal, 78
Imaginary unit
  as a Clifford number, 223
Indefinite mass, 18, 33
Induced gravity, 271, 283–284
Induced metric, 250
  on hypersurface, 129
Infinite-dimensional space, 107, 111
Inner product, 55, 58
  in infinite-dimensional space, 178
Interaction
  between the branes, 252
Interactive term
  for a system of branes, 257
Intersection
  of branes, 252
Intrinsic coordinates, 163
Invariant evolution parameter, 104
Invariant volume element
  in membrane space, 113
Kaluza–Klein theory, 69
Kinematically possible objects, 114
Kinetic momentum, 73
  for point particle, 13–14
Klein–Gordon equation, 6, 19
  in curved spacetime, 25
Local Lorentz frame, 27
Localization
  within a space like region, 291
Loop quantum gravity, 166, 202
Lorentz force, 14
Mach principle, 166
Many membrane universe, 161
Many worlds interpretation, 308, 316, 336
Mass
  constant of motion, 11, 14
  fixed, 4
Matter configurations
  on the brane, 263
Matter sheet, 273
Maxwell equations, 60–61, 152
Measure in momentum space, 217
Membrane action
  polyvector generalization of, 108
  unconstrained, 142
Membrane configuration, 161, 181, 249
  spacetime filling, 185
Membrane dynamics, 152
Membrane momentum, 139
Membrane space, 107–108, 130
  as an arena for physics, 165
Membrane velocity, 139
Membrane
  unconstrained, 203
Membranes
  dynamically possible, 114
  kinematically possible, 114
  unconstrained, 108–109, 222
Metric
  fixed, 145
  of $\mathcal{M}$-space, 113, 145
Minimal surface action, 272
'momentum' constraints, 123
Momentum operator, 288
  for membrane, 207
  for point particle, 16
  in curved spacetime, 24
Momentum polyvector, 61–62
Momentum representation, 38
Momentum space, 37
Momentum
  of the Stueckelberg field, 35
Motion
  in 3-space, 5
  in spacetime, 9, 18
Multi-particle states

superposition of, 228
Multivector, 57
Multiverse, 309, 316, 322, 336, 340
Negative energies, 51
Negative energy density, 101
Negative frequencies, 51
n-form field, 192
Negative norm states, 51, 97
Negative signature fields, 101
Null strings, 219
Ordering ambiguity, 207
Outer product, 55, 58
  in infinite-dimensional space, 178
Pandimensional continuum, 90, 103, 200
Parallel propagator
  prototype of, 164
Parameter of evolution, 298
Parametrized quantum field theory, 52
Passive diffeomorphisms, 107
Passive transformations, 109
Path integral, 272, 275
Pauli algebra, 77
Pauli matrices, 80
Pauli–Lubanski pseudo-vector, 65
p-brane, 119, 188
Phase space action
  for point particle, 12
Phase transformations, 41–42
Planck mass, 69
Planck units, 347
Point transformations, 22
Poisson brackets, 33–34, 95, 122
Polymomentum, 68
Polyvector action, 68, 133
  for p-branes, 192
  in $\mathcal{M}$-space, 195
Polyvector constraint, 76
Polyvector field, 177
Polyvector, 62
  in infinite-dimensional space, 178
Polyvectors, 57, 138
  in $\mathcal{M}$-space, 132, 197
Position operators
  for membrane, 205
Position polyvector field, 177
Position vector field, 177
Predictability, 104
Probability amplitude, 275
Probability current, 18, 208, 210
Probability density, 17, 208, 210, 219
  in 3-space, 18
  in spacetime, 18
Problem of time, 86, 138, 284, 298
Product
  of covariant derivatives, 30
  of operators, 27, 29
Propagator, 280

Proto-metric, 183
Proto-vectors, 183
Pseudo-Euclidean signature, 90
Pseudoscalar, 59
Pseudovector, 59
Quantization of gravity, 271
Quantization of the p-brane
  a geometric approach, 238
Quantized p-brane, 242
Quantum gravity, 293
Randall–Sundrum model, 264
Reduced action, 135, 137
Reduced density matrix, 325
Reference fluid, 165, 167, 186
Reference system, 165
Relative state, 307, 325
Relativity of signature, 87
Reparametrization invariance, 107
Reparametrizations, 111
Ricci scalar, 277
  in $\mathcal{M}$-space, 158
  in spacetime, 185
Ricci tensor, 277
  in $\mathcal{M}$-space, 158, 181
Riemann tensor, 277
r-vector, 57
Scalar field
  real, 229
Scalar fields
  in curved spacetime, 277
Scalar product, 66, 177
Schild action, 136, 215
Schrödinger equation, 17, 23, 31, 35, 71, 208–209, 215, 235, 289
Schrödinger equation
  for membrane
    stationary, 210
  in curved spacetime, 24
Schrödinger picture, 85
Schrödinger representation, 205
Schrödinger's cat, 316, 325
Second quantization, 31
  of the polyvector action, 83
Second quantized field, 280
Self-interaction, 258
Self-intersecting brane, 261, 263
Sharp localization, 297
Simultaneity surface, 285
Skeleton space, 164
Source term, 281
Spacetime polyvectors, 193
Spacetime sheet, 271–272, 284
  centroid, 289
  moving, 295
Spin angular momentum, 61
Spin foams, 202
Spin networks, 202

INDEX

Spin, 66
Spinor, 77–78
  conjugate, 81
Spinors, 90, 202
Static membrane, 140
Stationary state, 210, 217
Stress–energy tensor, 33, 35, 50, 100, 191, 279
  for dust of p-branes, 261
  for dust of point particles, 261
  for dust, 273
  of an extended object, 254
  of the membrane configuration, 187
  of the point particle, 254
String, 103, 107
  space-like, 341
  time-like, 341
Stueckelberg action, 70–71, 73–74, 136
  for point particle, 8
Stueckelberg field
  quantization of, 235
Submanifold, 171
Superluminal disturbances, 339
System of many membranes, 155
System of two scalar fields, 99
System of units
  dilatationally invariant, 347
System
  of many branes, 257, 260
  of many intersecting branes, 256
Tachyon, 14, 309–310, 339–340
Tangent vectors, 131, 171–172
Tangentially deformed membranes, 108
Target space, 156, 163, 195
Tensor calculus
  in $\mathcal{M}$-space, 111
Tensors
  in membrane space, 114
Test brane, 260
Test membrane, 153
Tetrad, 170
The worldsheet spinors, 195
Time machines, 101, 103, 340
Time slice, 286, 293
Unconstrained action, 135
Unconstrained membrane, 284, 286

Unconstrained theory, 297
Unitarity, 51
Vacuum energy density, 93, 282
Vacuum expectation value
  of the metric operator, 243
Vacuum state vector, 243
Vacuum state, 95, 97
Vacuum, 36, 46, 50, 100
Vector field, 174, 176
  in $\mathcal{M}$-space, 180
  in curved space, 176
  in spacetime, 185
Vector
  in membrane space, 130
  prototype of, 164
Vectors, 55
  in curved spaces, 168
Velocity polyvector, 62, 68
Vierbein, 30
Volume element in $\mathcal{M}$-space, 153
Warp drive, 101, 103
Wave function
  complexe-valued, 226
  of many particles, 37
  polyvector valued, 76
Wave functional, 86, 215, 275, 289, 291, 296
  packet, 289
Wave packet profile, 43, 228–229
Wave packet, 18, 26, 290, 296
  Gaussian, 219, 221, 290
  centre of, 291
  for membrane, 217
Wedge product
  in infinite-dimensional space, 178
Wheeler–DeWitt equation, 86
Wiggly membrane, 211, 284
World line
  as an intersection, 295
World sheet polyvector, 194
World sheet, 125, 188
  r-vector type, 200
  generalized, 199
World surface, 188
World volume, 188
Wormholes, 101, 340
Zero point energy, 93, 103

Fundamental Theories of Physics

46. P.P.J.M. Schram: *Kinetic Theory of Gases and Plasmas.* 1991 ISBN 0-7923-1392-5
47. A. Micali, R. Boudet and J. Helmstetter (eds.): *Clifford Algebras and their Applications in Mathematical Physics.* 1992 ISBN 0-7923-1623-1
48. E. Prugovečki: *Quantum Geometry. A Framework for Quantum General Relativity.* 1992 ISBN 0-7923-1640-1
49. M.H. Mac Gregor: *The Enigmatic Electron.* 1992 ISBN 0-7923-1982-6
50. C.R. Smith, G.J. Erickson and P.O. Neudorfer (eds.): *Maximum Entropy and Bayesian Methods.* Proceedings of the 11th International Workshop (Seattle, 1991). 1993 ISBN 0-7923-2031-X
51. D.J. Hoekzema: *The Quantum Labyrinth.* 1993 ISBN 0-7923-2066-2
52. Z. Oziewicz, B. Jancewicz and A. Borowiec (eds.): *Spinors, Twistors, Clifford Algebras and Quantum Deformations.* Proceedings of the Second Max Born Symposium (Wrocław, Poland, 1992). 1993 ISBN 0-7923-2251-7
53. A. Mohammad-Djafari and G. Demoment (eds.): *Maximum Entropy and Bayesian Methods.* Proceedings of the 12th International Workshop (Paris, France, 1992). 1993 ISBN 0-7923-2280-0
54. M. Riesz: *Clifford Numbers and Spinors* with Riesz' Private Lectures to E. Folke Bolinder and a Historical Review by Pertti Lounesto. E.F. Bolinder and P. Lounesto (eds.). 1993 ISBN 0-7923-2299-1
55. F. Brackx, R. Delanghe and H. Serras (eds.): *Clifford Algebras and their Applications in Mathematical Physics.* Proceedings of the Third Conference (Deinze, 1993) 1993 ISBN 0-7923-2347-5
56. J.R. Fanchi: *Parametrized Relativistic Quantum Theory.* 1993 ISBN 0-7923-2376-9
57. A. Peres: *Quantum Theory: Concepts and Methods.* 1993 ISBN 0-7923-2549-4
58. P.L. Antonelli, R.S. Ingarden and M. Matsumoto: *The Theory of Sprays and Finsler Spaces with Applications in Physics and Biology.* 1993 ISBN 0-7923-2577-X
59. R. Miron and M. Anastasiei: *The Geometry of Lagrange Spaces: Theory and Applications.* 1994 ISBN 0-7923-2591-5
60. G. Adomian: *Solving Frontier Problems of Physics: The Decomposition Method.* 1994 ISBN 0-7923-2644-X
61. B.S. Kerner and V.V. Osipov: *Autosolitons. A New Approach to Problems of Self-Organization and Turbulence.* 1994 ISBN 0-7923-2816-7
62. G.R. Heidbreder (ed.): *Maximum Entropy and Bayesian Methods.* Proceedings of the 13th International Workshop (Santa Barbara, USA, 1993) 1996 ISBN 0-7923-2851-5
63. J. Peřina, Z. Hradil and B. Jurčo: *Quantum Optics and Fundamentals of Physics.* 1994 ISBN 0-7923-3000-5
64. M. Evans and J.-P. Vigier: *The Enigmatic Photon.* Volume 1: The Field $B^{(3)}$. 1994 ISBN 0-7923-3049-8
65. C.K. Raju: *Time: Towards a Constistent Theory.* 1994 ISBN 0-7923-3103-6
66. A.K.T. Assis: *Weber's Electrodynamics.* 1994 ISBN 0-7923-3137-0
67. Yu. L. Klimontovich: *Statistical Theory of Open Systems.* Volume 1: A Unified Approach to Kinetic Description of Processes in Active Systems. 1995 ISBN 0-7923-3199-0; Pb: ISBN 0-7923-3242-3
68. M. Evans and J.-P. Vigier: *The Enigmatic Photon.* Volume 2: Non-Abelian Electrodynamics. 1995 ISBN 0-7923-3288-1
69. G. Esposito: *Complex General Relativity.* 1995 ISBN 0-7923-3340-3

Fundamental Theories of Physics

70. J. Skilling and S. Sibisi (eds.): *Maximum Entropy and Bayesian Methods*. Proceedings of the Fourteenth International Workshop on Maximum Entropy and Bayesian Methods. 1996
    ISBN 0-7923-3452-3
71. C. Garola and A. Rossi (eds.): *The Foundations of Quantum Mechanics Historical Analysis and Open Questions*. 1995
    ISBN 0-7923-3480-9
72. A. Peres: *Quantum Theory: Concepts and Methods*. 1995 (see for hardback edition, Vol. 57)
    ISBN Pb 0-7923-3632-1
73. M. Ferrero and A. van der Merwe (eds.): *Fundamental Problems in Quantum Physics*. 1995
    ISBN 0-7923-3670-4
74. F.E. Schroeck, Jr.: *Quantum Mechanics on Phase Space*. 1996   ISBN 0-7923-3794-8
75. L. de la Peña and A.M. Cetto: *The Quantum Dice*. An Introduction to Stochastic Electrodynamics. 1996
    ISBN 0-7923-3818-9
76. P.L. Antonelli and R. Miron (eds.): *Lagrange and Finsler Geometry*. Applications to Physics and Biology. 1996
    ISBN 0-7923-3873-1
77. M.W. Evans, J.-P. Vigier, S. Roy and S. Jeffers: *The Enigmatic Photon*. Volume 3: Theory and Practice of the $B^{(3)}$ Field. 1996
    ISBN 0-7923-4044-2
78. W.G.V. Rosser: *Interpretation of Classical Electromagnetism*. 1996   ISBN 0-7923-4187-2
79. K.M. Hanson and R.N. Silver (eds.): *Maximum Entropy and Bayesian Methods*. 1996
    ISBN 0-7923-4311-5
80. S. Jeffers, S. Roy, J.-P. Vigier and G. Hunter (eds.): *The Present Status of the Quantum Theory of Light*. Proceedings of a Symposium in Honour of Jean-Pierre Vigier. 1997
    ISBN 0-7923-4337-9
81. M. Ferrero and A. van der Merwe (eds.): *New Developments on Fundamental Problems in Quantum Physics*. 1997
    ISBN 0-7923-4374-3
82. R. Miron: *The Geometry of Higher-Order Lagrange Spaces*. Applications to Mechanics and Physics. 1997
    ISBN 0-7923-4393-X
83. T. Hakioğlu and A.S. Shumovsky (eds.): *Quantum Optics and the Spectroscopy of Solids*. Concepts and Advances. 1997
    ISBN 0-7923-4414-6
84. A. Sitenko and V. Tartakovskii: *Theory of Nucleus*. Nuclear Structure and Nuclear Interaction. 1997
    ISBN 0-7923-4423-5
85. G. Esposito, A.Yu. Kamenshchik and G. Pollifrone: *Euclidean Quantum Gravity on Manifolds with Boundary*. 1997
    ISBN 0-7923-4472-3
86. R.S. Ingarden, A. Kossakowski and M. Ohya: *Information Dynamics and Open Systems*. Classical and Quantum Approach. 1997
    ISBN 0-7923-4473-1
87. K. Nakamura: *Quantum versus Chaos*. Questions Emerging from Mesoscopic Cosmos. 1997
    ISBN 0-7923-4557-6
88. B.R. Iyer and C.V. Vishveshwara (eds.): *Geometry, Fields and Cosmology*. Techniques and Applications. 1997
    ISBN 0-7923-4725-0
89. G.A. Martynov: *Classical Statistical Mechanics*. 1997   ISBN 0-7923-4774-9
90. M.W. Evans, J.-P. Vigier, S. Roy and G. Hunter (eds.): *The Enigmatic Photon*. Volume 4: New Directions. 1998
    ISBN 0-7923-4826-5
91. M. Rédei: *Quantum Logic in Algebraic Approach*. 1998   ISBN 0-7923-4903-2
92. S. Roy: *Statistical Geometry and Applications to Microphysics and Cosmology*. 1998
    ISBN 0-7923-4907-5
93. B.C. Eu: *Nonequilibrium Statistical Mechanics*. Ensembled Method. 1998
    ISBN 0-7923-4980-6

Fundamental Theories of Physics

94. V. Dietrich, K. Habetha and G. Jank (eds.): *Clifford Algebras and Their Application in Mathematical Physics.* Aachen 1996. 1998   ISBN 0-7923-5037-5
95. J.P. Blaizot, X. Campi and M. Ploszajczak (eds.): *Nuclear Matter in Different Phases and Transitions.* 1999   ISBN 0-7923-5660-8
96. V.P. Frolov and I.D. Novikov: *Black Hole Physics.* Basic Concepts and New Developments. 1998   ISBN 0-7923-5145-2; Pb 0-7923-5146
97. G. Hunter, S. Jeffers and J-P. Vigier (eds.): *Causality and Locality in Modern Physics.* 1998   ISBN 0-7923-5227-0
98. G.J. Erickson, J.T. Rychert and C.R. Smith (eds.): *Maximum Entropy and Bayesian Methods.* 1998   ISBN 0-7923-5047-2
99. D. Hestenes: *New Foundations for Classical Mechanics (Second Edition).* 1999   ISBN 0-7923-5302-1; Pb ISBN 0-7923-5514-8
100. B.R. Iyer and B. Bhawal (eds.): *Black Holes, Gravitational Radiation and the Universe.* Essays in Honor of C. V. Vishveshwara. 1999   ISBN 0-7923-5308-0
101. P.L. Antonelli and T.J. Zastawniak: *Fundamentals of Finslerian Diffusion with Applications.* 1998   ISBN 0-7923-5511-3
102. H. Atmanspacher, A. Amann and U. Müller-Herold: *On Quanta, Mind and Matter Hans Primas in Context.* 1999   ISBN 0-7923-5696-9
103. M.A. Trump and W.C. Schieve: *Classical Relativistic Many-Body Dynamics.* 1999   ISBN 0-7923-5737-X
104. A.I. Maimistov and A.M. Basharov: *Nonlinear Optical Waves.* 1999   ISBN 0-7923-5752-3
105. W. von der Linden, V. Dose, R. Fischer and R. Preuss (eds.): *Maximum Entropy and Bayesian Methods Garching, Germany 1998.* 1999   ISBN 0-7923-5766-3
106. M.W. Evans: *The Enigmatic Photon Volume 5: O(3) Electrodynamics.* 1999   ISBN 0-7923-5792-2
107. G.N. Afanasiev: *Topological Effects in Quantum Mecvhanics.* 1999   ISBN 0-7923-5800-7
108. V. Devanathan: *Angular Momentum Techniques in Quantum Mechanics.* 1999   ISBN 0-7923-5866-X
109. P.L. Antonelli (ed.): *Finslerian Geometries A Meeting of Minds.* 1999   ISBN 0-7923-6115-6
110. M.B. Mensky: *Quantum Measurements and Decoherence Models and Phenomenology.* 2000   ISBN 0-7923-6227-6
111. B. Coecke, D. Moore and A. Wilce (eds.): *Current Research in Operation Quantum Logic.* Algebras, Categories, Languages. 2000   ISBN 0-7923-6258-6
112. G. Jumarie: *Maximum Entropy, Information Without Probability and Complex Fractals.* Classical and Quantum Approach. 2000   ISBN 0-7923-6330-2
113. B. Fain: *Irreversibilities in Quantum Mechanics.* 2000   ISBN 0-7923-6581-X
114. T. Borne, G. Lochak and H. Stumpf: *Nonperturbative Quantum Field Theory and the Structure of Matter.* 2001   ISBN 0-7923-6803-7
115. J. Keller: *Theory of the Electron.* A Theory of Matter from START. 2001   ISBN 0-7923-6819-3
116. M. Rivas: *Kinematical Theory of Spinning Particles.* Classical and Quantum Mechanical Formalism of Elementary Particles. 2001   ISBN 0-7923-6824-X
117. A.A. Ungar: *Beyond the Einstein Addition Law and its Gyroscopic Thomas Precession.* The Theory of Gyrogroups and Gyrovector Spaces. 2001   ISBN 0-7923-6909-2
118. R. Miron, D. Hrimiuc, H. Shimada and S.V. Sabau: *The Geometry of Hamilton and Lagrange Spaces.* 2001   ISBN 0-7923-6926-2

Fundamental Theories of Physics

119. M. Pavšič: *The Landscape of Theoretical Physics: A Global View*. From Point Particles to the Brane World and Beyond in Search of a Unifying Principle. 2001　　ISBN 0-7923-7006-6